Generation of Cosmological Large-Scale Structure

NATO ASI Series

Advanced Science Institutes Series

A Series presenting the results of activities sponsored by the NATO Science Committee, which aims at the dissemination of advanced scientific and technological knowledge, with a view to strengthening links between scientific communities.

The Series is published by an international board of publishers in conjunction with the NATO Scientific Affairs Division

A **Life Sciences**	Plenum Publishing Corporation
B **Physics**	London and New York
C **Mathematical and Physical Sciences**	Kluwer Academic Publishers
D **Behavioural and Social Sciences**	Dordrecht, Boston and London
E **Applied Sciences**	
F **Computer and Systems Sciences**	Springer-Verlag
G **Ecological Sciences**	Berlin, Heidelberg, New York, London,
H **Cell Biology**	Paris and Tokyo
I **Global Environmental Change**	

PARTNERSHIP SUB-SERIES

1. **Disarmament Technologies**	Kluwer Academic Publishers
2. **Environment**	Springer-Verlag / Kluwer Academic Publishers
3. **High Technology**	Kluwer Academic Publishers
4. **Science and Technology Policy**	Kluwer Academic Publishers
5. **Computer Networking**	Kluwer Academic Publishers

The Partnership Sub-Series incorporates activities undertaken in collaboration with NATO's Cooperation Partners, the countries of the CIS and Central and Eastern Europe, in Priority Areas of concern to those countries.

NATO-PCO-DATA BASE

The electronic index to the NATO ASI Series provides full bibliographical references (with keywords and/or abstracts) to more than 50000 contributions from international scientists published in all sections of the NATO ASI Series.
Access to the NATO-PCO-DATA BASE is possible in two ways:

– via online FILE 128 (NATO-PCO-DATA BASE) hosted by ESRIN,
Via Galileo Galilei, I-00044 Frascati, Italy.

– via CD-ROM "NATO-PCO-DATA BASE" with user-friendly retrieval software in English, French and German (© WTV GmbH and DATAWARE Technologies Inc. 1989).

The CD-ROM can be ordered through any member of the Board of Publishers or through NATO-PCO, Overijse, Belgium.

Series C: Mathematical and Physical Sciences – Vol. 503

Generation of Cosmological Large-Scale Structure

edited by

David N. Schramm

Department of Astronomy and Astrophysics,
University of Chicago,
Chicago, Illinois, U.S.A.

and

P. Galeotti

Istituto Nazionale di Fisica Nucleare,
Torino, Italy

Kluwer Academic Publishers

Dordrecht / Boston / London

Published in cooperation with NATO Scientific Affairs Division

Proceedings of the NATO Advanced Study Institute on
the Third Erice School on Particle Astrophysics:
Generation of Large-Scale Cosmological Structures
Erice, Italy
3–13 November 1996

A C.I.P. Catalogue record for this book is available from the Library of Congress.

ISBN 978-94-010-6513-9   e-ISBN-13: 978-94-009-0053-0
DOI: 10.1007/978-94-009-0053-0

Published by Kluwer Academic Publishers,
P.O. Box 17, 3300 AA Dordrecht, The Netherlands.

Sold and distributed in the U.S.A. and Canada
by Kluwer Academic Publishers,
101 Philip Drive, Norwell, MA 02061, U.S.A.

In all other countries, sold and distributed
by Kluwer Academic Publishers,
P.O. Box 322, 3300 AH Dordrecht, The Netherlands.

Printed on acid-free paper

Table of Contents

PREFACE

This volume is the proceedings of the third school in particle astrophysics that Schramm and Galeotti have organized at Erice. The focus of thirs third school was the Generation of Cosmological Large-Scale Structure. It was held in November of 1996. The first school in the series was on "Gauge Theory and the Early Universe" in May 1986, the second was on "Dark Matter in the Universe" in May 1988. All three schools have been successful under the auspices of the NATO Advanced Study Institute. This volume is thus the third in the series of the proceedings of these schools. The choice of the topic for this third school was natural, since the problem of generating a large-scale structure has become the most pressing problem in cosmology today. In particular, it is this generation of structure that is the interface between astronomical observations and particle models for the early universe. To date, all models for generating structures inevitably require new fundamental physics beyond the standard, $SU_3 \times SU_2 \times U_1$, model of high energy physics. The seeds for generating structures usually invoke unification physics, and the matter needed to clump and form them seems to require particle properties that have not been seen in laboratories to date. New astronomical observations seems to be finding very large structures with high peculiar veloicities, while the observations of the microwave background anisotropies by the COBE-satellite and by balloon and ground-based experiments are tightly constraining possible primordial density fluctuation scenarios. The symbiotic relationship of the seeds to form structures and "dark matter" (the topic of the second school) is obvious since the way that structures form is intimately related to the nature of the matter forming a structure. Particular excitement in the field today is due to the fact that is has shifted from theoretically driven work characterized by our first school to a field that is now very much driven by experiments and observations. This third school dramatically demonstrated this transition.

As in the case of the previous schools, our 1996 meeting at Erice provided a wonderful example of the interplay between astronomy and physics and among experiment, observation and theory. The speakers and students came from both traditional astronomical areas and traditional particle physics areas, as well as from the newly developed fields of experimental dark matter searches; and theorists came from the newly developed particle/cosmology interface. We feel that the proceedings of this meeting should be a useful compendium for anybody wishing to look at what is currently going on in the world of cosmology.

x

We would like to take this opportunity to thank the NATO Science Committee for supporting this school and to thank the staff of the Ettore Majorana Centre for Scientific Culture. We would also like to thank our co-organizers, especially Venya Berezinsky for his help in making sure that the representation of former Eastern Block scientists was maintained.

David N. Schramm
Piero Galeotti

Participants

Roberta Bernstein
Dept. of Astronomy and Astrophysics
AAC 140
University of Chicago
Chicago. IL 60637 USA
roberta@oddjob.uchicago.edu

Karlis Berzins
Astronomical Observatory
Bueks Bohr Institute
University of Copenhagen
The Rockefeller Institute
Juliane Maries Vej 30
DK-2100 Copenhagen, Denmark
kberzins@acad.latnet.lv

Richard Bond
CITA
McLennon Laboratories
University of Toronto
60 St. George St.
Toronto, ON M5S 1A1
Canada
bond@cita.utoronto.ca

Ari Buchalter
Dept. of Astronomy
Colombia University
538 W. 120th St.
New York, NY 10027 USA
ari@parsifal.phys.columbia.edu

Stephan Burns
Dept. of Physics and Astronomy
UCLA
405 Hildegard Ave.
Los Angeles, CA 90025 USA
burns@physics.ucla.edu

Xuelei Chen
Physics Dept.
Colombia University
538 W. 120th St.
New York, NY 10027 USA
xuelei@cuphy3.phys.columbia.edu

Guido Chincarini
Osservatorio Astronomico di Brera
Via Brera 28I,20121
Milano, Italy
guido@merate.mi.astro.it

Andrew Connolly
Dept. of Physics and Astronomy
Johns Hopkins University
3400 N. Charles St.
Baltimore, MD 21218 USA
ajc@tiamat.pha.jhu.edu

Craig Copi
Dept. of Astronomy and Astrophysics
University of Chicago
5640 S. Ellis Ave.
Chicago, IL 60637 USA
copi@oddjob.uchicago.edu

Catherine Cress
Astronomy Dept.
Colombia University
538 W. 120th St.
New York, NY 10027 USA
cress@astro.columbia.edu

Haakon Dahle
University of Toronto
Dept. of Astronomy
McLennan Labs, Rm. 1403
60 St. George St.
Toronto, Ont. M5S 3H8, Canada
dahle@cita.utoronto.ca

Julien Devriendt
Institut d'Astrophysique
98 bis Boulevard Arago
75014 Paris, France
devriend@iap.fr

Pasquale Di Bari
INFN
Laboratori Nazionale del Gran Sasso
67010 Assergi, Italy
dibari@lngs.infn.it

Valentina D'Odorico
SISSA/ISAS
via Beirut, 2-4
I-34014 Trieste, Italy
dodorico@sissa.it

Natalia Dounina-Barkovskaya
Dept. of Astrophysics
Inst. for Theoretical and Exper. Physics
B. Cheremushkinskaya 25
117259 Moscow, Russia
dunina@vitep2.itep.ru

Piero Galleotti
Istituto Nazionale di Fisica Nucleare
Via Pietro Giuria 1,I-10125
Torino, Italy
galeotti@tog3ax.to.infn.it

Massimo Galluccio
Via Luchino dal Verme, 90
00176 Roma, Italy
gallucci@oarhp1.rm.astro.it

Mario Gambera
Istituto di Astronomia dell'
Universita di Catania
Viale Andrea Doria n. 6
Catania, Italy
mgambera@alpha4.ct.astro.it

Jeffrey Gardner
Dept. of Astronomy, Box 35180
University of Washington
Seattle, WA 98195 USA
gardner@astro.washington.edu

Riccardo Giovanelli
Dept. of Astronomy
Space Sciences Bldg.
Cornell University
Ithaca, NY 14853
riccardo@astrosun.tn.cornell.edu

Thomas Goldbrunner
Fakultät für Physik der TU München
Physik-Dept. E15
James Franck Str.
85748 Garching , Germany
goldbrunner@e15.physik.tu-muenchen.de

Martin Götz
Physics Dept.
Brown University, Box 1843
Providence, RI 02912 USA
gotz@het.brown.edu

David Graff
Dept. of Physics
Univ. of Michigan
500 E. University Ave.
Ann Arbor, MI 48109-1120 USA
graff@umich.edu

Jeff Jewell
Dept. of Astronomy & Astrophysics
University of Chicago
5640 S. Ellis Ave.
Chicago, IL 60637, USA
jewell@oddjob.uchicago.edu

Roman Juszkiewicz
N. Copernicus Astronomical Center
Polish Academy of Sciences
Bartycka 18
00-716 WARSAW, Poland
juszkiewicz@iap.fr

Nick Kaiser
CITA
University of Toronto
60 St. George Street
Toronto, ON M5S 1A1
Canada
kaiser@cita.utoronto.ca

Pelin Karakaya
Orta Dogu Teknik Universities
Kimya Bolumu
06531 Ankara - Turkey
pelin@rorqual.cc.metu.edu.tr

Petteri Keranen
Dept. of Physics
P.O. Box 9
University of Helsinki
Helsinki, Finland
keranen@phcu.helsinki.fi

Peter Kernan
Physics Dept. , Rock 211
Case Western Reserve University
10900 Euclid Ave.
Cleveland, OH 44106-7079 USA
pjk6@po.cwru.edu

Martin Kerscher
Lehrstuhl Wagner
Universitaet Muenchen
Thereisenstrasse 37
D-80333 Meunchen, Germany
kerscher@stat.physik.uni-muenchen.de

Hong Liu
Dept. of Physics
Case Western Reserve University
10900 Euclid Avenue
Cleveland, OH 44106 USA
hxl20@po.cwru.edu

Robert Lopez
Dept. of Astronomy ad Astrophysics
University of Chicago
5640 S. Ellis Ave.
Chicago, IL 60637 USA
lopez@oddjob.uchicago.edu

Ariel Megevand
Instituto Balseiro
Centro Atomico Bariloche
8400 Bariloche, Argentina
megevand@cab.cnea.edu.ar

Pierluigi Monaco
SISSA
via Beirut 6-34013
Trieste, Italy
monaco@newton.sissa.it

Marco Montuori
Physics Dept.
University "La Sapienza"
Pl. Aldo Moro 2
00185 Roma, Italy
montouri@fermi.fis.unical.it

Marianne Neff
Physik Dept. E15
Technische Universitat Munchen
James Franck Straße
D-85748 Garching, Germany
neff@e15.physik.tu-muenchen.de

Kenneth Nollett
Dept. of Astronomy and Astrophysics
University of Chicago
5640 S. Ellis Ave.
Chicago, IL 60637 USA
nollett@oddjob.uchicago.edu

Siang Peng Oh
Princeton University Observatory
Peyton Hall
Ivy Lane
Princeton, NJ 08544, USA
peng@astro.princeton.edu

Lyman Page
Dept. of Physics -Jadwin Hall
Princeton University
P.O. Box 708
Princeton, NJ 08544 USA
page@pupgg.princeton.edu

Luciano Pietronero
Istituto di Fisica
Universita La Sapienza
Piazzale A. Moro 2
00185 Roma, Italy
pietronero@roma1.infn.it

Cristiano Porciani
SISSA/ISAS
via Beirut 2-4 , I-34014
Trieste, Italy
porciani@sissa.it

Nurur Rahman
139/1 South Mugda Para
Dhaka 1214
Bangladesh
sumon@bdcom.com

Martin Sawicki
Dept. of Astronomy
University of Toronto
Toronto, ON
M5S 3H8 Canada
sawicki@astro.utoronto.ca

Jens Schmalzing
Lehrstuhl Wagner
Universitaet Muenchen
Thereisenstrasse 37
D-80333 Meunchen, Germany
jens@stat.physik.uni-muenchen.de

D.N. Schramm
Dept. of Astronomy and Astrophysics
AAC 140
University of Chicago
Chicago. IL 60637 USA
dns@oddjob.uchicago.edu

Sean Scully
School of Physics and Astronomy
University of Minnesota
116 Church St. SE
Minneapolis, MN 55455 USA
sean@astro.spa.umn.edu

Sergei Shandarin
Department of Physics and Astronomy
University of Kansas
Lawrence, KS 66045 USA
sergei@kusmos.phsx.ukans.edu

Andrew Sonnenschein
Dept. of Physics
University of Calif. at Santa Barbara
Santa Barbara, CA 93106 USA
andrews@slac.stanford.edu

Mark SubbaRao
Dept. of Physics and Astronomy
Johns Hopkins University
Baltimore, MD 21207 USA
subbarao@pha.jhu.edu

Alex Szalay
Dept. of Physics and Astronomy
Johns Hopkins University
Homewood Campus
Baltimore, MD 21218 USA
szalay@pha.jhu.edu

Istvan Szapudi
Fermilab - MS 209
P.O. Box 500
Batavia, IL 60510-0500 USA
szapudi@traviata.fnal.gov

Hakan M. Taplamacioglu
Turkish Standards Institution
Emek Mah. 86. Sokak 19/1
06510 Ankara, Turkey
biotec@rorqual.cc.metu.edu.tr

Ramon Toldra
Grup de Fisica Teorica & Institut de
Fisica d'Altes Energies
Facultat de Ciencies. Edifici Cn
Universitat Autonoma de Barcelona
E-08193 Bellaterra, Barcelona, Spain
toldra@ifae.es

Michael S. Turner
University of Chicago
Dept. of Astronomy and Astrophysics
Ellis Avenue - AAC 142
Chicago, IL 60637 USA
mturner@oddjob.uchicago.edu

Neil Turok
DAMTP
Cambridge University
Silver Street
Cambridge CB3 9EW, United Kingdom
n.g.turok@damtp.cam.ac.uk

Augustine A. Ubachukwu
Dept. of Physics & Astronomy
University of Nigeria
Nsukka, Nigeria

Licia Verde
Institute for Astronomy
University of Edinburgh
Blackford Hill,
Edinburgh EH9 3HJ
United Kingdom
l.verde@roe.ac.uk

Peter Widerin
Theoretical Physics
ETH
Honggerberg
8093 Zurich, Switzerland
widerin@itp.phys.ethz.ch

Kazuhiro Yamamoto
Department of Physics
Hiroshima University
Kagamiyama
739 Hiroshima, Japan
yamamoto@astro.phys.sci.hiroshima-u.ac.jp

Capp Yess
Dept. of Physics and Astronomy
University of Kansas
Lawrence, KS 66045-2151 USA
yess@kusmos.phsx.ukans.edu

HIGH REDSHIFT GALAXIES

GUIDO CHINCARINI[1,2] AND PAOLO SARACCO[3]
1) Universita` degli Studi di Milano
2) Osservatorio Astronomico di Brera
3) Istituto di Fisica Cosmica e Tecnologie Relative, CNR

1.   Prologue

That we are in an evolving Cosmo where each single object, and the Universe as a whole, evolve in time has been accepted since a long time. The great excitement in current cosmology, as mentioned also by J. Peebles sometime before 1983, is that we have opened an observational and theoretical window in the world of mathematical models and we are now looking in them, confronting them with the observations with the feeling that the prospects are excellent to understand some of the details of the grand design of Nature. A variety of unforeseen facts are being, and expect to be, discovered. The advent of advanced detector technology, fast computer facilities and large telescopes are making all of this possible.
The evolution of galaxies is probably the key to understand the Universe as a whole: it is one of the most important ingredients in the various cosmological tests. The poor knowledge which we have of the evolution is the reason why such tests are still not able to give strong statement about the values of the main cosmological parameters. In this framework the comparison of the properties of galaxies at different redshifts represents the most powerful tool to understand their evolution: variations of their number density, their luminosity distribution, their morphology, their stellar, gas and mass contents may give fundamental insights on the acting mechanism(s) which determines the evolution of galaxies. Present day instrumentation is making possible to compare most of these properties among galaxies at very different epochs giving an unprecedented view of the Cosmo.

In these lectures we will deal with galaxy number counts to introduce the evolution of galaxies focusing the attention on the fundamental role which plays the Luminosity Function (LF; sec. 2 and 3). Through a review of the most recent results obtained in these research fields, we will explore the properties of the galaxies down to $z=1$-1.5. In this context we will also pay attention to some of the possible bias by which catalogues are affected and that may contribute to yield an unclear and sometime incoherent picture. In section 4 we will deal with the current search of distant galaxies reaching $z>4$. It is worth noting that while the samples used to study the evolution of galaxies down to $z\sim1$ are constitute of

1

D. N. Schramm and P. Galeotti (eds.), Generation of Cosmological Large-Scale Structure, 1–31.
© *1997 Kluwer Academic Publishers.*

hundreds to thousands galaxies, at higher redshifts the samples contain very few galaxies. Moreover while the selection effects affecting the local galaxy samples are fairly well known, those affecting high redshift galaxy samples (strictly related to the different search method used) are less evident. Thus, for the time being, the properties of the few observed high redshift galaxies could not represent the properties of the *population* of galaxies at high redshift. Finally, in section 5, we will discuss the searches of Clusters of Galaxies.

2.   Introduction

2.1   Time of Galaxy Formation

The current technology and telescopes allow to observe galaxies down to a redshift $z>4$. What we would expect to see and to test at these redshifts?
Assuming that galaxies form from gravitational instabilities, at the turn-around radius (radius of maximum expansion) the density of a spherical overdense region is given by

$$\approx \, {}^{9\pi^2}\!/\!_{16} \, \rho_b$$

where ρ_b is the mean background density.
By collapsing to form the galaxy, the density of the spherical perturbation increases by a factor f_c and the observed overdensity is of the order:

$$\rho_{obs} \cong f_c \frac{9\pi^2}{16} \rho_b (t_{max}) \approx 5.6 \Omega \rho_c (1 + z_{turn-around})^3 f_c^{\,3}$$

where ρ_c is the critical density Assuming $M_{galaxy} \sim 10^{11} \, M_\Theta$ *and* $r \sim 10 \, kpc$ we obtain:

$$(\rho_{obs}\!/\!\rho_c) \cong 10^5 \, and \, z_{turn-around} \cong \frac{1}{f_c}(\rho_{obs}\!/\!\rho_c)^{1\!/3} \approx \frac{30}{f_c \, \Omega^{1\!/3}}$$

Since $t_{collapse} \sim 2 \, t_{turn-around}$ and $\rho_b \propto t^{-2}$, the density contrast increases by a factor 2^2 and $z_{turn-around}$ decreases by a factor $2^{2/3}$. Using $f_c \sim 2$ for a dissipationless collapse

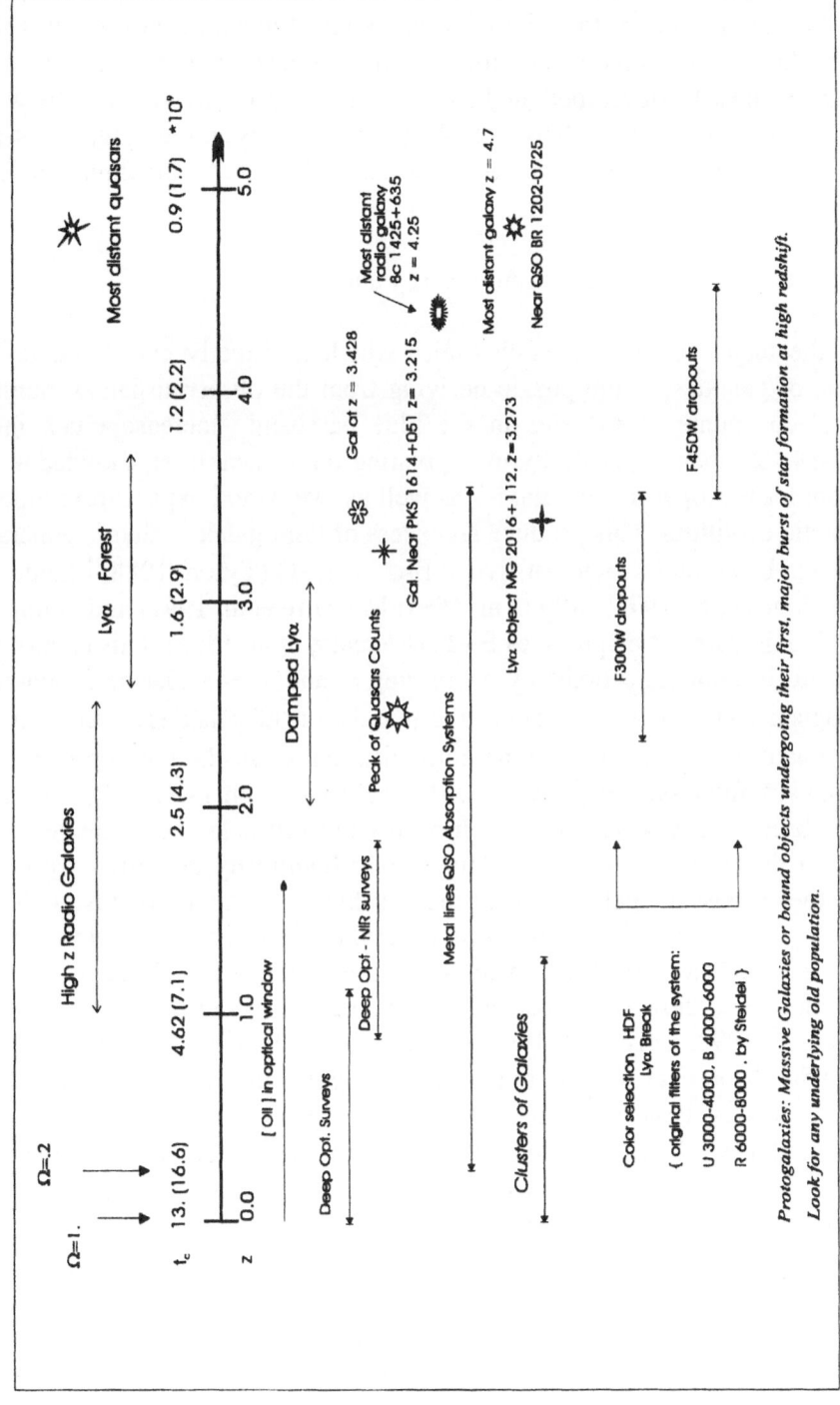

Protogalaxies: Massive Galaxies or bound objects undergoing their first, major burst of star formation at high redshift. Look for any underlying old population.

Figure 1. Some of the observed events as a function of redshift and age of the Universe.

we obtain $z_{\text{turn-around}} \leq 15\ \Omega^{-1/3}$ while for a disk and $f_c \sim 10$ we obtain $z_{\text{turn-around}} \leq 3\ \Omega^{-1/3}$. The point we want to make with this order of magnitude estimate is that we are in the redshift range where observations can test theories (many of these concepts are clearly developed in Padmanbhan, 1996). The relation between redshift and cosmic time, furthermore, shows that already in the range $0.5 < z < 1.0$ the look -back time is of the order of the stellar evolution time, see also Figure 1.

2.2   Counts of Galaxies and Luminosity Function

One of the major today's research topics which is directly connected to the evolution of galaxies, is the puzzle deriving from the observed galaxy number counts: deep counts of galaxies in the blue passband   increase much more rapidly at B>21 than expected by extrapolating the ensemble of knowledge we have from local properties of galaxies as well as we would expect from models w/o galactic evolution. This produce an excess of faint galaxies that is generally estimated to be around a factor of five at B≈24 (Ω_0=1) (Tyson, 1988, Maddox et al. 1990; Jones et al. 1991; Lilly et al. 1991; Metcalfe et al. 1991) and continues unabated to the faintest magnitudes B~28 (Metcalfe et al. 1995). This excess can be reduced (but not cancelled!) by decreasing q_0 and/or considering a non-zero cosmological constant Λ, by making the population of galaxies bluer than the local estimates and by increasing the normalisation ϕ^* or the faint-end slope of the luminosity function (see Lilly et al. 1991; Koo & Kron 1992). The question is: such observed excess is due to high or low redshift galaxies? Galaxies were brighter in the past than now. Invoking strong luminosity evolution we could explain the observed galaxy counts excess but run into problems with the spectroscopic observations: despite this excess, the redshift distributions of the optically selected samples down to B<24 look remarkably like that expected for no evolution, and show that most of the galaxies are at redshifts $<z>\sim0.4$-0.5 (Colless et al. 1990; Lilly et al. 1991; Cowie et al. 1991; Ellis 1993; Glazebrook et al. 1995a). Moreover they do not show any significant population of " blue objects " at very large distances. On the other hand counts in the K' band, Gardner et al. (1993), Djorgovski et al. (1995), McLeod et al. (1995) Saracco et al. (1997), are not as steep as those in the B band showing that somewhere an excess of blue objects might exist. This is, in a simple form, the statement of the problem. The understanding of these observational facts forms one of the major today's research topics. Taking into account that in the observed redshift distributions most of the galaxies with B<24 are at redshift $z\sim0.4$-0.5, this implies that the galaxies responsible of the counts excess at these magnitudes should have slightly sub-L^* luminosities in this passband (Lilly 1993). Considering that the observed surface density of galaxies brighter than B=27 is

$N_{obs} \sim 4.5 \ 10^5$/sqdegree and that $\int_{L_{min}}^{\infty} \varphi(L)VdL = N_{obs}$, where $\varphi(L)$ is the LF of galaxies, this indicates that L_{min} is rather faint: about 0.1 L^* for $\Omega = 0.1$ and 0.01 for $\Omega = 1$. Such considerations shows the big role played in this field by the estimate of the local (and possibly at high z) luminosity function and, in particular, of its faint end slope. The strong dependence of the interpretation of the observed galaxy counts on the shape and slope of the LF will be discussed in the next section in addition to the most recent data and results.

3.   The Luminosity Function

3.1   Local Luminosity Function

The Luminosity Function as derived in the Virgo Cluster, Sandage et al. (1985), has been (and it is) a very educating guideline for any estimate in other samples and different environment. In that sample it has been found that the L.F.s derived for each morphological type has a Gaussian distribution with the exception of the dwarf galaxies whose L.F. show an exponential rise toward the faintest magnitudes. Faint irregular galaxies are also abundant The envelope of the distributions for the different morphological types, that is the total LF, is well fitted by a Schechter approximation, Schechter (1976), with a steep faint end. As we have seen in the previous section the percentage of faint galaxies is relevant in the interpretation of galaxy counts, and the result obtained by Sandage et al. (1985) has been the first indication that the population of intrinsically faint galaxies could be much larger than what we thought. However this result is obtained in a cluster which is a peculiar and a quite different environment from the field. The first evidence that also the local field L.F. shows an upturn at faint magnitudes came from the CfA survey, Marzke et al. (1994a,b), where the change of slope beginns at about $M_{zw} \sim -17$ with the main contribution due to Sm - Im galaxies. A similar tendency has been later observed in the SSRS2 survey, da Costa et al. (1994).

A detailed analysis of the Local Luminosity Function has been recently derived from the ESP sample, Zucca et al. (1996), where evidence is presented that the L.F. upturns at $M_B = -17 + 5 \log h$. The ESP sample, Vettolani et al. (1996), goes to the limiting magnitude $b_j = 19.4$ and covers a volume of $\sim 5 \ 10^4 \ h^{-3} \ Mpc^3$ at the sensitivity peak of $z \sim 0.1$ and consists of 3342 galaxies. In about 50% of these emission lines have been detected. The total L.F. is well represented by a

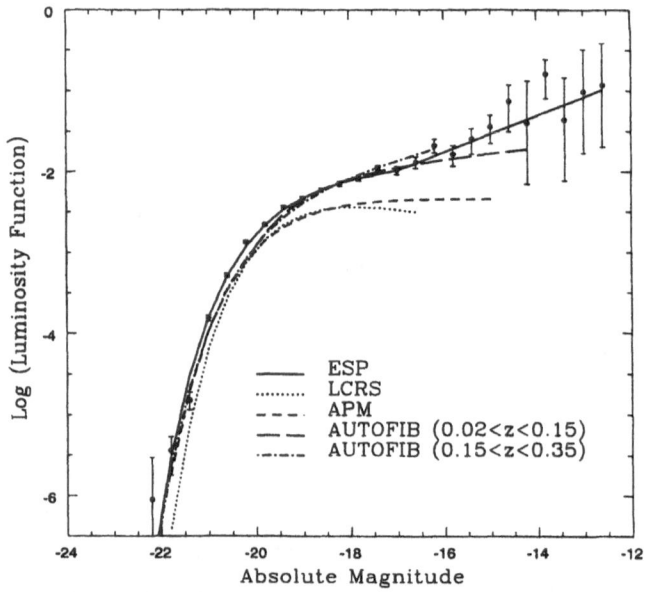

Figure 2. The ESP Luminosity Function in comparison with previous results. For sake of clarity, the two shallow surveys (CfA2 and SSRS2) are not plotted.

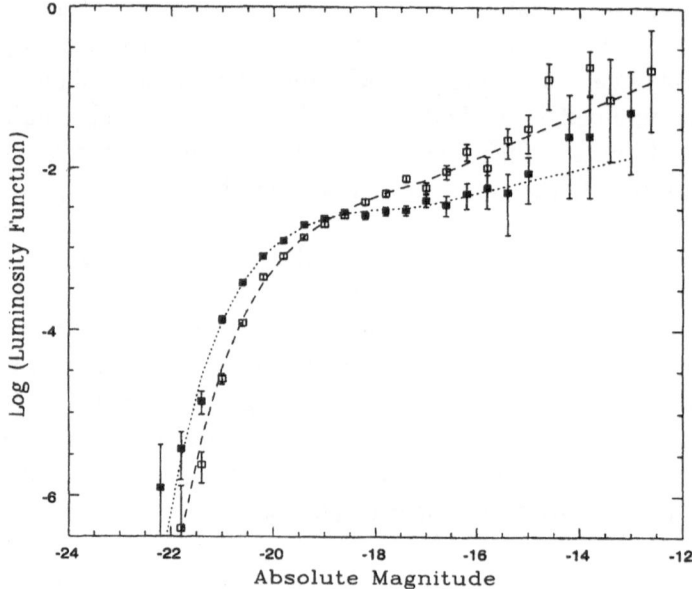

Figure 3. Normalised Luminosity Function for the ESP galaxies. Open squares and dashed line refer to galaxies with emission line while filled squares and

dotted line refer to galaxies with emission line. Fit with Schechter function and power law.

Schechter function to which we added a power law to fit the faint end, which is defined by about 40 galaxies fainter than -16 and with a redshift smaller than about 10 000 km/s. Most of them are emission lines galaxies, Figure 2 and Figure 3 . Such results show that the steeping of the LF at faint magnitudes is a fact and it is not dependent on the environment. In other words there is an increasing number of irregular and intrinsically faint galaxies going to faint magnitudes. In this context Saracco et al. (1996) show that a steeping at the faint end of the LF could explain the observed blue faint galaxy counts excess whithout invoking strong luminosity and/or density evolution. On the other hand the agreement with the K counts, which do not show excess, is obtained hypothesising a flattening of the LF from B to K band, i.e. the faint galaxies that contribute to the rising of the B-band LF have to be bluer than the canonical value (B<2.5).

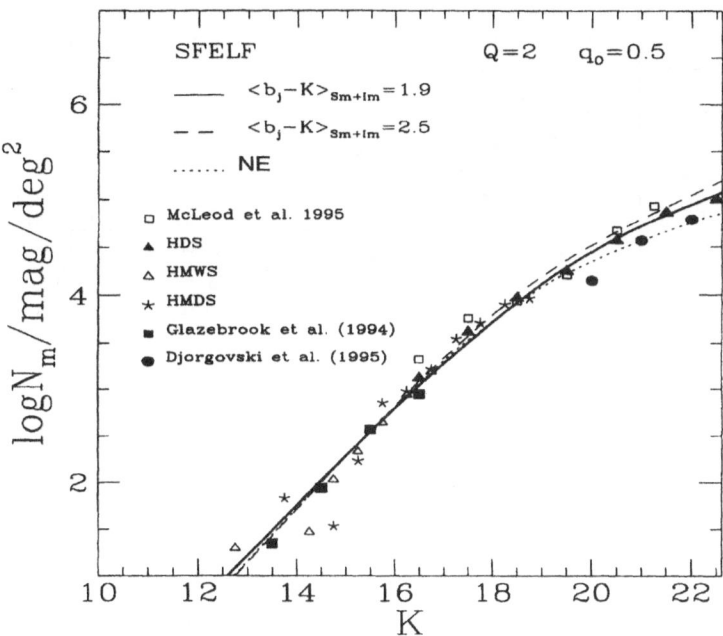

Figure 4. K-band number counts are compared with models. The dotted line refers to no density evolution. Both dashed and dotted lines accounts for a merging rate of Q=2; in one case (dashed) we use $<b_j - k> = 2.5$ for Sm + Im galaxies, while in the second case (dotted) $<b_j - k> = 1.9$

8

This hypothesis is strongly supported by the correlation which they found between the colour (B-K) and the absolute infrared magnitude (M_k): galaxies tend to be bluer with decreasing IR luminosity , Figure 4 and Figure 5. Following the authors: " The correlation found [here M_k was taken as a measure of the mass] implies also a dependence of the evolution of galaxies from their mass. It suggests that low mass galaxies are younger than the high mass ones and that the low mass red stellar population, with eventually some contribution by evolved giants, tends to dominate at larger galaxy masses". That is in agreement with the demonstration by Cowie of the downsizing effect, see section 4.1.

However there is no doubt that a full understanding of the faint galaxy counts is related to an accurate knowledge of the local LF at various wavelengths Garilli et al., (1997).

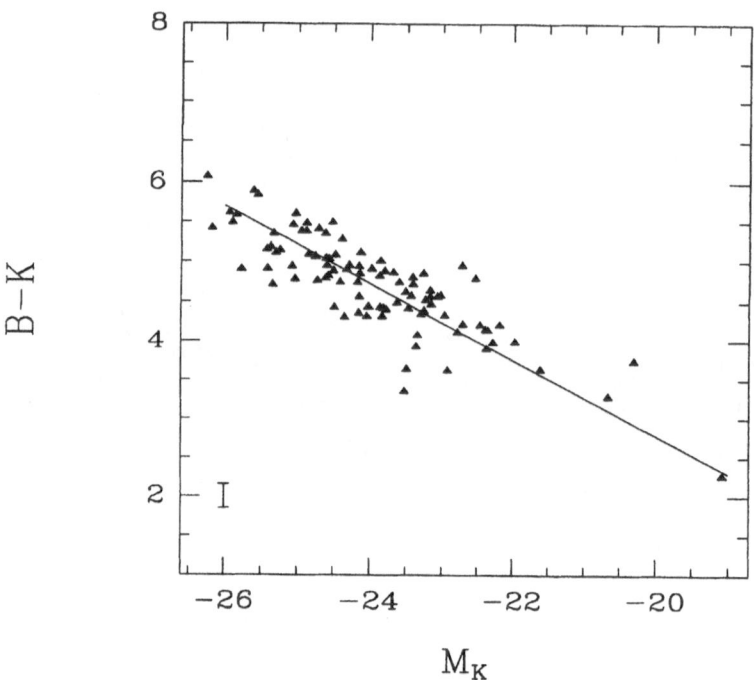

Figure 5. Colour Luminosity plot for the complete IR subsample extracted from the optically selected sample by Bershady et al. (1994).

3.2   Evolution of the Luminosity Function

Deep galaxy surveys, first among these the Canada-France Redshift Survey (CFRS; Lilly et al. 1995), evidence evolution in the Luminosity Function. The CFRS has·produced a complete sample of 591 galaxies in the magnitudes range

$17 < I_{AB}^{1} < 22.5$, and in the redshift range $0.0 < z < 1.3$. The median redshift of the sample is $<z>=0.55$. In our context the main results derived by the analysis of the sample are: a) a continuos change of population with redshift, as shown by the increase of the percentage of emission line galaxies with increasing redshifts, b) the LF of red galaxies does not change all the way back to $z \sim 1$, which implies that such a population of galaxies was to a large degree in place at that epoch, c) the LF of bluer galaxies shows strong signs of evolution: at $z > 0.5$ the number of galaxies with approximately present day L^* ($M_{AB}(B) \sim -21$) increases considerably. This result is in agreement with the steeping of the galaxy counts down to $B \sim 24$, however it is not clear whether such an increase is due to a brightening of blue galaxies, to a real increase in the number density (i.e. merging) or to a mixture of both effects. HST observations of 32 galaxies in the redshift range $0.5 < z < 1.2$ of the CFRS sample (Schade et al. 1996) show that these galaxies exhibit the same range of morphological types as seen locally, in particular 70% form a regular Hubble sequence and can be decomposed in a red bulge and a blue disk.

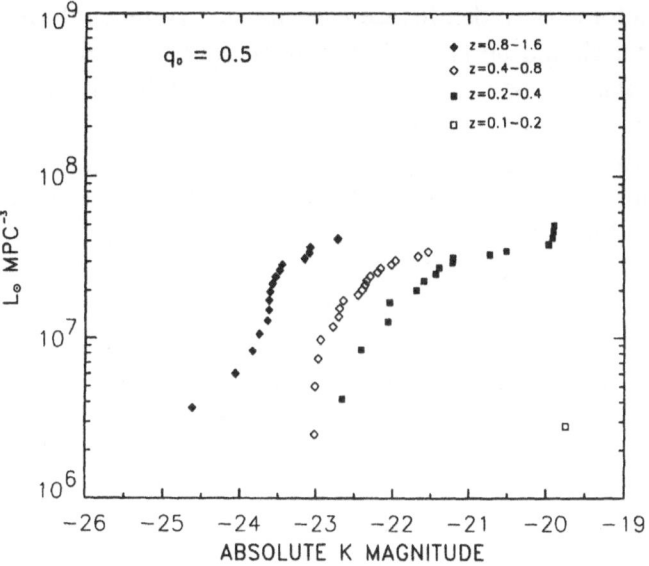

Figure 6.   For caption see next page

---

[1]

$m_{AB} = -2.5 \log (f_v) - 48.595$ (Oke and Gunn 1993). The AB magnitude system (Oke 1974) is defined so that a flux of $3.63 \ 10^{-20}$ erg^{-2} cm s^{-1} Hz^{-1} corresponds to $m = 0.0$ regardless of the wavelength of observation. To convert to Johnson UBVRI system: U (-0.8 mag), B(+0.2), V (0.0), R (-0.3), I (-0.5).

Figure 6. Rest frame K-band light density of blue galaxies as a function of redshift interval for a K ≤ 20.5 sample. The smooth increase in the luminosity of the most luminous forming galaxies with increasing redshift is readily apparent. (Courtesy of Cowie et al., 1996)

The other galaxies, blue nucleated galaxies, have an asymmetric and peculiar morphology and, some of them, appear to be interacting.

Of paramount importance are the spectroscopic studies of the Hawaii Deep Fields (Cowie et al. 1996). These samples definitely rule out the Gronwall and Koo (1995) LF, which is also in disagreement with that by Zucca et al. (1996) mentioned above, and confirm on solid statistical grounds the relatively rapid evolution in the blue selected samples compared to the red selected samples, as previously found in the CFRS, Figure 6.

Recently various authors improved the technique of estimating photometric redshifts using broadband photometry. Such methods are of paramount importance since they allow to estimate redshifts also for galaxies which are beyond the reach of the 8 meter class telescopes (these limits will be overcome, however, as soon as superconducting tunnel junctions will become a fact for observers). Photometric redshifts are essential for getting information of very faint objects but should be taken " cum grano salis" especially in those cases where extrapolation beyond spectroscopic redshifts is based only on evolutionary models.

Among the various excellent recent results it is worth mentioning the work by Connolly (1996), Subbarao et al. (1996), and Sawicki et al. (1996). In this work the correlation between the photometric and spectroscopic redshift is quite good, $\sigma = 0.1 - 0.3$, up to $z \sim 3.5$. It is possible using the HDF to estimate the evolution of the L.F. up to that redshift, Figure 7.

The Luminosity Function fades and flattens at lower redshifts. This flattening accompanied by a moderate fading of the bright end could be an indication of merging and luminosity evolution. Assuming, as it seems to be, that the majority of the HDF galaxies are star forming galaxies, the evolution of the luminosity function could be interpreted as a migration of star-forming galaxies from brighter at high z to fainter at lower z. The less massive the galaxy, the more recently it is undergoing a period of strong star formation. That is the 'downsizing' detected by Cowie et al. in the range $0.2 < z < 1.7$ could cover the whole range of redshifts up to $z \sim 3.5$.

An obvious way to express the evolution of the Luminosity Function as a function of z is, as done by Heyl et al. (1996) to write the parameters as:

$$\varphi^*(z) = \varphi_o(1+z)^{\gamma_\varphi}; \quad L^* = L_o^*(1+z)^{\gamma_L}; \quad \alpha(z) = \alpha_o + \gamma_\alpha z$$

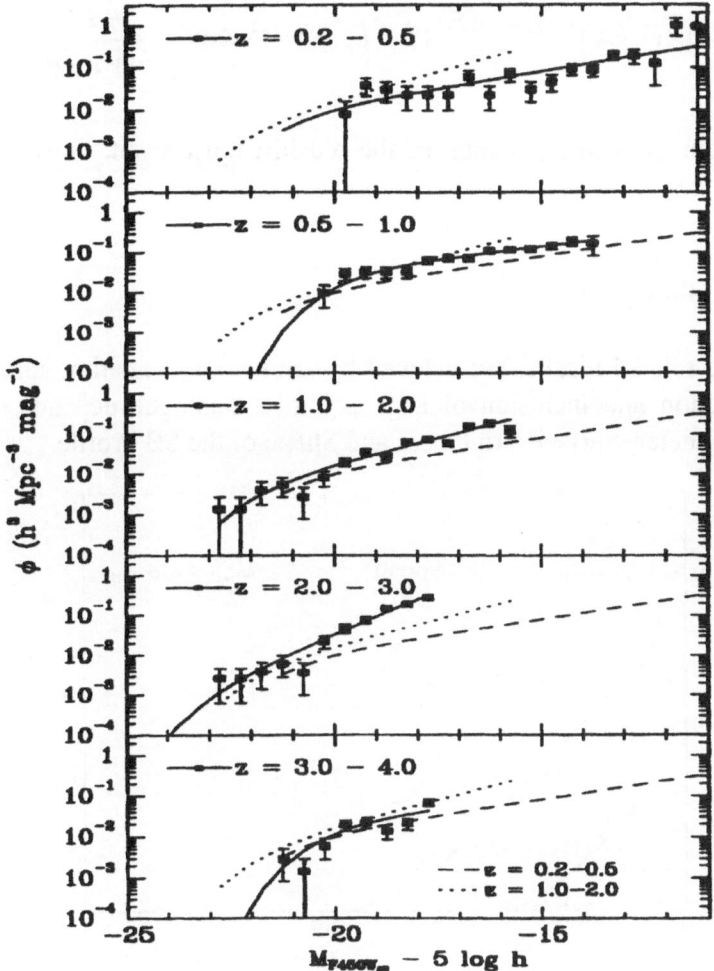

Figure 7. Evolution of the Luminosity Function with redshift (Courtesy of Sawicki,Lind and Yee, 1996). Solid lines are fits to the data using a Schechter function, dotted line is fiducial HDF LF in the range 1 < z < 2 and the dashed line is the HDF LF in the range 0.2 < z < 0.5.

which results in the following analytical expression for the L.F.:

$$\varphi(L,z) = \varphi_o^* (1+z)^{\gamma_\varphi - \gamma_L (\alpha_o + \gamma_a z)} (L/L_o^*)^{\alpha_o + \gamma_a z} \exp(\frac{-L}{L_o^* (1+z)^{\gamma_L}})$$

A firm derivation of such a function of the redshift must await, however, for larger and deeper samples.

3.3   Bias in Catalogues

Catalogues, and related results, are plagued by known and sometime unknown bias. The detection and inclusion of a galaxy in a catalogue depends on its Luminosity, Diameter, Surface Brightness and Shape of the SB Profile.

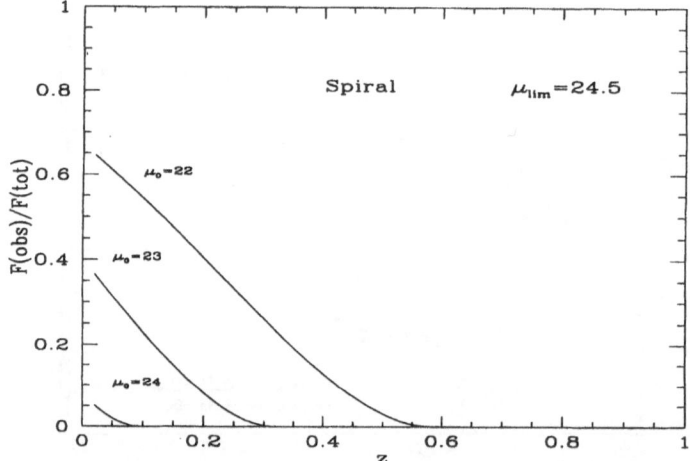

Figure 8. The ratio between the measured and the total flux versus redshift for galaxies of various central S.B. and for a Surface Brightness limit of $\mu = 24.5$

It can be shown, in particular, that the volume of a diameter limited sample is related to the above parameters by the relation, McGaugh (1994), Mc Gaugh et al. (1995):

$$V \propto d^{3_{max}} \propto (a/\theta_{min})^3 (\mu_{lim} - \mu_o)^3$$

where a is the scale length for a disk dominated galaxy, μ_o is the central surface brightness, θ is the minimum diameter of the survey and μ_{lim} the limiting SB of the survey. This is generally B ~ 24 mag/sqarcsec or, for deep surveys, B ~ 28 mag/sqarcsec. For a Flux limited sample (m_{lim}) the maximum Volume of the survey can be written as:

$$V \propto d^3{}_{max} \propto L^{\frac{3}{2}} \propto a^3 \, 10^{-0.6\mu_o} \, f(x_{min})^{\frac{3}{2}} \; ; \; x = \frac{r}{\alpha} = (\mu_{lim} - \mu_o) \frac{1}{1.086}$$

where α is the angular scale length, μ the isophotal limit above which we measure the flux and the other symbols as for the above relation. In Figure 8 we show how the ratio between the measured and the total flux of galaxies changes as a function of z. In Figure 9 we show the highest redshift at which an E or S galaxy with central surface brightness (SB) μ_o can be detected in a survey with limiting SB $\mu = 24.5$ mag/sqrcsec.

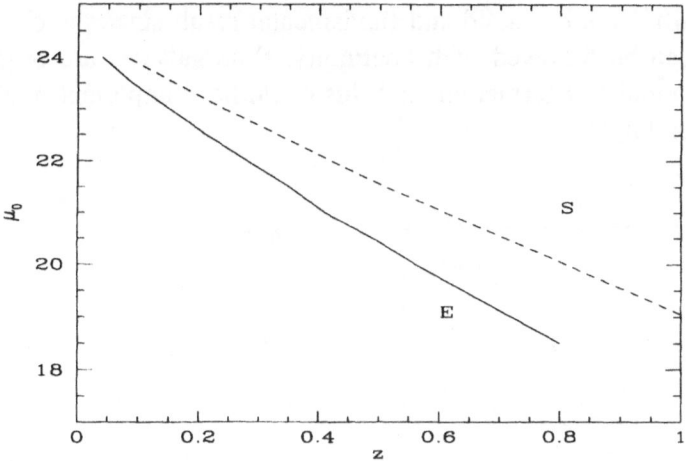

Figure 9. ' Visibility' for E (continuos line) and S (dotted line) galaxies of the same central surface brightness for a sample with $\mu_{lim} = 24.5$.

Taking into account that any survey is limited also in surface brightness we can draw the following conclusions: a) really complete catalogues of galaxies limited by total flux hardly exist, these are at most a reasonable compromise, b) in general magnitude limited samples detect more galaxies than diameter limited samples because they include very distant and intrinsically luminous objects. These catalogues are strongly dominated by L^* galaxies, c) diameter selection should yield local samples which are less biased and more representative of the general field population, d) at high redshift we have an higher probability of detecting spiral galaxies.

4. Search for Distant Galaxies

4.1   The Fairly Distant Galaxies

As shown in Figure 1, recent surveys of galaxies in the optical and near infrared allow detection of objects up to a redshift z ~ 1.7. The Hawaii Deep Fields show an extended tail, up to z ~ 1.7, in the redshift distribution where blue (B-I) galaxies show strong [OII] emission. Selecting subsamples at progressively higher redshift Cowie et al. observe that brighter galaxies tend to show large [OII] equivalent width, an indication that by going back in time brighter galaxies show sign of star formation. It appears that at the present time there is almost no galaxy formation, but that as recently as z ~ 0.2, low mass galaxies were forming, Figure 10. Progressively higher luminosity galaxies are seen with strong star formation at higher redshifts, so that more massive galaxies form at higher redshift. This is a solid and fundamental result since the changes in the population can be followed with continuity. Blue galaxies are brighter in the past, have a rapid star formation, and this could have important implication for the background light.

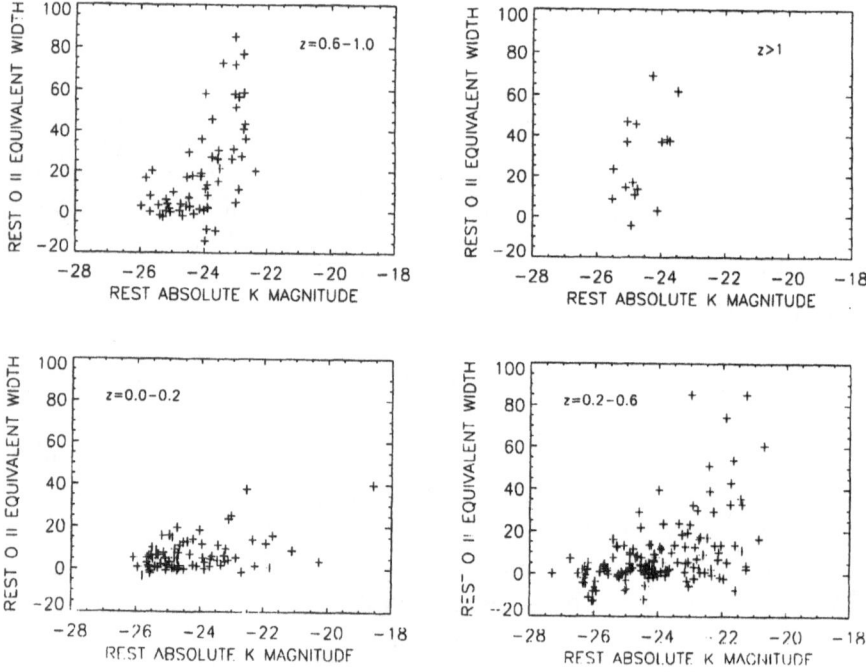

Figure 10. Rest - frame [OII] equivalent width versus rest absolute K magnitude and redshift for the Hawaii K < 20 sample. Progressively, from low to high

redshifts, more massive galaxies are undergoing rapid star formation
$(E\bar{W}([0II]): \geq 25 \, A \, ?)$. *Courtesy by Cowie et al. (1996).*

4.2   The Medium and High Redshift Galaxies

The sample selection uses mainly three types of approach:
- search of objects with emission in Lyα. The guiding idea is that a distant and therefore rather young, galaxy should present high rate of star formation and look, in its spectroscopic characteristics, rather similar to an HII region. The selected fields are often chosen about a QSO or Radio-Galaxy of known redshift. In addition to the reasonable assumption of correlation of clusters with QSO or Radio - galaxies, this approach will obviously select only Lyα galaxies with little or none internal absorption, that is a particular type of object will be selected rather than a complete population at high redshift,
- search of objects absorbing the light of a background quasar. The method, which has been shown to be very efficient, select only objects which are along the line of sight of quasars and allow, in spite of its limitations, very important information both on the halo or disk of the intervening galaxies and, at the same time, allow statistical studies on the Large Scale distribution of galaxies;
- photometric searches using either high resolution imaging, HDF for instance, or colour selection followed by spectroscopic follow up of the candidates. This has been particularly efficient when the colour selection has been based on the redshift of the Ly break and has the big advantage of selecting a population of galaxies in a predefined range of redshift.

4.2.1   Lyα Galaxies

The method is not very effective due to dust suppression. However it is an excellent complement for detecting objects with faint continuum and without much dust. To name only one among the most recent results we refer to the detection of galaxies at a redshift of z ~ 4.55 by Hu and McMahon, (1996) and Hu et al. (1996), Figure 11, (see also Fontana et al., 1995).

4.2.2.   QSO Line Absorption Systems

For an ecellent review on the subject see Lanzetta (1993). Bergeron and Boisse (1991) showed that the use of the Mg II doublet (2796,2803) was an excellent probe for detecting galaxies on the line of sight of the quasars. Out of 13 objects examined they were able to identify 11 objects with a mean redshift <z> ~ 0.4. The high success rate established the validity of the method and supported a large cross section of galaxies: $R_{gal} \sim 30$ h^{-1} kpc, h=H/100. Selection criteria have been based also on the detection of galaxies at the redshift of the high ionisation lines of quasars and damped Lyα. Djorgowsky et al. (1996), in such a way, detects a galaxy at z = 3.153 whose total mass and estimated mass in gas are that of a typical normal disk galaxy today, Figure 12.

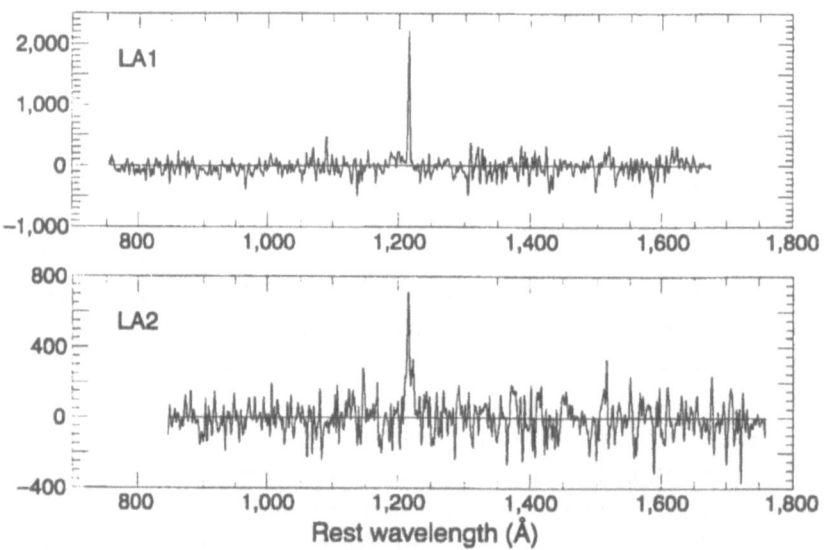

Figure 11. Galaxies detected by Lyα emission at a redshift z ~ 4.55 using the LRIS spectrograph on the Keck telescope. Courtesy of Hu and McMahon (1996).

4.2.3   High Resolution Imaging and Colour Selection

Deep Imaging allows selection of faint galaxies, intrinsically faint objects which are nearby and bright objects which are very distant; the expected distribution in redshift can be easily derived assuming the Luminosity Function is well known. The HST, with its excellent resolving power, has the capability to resolve a normal galaxy at any redshift.

The HST Medium Deep Survey (MDS), Phillips et al. (1995), bright sample deals with about 100 faint galaxies with I < 20.5, corresponding to B ~ 22-23, and shows that the apparent size and distribution of the galaxies is roughly consistent with a non-evolving galaxy population observed at a median redshift of z ~ 0.3. Size and luminosity are correlated as observed in local samples. This result implies that there is no evidence for substantial size and luminosity evolution of galaxies brighter than about on tenth of L* to redshift of about z ~ 0.4.

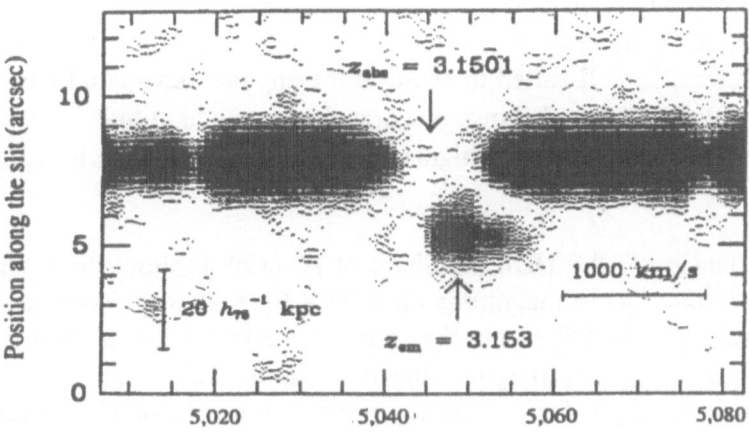

Figure 12. Reproduction of the spectroscopic frame showing the Ly$_\alpha$ line emission from the DLA 2233+131 galaxy and the associated absorption in the quasar spectrum. (Courtesy of Djorgovski et al., 1996)

From the more extensive faint sample, Casertano et al. (1995), which consists of 13 500 objects in the magnitude range $18 < I < 25$, 70 % of which are galaxies, the authors find that : a) in the range $18 < I_{785} < 22$ mag the number of galaxies exceed by about 50% the counts based on ground observations, b) a large fraction of small angular size galaxies is present when compared to conventional evolutionary model. The median half light radius is of 0.27" for 22 $< I < 23$ galaxies, c) Ellipticals and Spirals bifurcate in color. Elliptical tend to be redder at fainter apparent magnitude, d) the angular size of disklike galaxies in the magnitude range $19 < I < 22$ is about a factor two larger than the angular size of bulgelike galaxies, e) small angular size objects do not show a bluing, V_{555} - I_{785} color, trend, f) the correlation function, w(θ), is well represented by a power law with a significant excess of galaxy pairs on scales < 4", or 12/h kpc at a median redshift $<z> \sim 0.5$ of the sample. Also these data rule out extreme merging models, g) the amplitude of the correlation function increases moderately rapidly with decreasing median redshift which could be due to weak clustering of dwarf galaxies which are dominant at $I > 22$, or to a mild galaxy evolution in a low density Universe with strong clustering, h) bulge dominated galaxies correlate stronger than later type galaxies. This is also is in agreement with what happens in the local Universe, Giovanelli et al. (1986).

Similar results on clustering evolution have been obtained by Iovino et al. (1995). These authors deal with a sample of 200 000 objects distributed over a field of about 150 sqdegrees. About 7000 distant early galaxies are selected by color, $m_j < 18.9$, $B_j-R > 2.15$ and $B_j-I > 2.95$. A first set of spectroscopic observations, about 200 objects, shows that the contamination by stars is smaller than 5%, while the contamination by late type galaxies is of about 10%; the redshifts are in the range $0.3 < z < 0.55$ with a median of $<z> \sim 0.4$. The sample, therefore, is well suited to estimate the angular and spatial correlation functions, $w(\theta)$ and $\xi(r)$, and shows that $\xi(r)$ changes in amplitude by a factor $\sim 2.6 \pm 0.5$ (smaller amplitudes at higher redshift) and maintains, however, the same slope.

The superb images of the HDF, Williams et al. (1996), allow fairly accurate morphology to the faintest magnitude limit, $21 < I_{814} < 25$ mag. The analysis by van den Bergh et al. (1996) shows that, (see also Abraham et al. 1996): a) the fraction of interacting and merging objects is significantly higher in the HDF that it is among nearby galaxies. Spectroscopic data will be however needed in order to disentangle merging phenomena from projection effects. These statistics, however, should be taken with some caution and corrected for possible bias. Giavalisco et al. (1996), in fact, using simulations and Ultraviolet Imaging Telescope data show that local galaxies observed in the UV and transferred to cosmological distances have morphologies that appear to be of later type or more irregular than their local (optical) counterparts. Some of these are in qualitative agreement with images revealed by the faint HST survey. Thus in order to fully understand the morphologies at high z , we should know the UV morphology of galaxies in the local Universe. This would be an excellent goal for a small UV (< 3000 A°) telescope to be flown on a small medium satellite program. An other word of caution comes from the work by Colley et al. (1996). These authors point out that some of the faint galaxies observed could be rather knots of the same and larger galaxies. The difference is only partly semantic since the interpretation would affect the estimate of galaxy counts. Going back to the work of van den Bergh et al. (1996), their results show also that: b) Early type galaxies are as frequent as in the Shapley - Ames catalogue, that is to say that elliptical galaxies were already well in place at the median depth of the HDF imaging, c) galaxies resembling the archetypal grand-design late type spirals are very rare. Barred spirals are essentially absent from the deep field except for a possible detection. Taken at face value it means that such galaxies form late, d) number counts increase considerably, at least up to I

= 25, for irregular, peculiar and merging galaxies and the distribution of objects is strongly skewed toward highly asymmetric objects. This confirm also previous results, Glazebrook et al., (1995b) and many others.

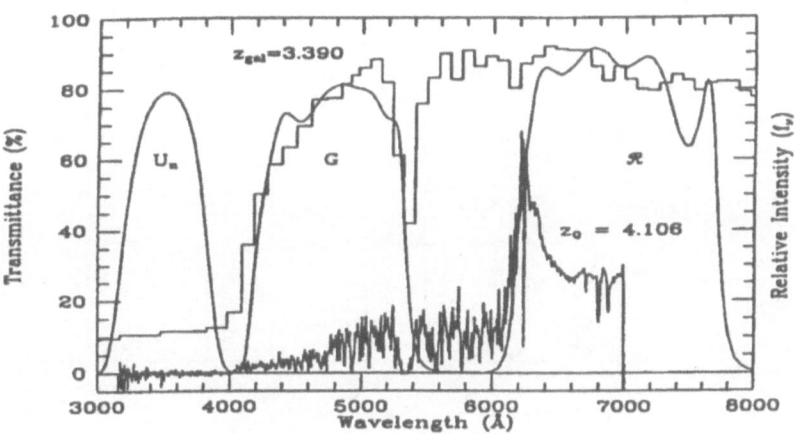

Figure 13. Plot illustrating the colour selection of high redshift objects using three broad bandpasses. The low dispersion spectrum of the quasar is due to Sargent et al. 1989. Courtesy of Steidel and Hamilton (1992).

Given the high surface density of galaxies on deep exposures it is essential, before attempting spectroscopic observations, to select subsamples of galaxies which optimize the science goal to be pursued. Steidel and Hamilton (1992) showed that broad band photometry is a powerful tool in detecting galaxies in the redshift range where the colors enable detection of the flux decrement across the rest frame of the Lyman limit of the galaxy, Figure 13.

A refinement and detailed discussion of the method is given by Madau et al. (1996). The Lyman break color criterion is applied to the HDF which had been observed in four passbands: F300W, F450W, F606W and F814W. These wavelengths allow detection in the redshift range $2 < z < 3.5$ and in the redshift range $3.5 < z < 4.5$. A large sample of candidate galaxies in the first redshift bin has been confirmed spectroscopically with the Keck telescope. The candidates sample is not affected by contamination of galaxies at low redshift. We must always remember, however, wysiwyg (what you select is what you get).

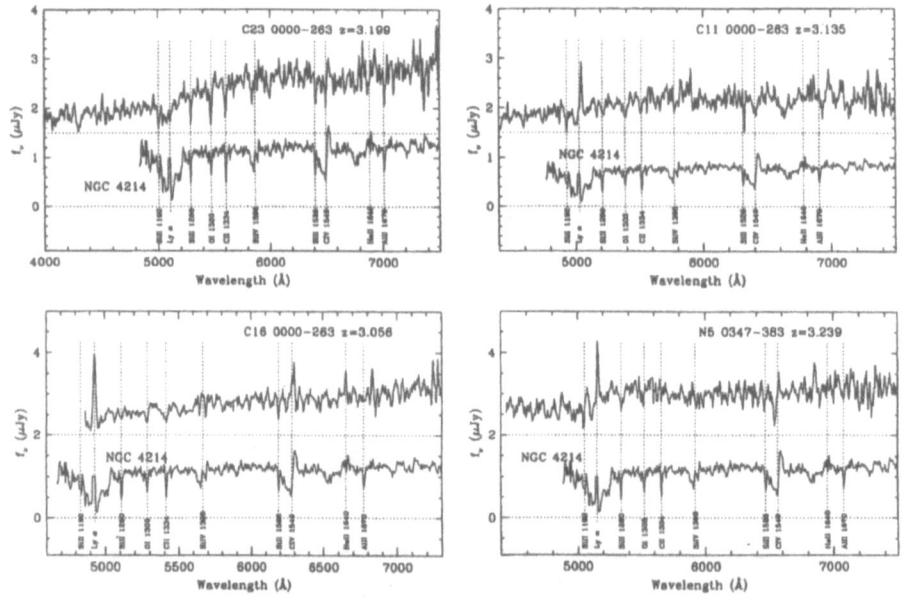

Figure 14. Sequence of Keck spectra for galaxies with z > 3. For comparison the spectrum of the nearby star forming galaxy NGC 4214 is plotted at the bottom of each panel. Courtesy of Steidel et al. (1996).

The high redshift galaxies, $z \sim 3$, detected by Steidel et al. (1996) have in general compact cores, 0.2" - 0.4" i.e. 1.5 - 3 h_{50}^{-1} kpc for $q_0 = 0.5$ with scale comparable to the cores of present-day luminous galaxies. In most cases they are relatively "regular " in their appearance but surrounding by irregular "halos". The bulk of the star formation at high z is occurring in very compact regions of very high surface brightness. The spectra are similar to local examples of starburst galaxies. They present a flat continuum, rather strong interstellar absorption and fairly low dust content. Lyα is weak or absent and the prominent high ionisation lines, HeII, CIV, SiIV, NV, CIV 1549, SiIV 1393,1402 are observed to be somewhat weaker than in present day starburst galaxies, Figure 14. This could be due to a smaller metal abundance in the high redshift galaxies, and needs to be carefully estimated in different samples. The far UV continuum is equivalent to that produced by about 2 - 6 10^5 O7 stars. The approximate internal velocity dispersion are $\sigma \sim 180$ - 320, this means that if the motion is due to gravitation then the mass implied are comparable to those of present day luminous galaxies. This is a result similar to that of Djorgovski et al., (section 4.2.2) and shows that galaxies were already formed, or in advanced phase of formation, at $z \sim 3$.

The surface density of Lyα break candidate is 0.40 ± 0.07 arcmin. That is for $3.0 < z < 3.5$ these consist of 2% of the objects with magnitude $23.5 < R < 25.0$. The star formation rate estimated is of about 4 to 25 h_{50}^{-2} $\mathcal{M}_o$ yr^{-1} (q = 0.5) and a mean value of about 8.5 h_{50}^{-2} $\mathcal{M}_o$ yr^{-1}. The comoving density of star forming galaxies is estimated to be 3.6 10^{-4} h_{50}^{3} Mpc^{-3} (q = 0.5) or about 1/2 (1/10) the space density of present day galaxies with $L > L*$ so that we can derive a total formation rate per comoving volume (at z > 3) of 3.1 10^{-3} h_{50} $\mathcal{M}_o$ yr^{-1} Mpc^{-3} (q = 0.5).

Again the most immediate and important conclusion is that galaxy formation was well underway by $z \sim 3.5$.

4.2.4   Distant Galaxies: Summary

While essential details are needed, the most recent observations and related simulations and modelling, are evidencing the gross picture. Star formation in galaxies is very active since $z \sim 4$, the more massive the galaxy the earlier, high z, the star formation; dwarf galaxies are bursting stars now. The Luminosity Function evolves due to both luminosity and number density galaxy evolution and the number counts at different wavelenghths can not be complitely accounted for invoking a rising faint tail luminosity function.

Red galaxies are well in place even at high redshift and do not show sign of evolution. Indeed it is very clear from the work by Cowie et al. that the luminous density of blue objects change considerably in the range $0.1 < z < 1.6$ while the luminosity function does not change much for red objects. Both Lyα and Ly break selected objects evidence that up to a redshift $z \sim 3$ we detect objects with a mass and spectroscopic characteristics which are very similar to those of present time normal galaxies. Finally galaxies are detected at $z \sim 4.5$ which means that the real action started at $z \geq 5$. At some point we should see only star forming galaxies and clearly identify the progenitors of Ellipticals and of the red population of objects. The cosmic time at $z \sim 5$ is of the order of the star evolution time. A census of the population which is present at high redshift is needed.

Even if Elliptical galaxies have a different history than Spirals we expect some evolution in morphology, indeed we witness it in the HDF even if we do not have yet enough clues about the sequence. Morphological changes and soft merging, and/or relaxation and merging of subclumps, must affect counts and the number density. These effects need to be quantified in conjunction with an accurate determination of the abundances and of the role played by the SN events as a function of cosmic time. In this context the finding by Songaila and Cowie (1996) on the metallicity of the high redshift absorbing clouds (Lyα Forest) are of primary importance. Finally the big picture should explain the

22

relation, if any, between the evolution of quasars, density peak at z ~ 2.5, and galaxies.

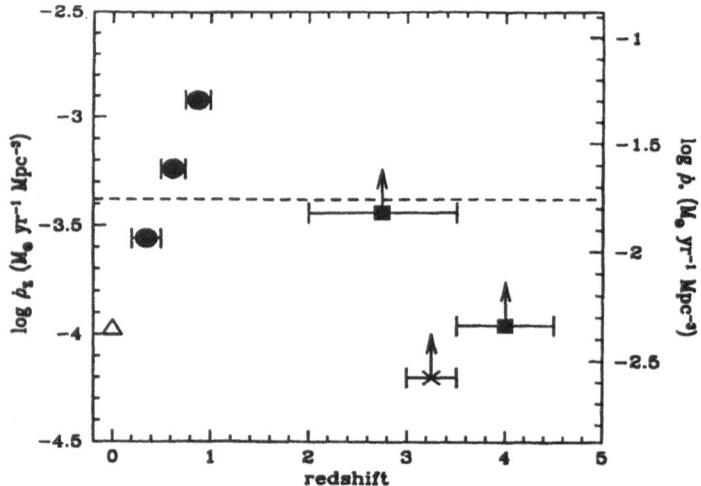

Figure 15. Element and star formation history of the Universe. $\dot{\rho}_z$ is a measure of the universal metal ejection density as a function of redshift. Courtesy of Madau et al. (1996).

The knowledge of the star formation as a function of cosmic time will shed light on global effects: the eventual contribution to the ionisation of the intergalactic medium , the contribution to the UV sky surface brightness and the universal metal ejection rate. Cowie (1996) shows that " forming ' galaxies contribute largely to the extragalactic background light and that we are observing the bulk of galaxy formation. As for the star formation history of field galaxies we must refer to the perused figure by Madau et al. (1996), Figure 15 (see also Figure 10 in Sawicki et al. 1996), which clearly shows that the bulk of the present day stellar population was assembled in the epoch 1 < z < 2.3 as well as most of the metals observed today. This is a range in redshifts where detailed information are still lacking and must await both new near infrared instrumentation coadiuvated by optical-uv imaging. In a more general context the history of star formation would greatly benefit from far UV observations of a local sample.

5.   Clusters of Galaxies

5.1   Preliminaries

Evidence that galaxies evolve came first from observation of clusters of galaxies. Butcher and Oemler (1978) and later developments, as in Dressler and

Gunn (1982,1983), detected a progressively higher percentage of blue galaxies, active and star- forming or post-starburst galaxies in increasingly distant clusters. An account of cluster evolution, mainly from the point of view of member galaxies, was recently given by Dressler in the 37th Herstmonceux Conference (1996). On the basis of a sample of about 2000 galaxies in 11 rich clusters in the redshift range $0.36 < z < 0.55$ it is remarked that:

- The colour dispersion of bona-fide elliptical galaxies show a rather small spread suggesting an early formation epoch for cluster ellipticals.
- Ellipticals are as abundant at $z \sim 0.5$ as today and well located in the dense regions.
- SO galaxies are less plentiful and less concentrated when compared to the present epoch
- Spirals are everywhere more abundant.
- Very actively star forming galaxies are often found to be late type disk galaxies with disturbed or peculiar forms.
- post-starburst galaxies are more frequently disky as opposed to spheroidally dominated.

The suggestion is that the observed rapidly evolving systems are not likely to become ellipticals further supporting the model of an early genesis of elliptical galaxies. The problem of understanding how a spiral could turn into an SO, or be destroyed, is still open.

Any study of clusters of galaxies must obviously be based on catalogues as complete as possible. These are needed to carry out statistical studies on the distribution of clusters at various z, to estimate how global cluster properties change as a function of cosmic time and to allow detailed studies on cluster members on the base of complete cluster samples.

These catalogues, however, are very hard to compile and the information, especially at moderately high redshifts, are largely lacking. Cluster surveys are extremely important also in view of some of the results which came out, at a rather low statistical significance however, of the Einstein Medium Sensitivity Survey (EMSS). It seemed that cluster at redshift higher than about 0.4 - 0.5 were practically absent and that was invoked as due to very fast evolution at least in the X-Ray band. This uncertain result has been later either accepted at face value or " confirmed " on rather ill defined, or preliminary, cluster samples.

We will briefly describe the present situation on catalogues and mention some of the surveys which are presently going on. In doing this we will in particular stress some of the limitations of the EMSS at faint fluxes not to deny the possibility of evolution, but to evidence the limitations of the data and encourage researcher to further work in surveys capable of detecting distant clusters.

5.2   Catalogues

5.2.1   Radio

Following a suggestion by Burns et al. (1993) Blanton et al. (1996) are searching clusters at the location of radio galaxies presenting a bent double morphology. Indeed a dense intracluster medium, as present in clusters of galaxies, make the lobes of an embedded radio galaxy appearing bent or pushed back as the galaxy moves through the medium. Out of 138 000 radio sources detected in the FIRST survey, more than 12 000 are double or multiple sources and the authors have selected 450 objects with bent double morphology where to look for clusters. The work is in progress and the two brightest cluster candidates are indeed clusters at a redshift of $z = 0.26$. As soon as the catalogue will be complete it will be very interesting to correlate it with optical and X-ray catalogues since in this survey the requirements are: a) presence of an intracluster medium and b) presence of a radio galaxy.

5.2.2   Optical

It is well known that the main limitation in the optical detection of clusters of galaxies is the contrast between the cluster, especially if it is a poor cluster, and the background/foreground field galaxies. A magnitude limited sample will detect, in distant clusters, only the luminous tip of the luminosity function which is progressively embedded in an ever increasing background of objects $\propto 10^{0.6\,m}$, see Cappi et al. (1989) and Postman et al. (1996) for a detailed discussion. Catalogues are then plagued by projection effects, completeness which differs from catalogue to catalogue and depends mainly from the method and algorithm used to detect clusters and often from the background subtraction. As a consequence the cluster density, which is obviously an important cosmological parameter which we should try to estimate at various z, is poorly known. It is uncertain even locally. Using clusters with R > 0 of the Abell and ACO catalogues Scaramella et al. (1991) estimate $8.7*10^{-6}$ (h^{-1} Mpc^{-3}) and $12.5*10^{-6}$ (h^{-1} Mpc^{-3}) respectively, Postman et al. (1992) $12*10^{-6}$ (h^{-1} Mpc^{-3}) for R $\geq$ 0, Bahcall (1979, 1988) and Bahcall and Soneira (1983) 11.3 (R=0), 4.04 (R=1),

1.28 (R=2) and 0.274 (R≥3) $*10^{-6}$ (h^{-1} Mpc^{-3}). The EDCC estimate, Lumsden et al. (1992), which is based on different data gives, for $R \geq 0$ 10 - 15 $*10^{-6}$ (h^{-1} Mpc^{-3}). Note that such different estimates, however not too largely different among them, are based on the same catalogues and should possibly be confronted with estimates obtained in other wavelengths, especially in the X-Ray band and possibly using weak lensing surveys, obtained in a completely independent way.

A cluster Luminosity Function is practically unknown at optical wavelengths.

Among the deepest surveys we refer to that by Gunn et al. (1986) detect clusters up to $z \sim 0.5$ with an estimated density of 11 clusters/sqdegree, and to that by Couch et al. (1991) and Postman et al. (1996). Clusters of galaxies at a redshift somewhat larger than $z \sim 1$ have been detected in this last survey and spectroscopic follows up is in progress.

In collaboration with ESO the European astronomical community decided, in order to overcome some of this deficiencies and have a catalogue to point the VLT, to survey an area of about 30 square degrees with the main goal of searching clusters of galaxies at a redshift of $z \sim 1$.

At very high redshift we have practically no information except for a few indications near quasar fields. Malkan et al. (1996), for instance, using narrowband infrared imaging, see also Djorgovski et al. (1996), detect a pair of galaxies at $z \sim 2.5$ near the quasar SBS 0953+545. From the small field searched and the I-K color of the other galaxies in this field they conclude that a cluster at that redshift could have been detected. Similarities exist, according to the authors, with the cluster of subgalactic clumps detected by Pascarelle et al. (1996) near the radio galaxy 53W002 at $z \sim 2.39$. While future observations will certainly show whether we are detecting clusters at the time of formation, we certainly need to have informations at intermediate z in order to observe the various phases of evolution.

The cluster X-ray emission is easily detectable over the x ray background emission which in the soft band (0.5 - 2 KeV) amounts to $\sim 1.62 \ 10^{-15}$ erg s^{-1} cm^{-2} KeV arcmin^{-2} . However in order to carry out a deep survey over a reasonably large area of the sky the telescope must have: 1) rather large field of view, about 1 degree, 2) good resolution over the whole field of view, about 5 - 10 arcsec at half power diameter and 3) reasonable high sensitivity, effective area about 300 cm^2. These requirements have not been met yet in any flown telescope and this is the reason why we do not have yet a proper catalogue. Hopefully we will be able to fill this urgent need soon since now the technology for such a survey is at hand.

The Einstein and the ROSAT satellites, however, allowed fundamental progress in this field. Most of the ROSAT based survey are still in progress and we will refer briefly to some of them later. As for the Einstein data the perused cluster catalogue, Henry et al. (1992), was generated by the EMSS (Einstein Medium Sensitivity Survey). The redshift distribution of clusters is given in Figure 16.

The deficiency of clusters in the range $0.4 < z < 0.7$ could be simply due to low statistics. The sky coverage at low fluxes is very small, about a factor 10 smaller than that at large fluxes and it could cause relatively high statistical fluctuations.

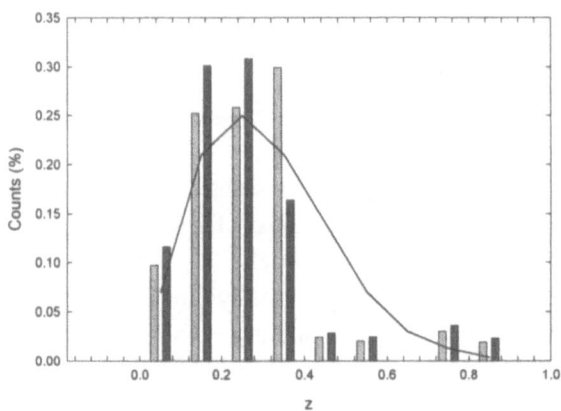

Figure 16. Distribution in redshifts of the EMSS sample. The continuous line shows the expected distribution according to the Luminosity Function used. Counts are here given in percentage to ease a detail account of the sky coverage. See text.

For instance by cutting the limiting flux of the catalogue slightly higher, we simply eliminate the last bin and pass from $f_{lim} = 1.33 \bullet 10^{-13}$ to $f_{lim} = 1.61 \bullet 10^{-13}$ erg cm^{-2} s^{-1}, obtaining the dark histogram of Figure 16 which has, compared to the grey histogram where all the data have been graphed, a difference of 50% in the counts in the bin at $z = 0.3$.

The distribution of EMSS clusters of galaxies in the redshift - Luminosity plane is represented in Figure 17. The number of objects in the interval ΔL, Δz is given by the relation:

$$N(L_i, L_j; z_k, z_l) = \int_{L_i}^{L_j} \varphi(L)\Omega(f = L/{4\pi D_L^2})dL \int_{z_k}^{z_l} dV/{4\pi}$$

where: N is the number of objects, $\varphi(L)$ the Luminosity Function, $\Omega(f)$ the sky coverage at different flux limit, D the luminosity distance and dV the volume in the interval z over 4π.

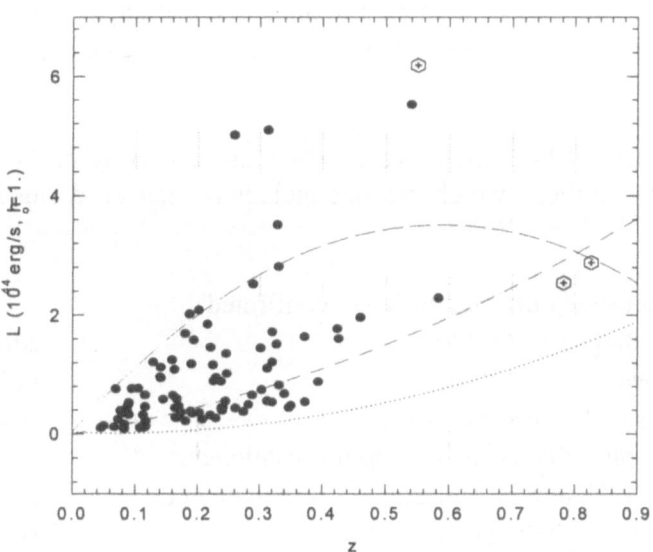

Figure 17 . Distribution of the EMSS clusters of galaxies in the Luminosity - redshift plane. Open circles with central cross refer to the clustres for which the redshift has been obtained later by Lupino and Gioia (1995). The bottom dotted line marks the sensitivity limit of the survey as a function of redshift, the top line, long dashed, marks the probability of detecting a cluster with a Luminosity higher than that defined by the line in that redshift bin, the line in between the two, dash-dot-dot, define the luminosity above which, for each redshift bin, 50% of the clusters are expected or, equally, 50% of the clusters are expected to be between this line and that defining the sensitivity limit. The number of clusters expected in each redshift bin are: 19 (bin 0.1-0.2), 22 (bin 0.2-0.3), 19 (bin 0.3-0.4), 14 (bin 0.4-0.5), 9 (bin 0.5-0.6), 5.4 (bin 0.6-0.7), 2.9 (bin 0.7-0.8), 1.5 (bin 0.8-0.9). For this graph the Luminosity function used has been normalised to the number of clusters in the bin 0.2 - 0.3. That is the Luminosity Function used is: $6.0*10^{-7}$ Exp (L (units 10^{44})/1.34) (L (units 10^{44})/1.34)$^{-1.11}$.

Here again is evidenced that the number of clusters detected at z > 0.4 is less than those expected and however we must keep in mind a) the rather small statistics, b) the presence of possible bias related to the limited space resolution of the telescope and c) the uncertainties in the expectation due to a yet poor knowledge of the local cluster luminosity function. In conclusion:

- The lack of clusters can not be taken as evidence of evolution and could be a fluke of the statistics. The eventual lack of X-ray clusters at medium-high redshifts has yet to be proved with reasonable confidence. Better samples are needed.
- More interesting perhaps is the evidence, Figure 17, that unusually bright clusters, have been detected at rather high redshift. Extremely bright clusters, not accounted for properly by the Luminosity Function, exist also at medium high redshift, Shindler et al. (1995) , and these might give important clues on evolution.

Following Rosati (1996), the surveys which are underway in the X-ray band, (we apologise for those which are not included) can be divided, into three categories:

1. Optically selected clusters and X-ray confirmed:
    - Ebeling et al. (1995)   survey data
    - Bower et al. (1994)   pointed data
    - Castander et al. (1994)   pointed data
2. Hybrid damples, that is X-ray + optical catalogues:
    - ESO KP & SGP subsample   Boeringer et al.   survey data
    - XBCS , Ebeling et al. (1996)   survey data
    - SGP, Cruddance et al. (1995)   survey data
    - Crawford et al. (1996)   survey data
3. Pure X-ray selected
    - NRASS, Giacconi et al.   survey data
    - NEP, Henry et al.   survey data
    - RIXOS, Castander et al.   pointed data
    - RDCS, Rosati et al.   pointed data
    - etc.

It is out of the scope of this work to discuss the preliminary results of these surveys. We simply mention that the analysis of a bright ESO KP subsample (SGP) is now completed and the local luminosity function is well determined and will be soon available, De Grandi et al. (1997) and De Grandi (1995). Excellent results are coming out of the RDCS sample which, thanks to the wavelet technique of detection, reach the flux limit, in the range 0.5 - 2.0 KeV, of about $1*10^{-14}$ erg cm^{-2} s^{-1}.

The conclusion is that a massive investment has been done in order to understand clusters of galaxies, their evolution and the evolution of member galaxies. The theory in this field, and especially numerical and hydrodynamics simulations, give extremely important guideline showing a) how dependent is cluster evolution on Ω, b) how substructures are expected to evolve with z and c) how the Luminosity or Temperature function is expected to evolve as a function of Ω and their relation to primordial fluctuations. In particular the theory and the simulations point out to a fast evolution and indicate that at high $z \sim 2$ we probably will be able to detect only subclumps of forming clusters. It is because of these indication that the only way to understand what is going on is to fly an X ray telescope with the resolution and sensitivity needed to detect clusters in their various phases of evolution. This is now feasible and of paramount cosmological importance.

Acknowledgements: We would like to thank Gill Harrison for making available copy of the Conference Guide of the 37th Herstmonceux Conference " HST and the high redshift Universe".

6. References

Abraham R. G., Tanvir N. R., Santiago B. X., Ellis R. S., Glazebrook K. & van den Bergh S. 1996, MNRAS 279, L47

Bahcall N. A. 1979, ApJ 232, 83

Bahcall N. A. 1988, ARA&A 26, 631

Bahcall N. A. & Soneira R. M. 1983, ApJ 270, 20

Bergeron J. & Boissè P. 1991, AA 243, 344

Blanton E.L., Helfand D.J., Becker R.H., Gregg M.D., White R.L., 1996, 187th AAS Meeting, San Antonio, Texas

Butcher H. and Oemler A., 1978, Ap.J. 219, 18

Cappi A., Chincarini G., Conconi P., Vettolani G., 1989, A&A 223, 1

Casertano S., Ratnatunga K.U., Griffiths R.E., Neuschaefer M. Im and L. W., 1995, Ap.J. 453 599

Colless M., Ellis R. S., Taylor K. & Hook R. N. 1990, MNRAS 244, 408

Colley W. N., Rhoads J. E., Ostriker J. P. & Spergel D. N. 1996, preprint

Couch W. J., Ellis R. S., MacLaren I. & Malin D. F. 1991, MNRAS 249, 606

Cowie L. L., Songaila A. and Hu E. M. 1991, Nat. 354, 460

Cowie L. L., Songaila A. & Hu E. M. 1996, AJ 112, 839

da Costa L.N., Geller M.J., Pellegrini P.S., Latham D.W., Fairall A.P., Marzke, R.O., Wilmer C.N.A., Huchra J.P., Calderon J.H., Ramella M., Kurtz M.J., 1994, Ap.J. 424, L1

De Grandi S., 1995, PhD Dissertation at the Universita' degli Studi di Milano.

Djorgovski et al. 1995, ApJ 438, L13

Djorgovski S.G., Pahre M. A., Bechtold J. & Elston R., 1996, Nature, 382, 234

Dressler A. and Gunn J.E., 1982, Ap.J. 263, 533, 1983, Ap. J., 270, 7

30

Ellis R. 1993, in'The Environment and Evolution of Galaxies', eds. J.M. Shull and H. A. Thronson, Kluwer Academic Publishers

Fontana A., Cristiani S., D'Odorico S., Giallongo E. and Savaglio S., 1996, M.N.R.A.S. 279, L27

Gardner J. P., Cowie L. L. & Wainscoat R. J. 1993, ApJ 415, L9

Garilli B., Saracco P., Chincarini G. & Iovino A. 1995, Sesto- Pusteria workshop. (1997 Astrophysical Letter and Communications, in press)

Giovalisco M., Livio M., Bohlin R. C., Duccio Macchetto F. & Stecher T. P. 1996, AJ 112, 369

Giovanelli R., Haynes M.P. and Chincarini G., 1986, Ap.J. 300, 77

Glazebrook K., Ellis R., Colless M., Broadhurst T., Allington-Smith J. & Tanvir N. 1995a, MNRAS 273, 157

Glazebrook K., Ellis R., Santiago B., Griffiths R., 1995b, MNRAS 275, L19

Gronwall C. & Koo D. C. 1995, ApJ 440, L1

Gunn J. E., Hoessel J. G. & Oke J. B. 1986, ApJ 306, 30

Heyl J., Colless M., Ellis R. S. & Broadhurst T. 1997, MNRAS in press

Henry J.P., Gioia I.M., Maccacaro T., Morris S.L., Stocke J.T. and Wolter A., 1992, Ap. J. 386, 408

Hu E.M., McMahon R.G., 1996, Nature 382, 231

Hu E. M., McMahon R. G. & Egami E. 1996, ApJ 459, L53

Iovino A., Warren S.J., Hewett P., Shaver P.A., 1995, Sesto- Pusteria workshop. (1997 Astrophysical Letter and Communications, in press)

Koo D. C. and Kron R. G. 1992, ARAA 30, 613

Jones L. R., Fong R., Shanks T., Ellis R. S. & Peterson B. A. 1991, MNRAS 249, 481

Lanzetta K. M. 1993, in'The Environment and Evolution of Galaxies', eds. J.M. Shull and H. A. Thronson, Kluwer Academic Publishers

Lilly S. J. 1993, ApJ 411, 501

Lilly S. J., Cowie L. L. and Gardner J. P. 1991, Ap.J. 369, 79

Lilly S. J., Tresse L., Hammer F., Crampton D. & Le Fevre O. 1995, ApJ 455, 108

Lumsden S. L., Nichol R. C., Collins C. A. & Guzzo L. 1992, MNRAS 258, 1

Lupino G. A. and Gioia I.M., 1995, Ap.J. Letters, 445, L77

Maddox S. J., Sutherland W. J., Efstathiou G., Loveday J. & Peterson B. A. 1990, MNRAS 247, 1

Madau P., Ferguson H. C., Dickinson M., Giavalisco M., Steidel C. C. & Fruchter A. S. 1996, AJ submitted

Malkan M. A., Teplitz H., McLean I. S. 1996, ApJ 468, 9

Marzke R. O., Geller M. J., Huchra J. P., et al. 1994b, A.J. 108, 437

Marzke R. O., Huchra J. P., Geller M. J. 1994a, Ap.J. 428, 43

McGaugh S. S. 1994, Nature 367, 538

McGaugh S. S., Bothun G. D. & Schombert J. M. 1995, AJ 110, 573

McLeod B. A., Bernstein G. M., Reike M. J., Tollestrup E. V. & Fazio G. G. 1995, ApJS 96, 117

Metcalfe N., Shanks T., Fong R. & Jones L. R. 1991, MNRAS 249, 498

Metcalfe N., Shanks T., Fong R. & Roche N. 1995, MNRAS 273, 257

Padmanbhan T., 1993, " The structure formation in the Universe", Pub. Cambridge University Press

Pascarelle, S.M., Windhorst R.A., Kee W.C. & Odewahn S.C., 1996, 383, 45

Phillips, A.C., Bershady, M.A., Forbes D.A., Koo D. C., Illingworth G.D. and Reitzel D.B., 1995, Ap.J. 444, 21

Postman M., Lubin L. M., Gunn J. E., Oke J. B., Hoessel J. G., Schneider D. P. & Christensen J. A. 1996, AJ 111, 615

Postman M., Huchra J. P. & Geller M. J. 1992, ApJ 384, 404

Rosati P., 1996 Private communication

Sandage A., Binggeli B. & Tamman G. A. 1985, AJ 90, 1759

Saracco P., Chincarini G. and Iovino A., 1996, MNRAS 283, 865

Saracco P., Iovino A., Garilli B., Maccagni D. & Chincarini G. 1997, AJ submitted

Sawicki M.J., Lin H. and Yee H.K.C., 1996, submitted to A.J.

Scaramella R., Zamorani G., Vettolani G., Chincarini G., 1991, AJ 101, 342

Schade D., Lilly S. J., Crampton D., Hammer F., Le Fevre O. & Tresse L. 1996, preprint

Schechter P. 1976, ApJ 203, 297

Schindler, S., Guzzo L., Ebeling H., Boeringer H., Chincarini G., Collins C. A., De Grandi S., Neumann D. M., Briel U. G., Shaver P. & Vettolani G., 1995 Astron. and Astroph., 299, L9

Songaila A and Cowie L.L., 1996, A.J. 112, 335

Steidel C. C. & Hamilton D. 1992, AJ 104, 941

Steidel C. C., Giavalisco M., Dickinson M. & Adelberger K. L. 1996, AJ 112, 352

SubbaRao M. U., Connolly A. J., Szalay A. S. & Koo D. C. 1996, AJ 112, 929

Tyson J. A. 1988, A.J. 96, 1

van den Bergh S., Abraham R. G., Ellis R. S., Tanvir N. R., Santiago B. X. & Glazebrook K. G. 1996, AJ 112, 359

Vettolani et al. (co-authors as in Zucca ...), 1996, A.J. submitted

Williams R. et al. 1996, in Science with the Hubble Space Telescope - II, eds. P. Benvenuti, F. D. Macchetto & E. J. Schreier (Baltimore: STScI)

Zucca E., Zamorani G., Vettolani G., Cappi A., Merighi R., Mignoli M., Stirpe G.M., MacGilliwray H., Collins c., Balkowski C., Cayatte V., Maurogordato S., Proust D., Chincarini G., Guzzo L., Maccagni D., Scaramella R., Blanchard A., and Ramella M., 1996, A.J. submitted

DYNAMICS OF THE LARGE-SCALE STRUCTURE

S. F. SHANDARIN

Department of Physics and Astronomy, University of Kansas
Lawrence, KS, 66045, USA

1. Introduction

The observations of distant objects provide strong evidences that the universe is nearly isotropic and homogeneous on very large scales $l \geq 300 Mpc$. On the other hand, the universe is extremely clumpy on small scales $l \leq 1 Mpc$. The inhomogeneities in the intermediate range of scales – groups and clusters of galaxies, superclusters and voids – that are collectively termed as the large scale structure of the universe (hereafter LSS) show a variety of shapes and sizes.

Quantitatively the degree of inhomogeneity in this range of scales is measured primarily by the inhomogeneities in galaxy distribution or by the galaxy large-scale motions (Giovanelli, this volume), (Szalay, this volume), see also (Strauss & Willick 1995), (Dekel 1994). It depends crucially on the statistics employed to measure the inhomogeneity. For instance, the two-point correlation function of galaxies falls significantly bellow unity on scales greater than $10h^{-1}Mpc$ ($h = H_0/(100 \ km \ s^{-1} \ Mpc^{-1})$). This fact was often used as an argument against the reality of superclusters extending over $20 - 30h^{-1}Mpc$. Other statistics, the two-point correlation function of clusters of galaxies, the sizes of voids of galaxies, the large-scale streaming flows suggest that the universe is significantly inhomogeneous up to $50h^{-1}Mpc$ or greater.

Understanding the formation of LSS is one of the most important problems in modern cosmology. Many fundamental questions ranging from the physical nature of dark matter (hereafter DM), to the temperature anisotropies of the cosmic microwave background radiation (hereafter CMBR) and the determination of the epoch of galaxy formation join together here (Peebles 1993).

D. N. Schramm and P. Galeotti (eds.), Generation of Cosmological Large-Scale Structure, 33–62.

If compared with the problem of the origin of galaxies the formation of LSS is a much easier problem. A common feature of the both is nonlinear gravitational clustering. However, the physics of galaxy formation also includes gas dynamics, various thermal processes, star formation, supernova explosions, etc, see e.g. (Blumenthal et al 1984). The complexity of these processes is so great that there is little hope that it can be properly modeled and computed in feasible future. It is most likely that the galaxy formation will be a phenomenological theory based on numerous empirical relations to be obtained observationally. In contrast to this the properties of LSS are essentially determined by gravitational clustering which is a much simpler process. It makes the theoretical study of the large-scale structure much more easier and therefore allows a much more fundamental (as opposite to phenomenological) approach.

Here I discuss the formation of the large-scale structure as a result of gravitational instability of the uniform universe. The gravitational instability is an amplifying mechanism that requires the seed density fluctuations (Lifshitz 1946), (Novikov 1964). For this reason the discovery of the anisotropy in the temperature of the CMBR by the COBE team has played such a fundamental role. The temperature anisotropies of the CMBR ($\delta T/T$)) have been interpreted as the primordial perturbations of the universe observed at the decoupling epoch. If so it provides a very strong argument for the gravitational instability paradigm for the formation of cosmic. Gravitational clustering is not a purely theoretical speculation anymore but a model for the real universe.

It is generally accepted that the seed density fluctuations were probably generated at the inflationary stage when the universe underwent a period of accelerated expansion [1]. The primordial perturbations generated during the inflationary stage are usually assumed to be Gaussian random fields although other possibilities are not excluded. As the universe evolved, the density perturbations grew by the amplitude and also underwent dissipative damping. The dissipation strongly depends on the scale of perturbations and thus determines the shape of the spectrum immediately after decoupling. The COBE detectors had no sufficient angular resolution for measuring the temperature fluctuations on the scale of galaxies or even on the scale of superclusters. However, there is little doubt that the ongoing and future microwave background experiments having better angular resolution will eventually measure the primordial fluctuations on the scales of LSS (Page, this volume). On the other hand the upcoming deep redshift

---

[1] Various aspects of the epoch of inflation are discussed in (Turner, this volume); an alternative view assuming that the phase transitions in the early universe could play an important role in the generation of the seed perturbations is presented by (Turok, this volume)

surveys (in particular the Sloan Digital Sky (SDSS) and 2 degree Field (2dF) surveys) will significantly enlarge the scales probed by the galaxy surveys.

In this lectures I will discuss the evolution of density inhomogeneities after decoupling of baryonic matter and radiation at the redshift $z \approx 1400$. The most of the discussion will concentrate on on the processes that happened in the relatively recent epoch ($z \leq 10$) when the density fluctuations on the scales of LSS began to approach or reached the nonlinear regime. A simple estimate based on the linear theory shows that a significant amount of dark matter is needed for reconciling the limits of the $\delta T/T$ on the scales of the LSS and the existence of the structures in the universe. Thus, in addition to substantial astronomical evidences for the dark matter (Schramm, this volume) and (Bond, this volume) there is a strong theoretical necessity for some form of dark matter if one wants to explain the cosmic structures by gravitational clustering (see e.g. (Shandarin, Doroshkevich, & Zel'dovich 1983)). Thus I will assume that the most of mass in the universe comprises weakly interacting particles like axions and/or massive neutrinos.

Observationally LSS is defined as the galaxy distribution in space, but theoretically one can reliably calculate only the mass distribution. The physics of the galaxy formation is prohibitively complex. A priori, it is not obvious that the light should trace mass. Moreover, there are strong evidences that it does not. However, the greater the scale the closer the two distributions should be. The formation and evolution of the largest inhomogeneities is determined mainly by gravity that does not distinguish between luminous and dark matter. On scales of superclusters the distribution of galaxies is probably similar to the mass distribution although they are not yet identical. This poses an additional problem in studying LSS. Often the lack of understanding the differences in galaxy and mass distributions is concealed in a phenomenological parameter called a biasing factor b. The exact definition of b varies from author to author but roughly it is the ratio of galaxy density contrast to the mass density contrast when both are smoothed with the same window function.

The easiest part of the structure formation problem – gravitational clustering – has not completely solved yet. However, I will try to show that we already understand it quite well, at least on the qualitative level. In the currently popular (CDM based) models the structure begins to form when the perturbations on the smallest scales $M \sim 10^6 M_{\odot}$, allowed by Jeans' criterion to grow, reach the nonlinear stage and start off the hierarchical clustering process. It is assumed that by the present time nonlinear clustering has reached the scales of superclusters $M \sim 10^{14} M_{\odot}$. The straightforward analysis of the dynamics in the full range of scales is possible neither analytically nor numerically. The analysis of a part of this range requires an assumption on what happens on smaller scales. Typically the small scale inhomogeneities are either smoothed out or modeled by a random distribution of small compact objects (particles). The former is typical for analytical studies and the latter for N-body simulations. We begin the discussion with the theoretical models assuming that the initial state is a slightly perturbed continuous medium of collision-less matter. The baryonic component is assumed to follow the DM component because its mean density is supposed to exceed the mean density of baryons roughly by an order of magnitude (Bond, this volume).

2. Basic Dynamic Model

Describing the evolution of density inhomogeneities and motions in an expanding Universe it is convenient in the comoving coordinates $\mathbf{x}$ in terms of the peculiar velocities $\mathbf{v}_p$

$$\mathbf{x} = \frac{1}{a(t)}\mathbf{r}, \quad \mathbf{v}_p = \frac{d\mathbf{r}}{dt} - H(t) \cdot \mathbf{r} = a(t)\frac{d\mathbf{x}}{dt}, \tag{1}$$

where $a = a(t)$ is a scalar factor describing the uniform expansion of the Universe, $\mathbf{r}$ is the physical coordinate, and $H(t) = \dot{a}/a$ is the Hubble parameter ($H(present\ time) = H_0 = 100\,h\ km\,s^{-1}\,Mpc^{-1}$; $0.5 \le h \le 1$). In order to simplify the discussion we will assume the cosmological term $\Lambda = 0$ [2].

In a homogeneous Universe the comoving coordinates of a particle remain constant ($\mathbf{x} = \mathbf{q} = const$ and the peculiar velocities equal zero $\mathbf{v}_p = 0$.

Density $\rho(\mathbf{x}, t)$, peculiar velocities $\mathbf{v}_p(\mathbf{x}, t)$, and gravitational potential perturbations $\phi(\mathbf{x}, t)$ are coupled by three equations (see e.g. (Peebles 1980)):
the continuity equation

$$\frac{\partial \rho}{\partial t} + 3H\rho + \frac{1}{a}\nabla \cdot (\rho \mathbf{v}_p) = 0, \tag{2}$$

the Euler equation

$$\frac{\partial \mathbf{v}_p}{\partial t} + \frac{1}{a}(\mathbf{v}_p \cdot \nabla)\mathbf{v}_p + H\mathbf{v}_p = -\frac{1}{a}\nabla\phi, \tag{3}$$

and the Poisson equation

$$\frac{1}{a^2}\nabla^2\phi = 4\pi G(\rho - \bar{\rho}), \tag{4}$$

[2] In a typical model with the non-zero cosmological term the qualitative behavior of the perturbation is similar to open Friedmann models. The linear theory of gravitational instability in the models with the Λ-term is given by (Kofman & Starobinsky 1985)

where $\bar{\rho} = \bar{\rho}(t)$ is the mean density of the universe at time t.

For the further discussion it is convenient to rewrite the above equations (2 - 4) in a different form (Gurbatov, Saichev, & Shandarin 1989). Changing the time variable from t to $D \equiv D_g(t)$ ($D_g(t)$ is the growing mode of the linear theory) and rescaling the density

$$\eta = \rho \cdot a^3, \tag{5}$$

peculiar velocity

$$\mathbf{v} = \frac{1}{a \cdot D} \mathbf{v}_p = \frac{d\mathbf{x}}{dD}, \tag{6}$$

and the perturbation of the gravitational potential

$$\varphi = (\frac{3}{2}\Omega_0 \dot{a}^2 D)^{-1}\phi. \tag{7}$$

one easily obtains

$$\frac{\partial \eta}{\partial D} + \frac{\partial(\eta \cdot v_i)}{\partial x_i} = 0, \tag{8}$$

$$\frac{\partial v_i}{\partial D} + v_k \frac{\partial v_i}{\partial x_k} = -\frac{3}{2} \frac{\Omega_0}{D \cdot f^2} (\frac{\partial \varphi}{\partial x_i} + v_i), \tag{9}$$

$$\frac{\partial^2 \varphi}{\partial x_i^2} = \frac{\delta}{D}, \tag{10}$$

where $f(t) = d\ln D/d\ln a$, $\Omega_0 = 8\pi G\bar{\rho}_0/3H_0^2$ is the dimensionless mean density, $\delta = (\eta - \bar{\eta})/\bar{\eta} = (\rho - \bar{\rho})/\bar{\rho}$, and summation over dummy indices is assumed. In the Einstein - De Sitter universe ($\Omega(t) = 1$) $f \equiv 1$ and $\varphi = (\frac{3}{2}\dot{a}^2 a)^{-1}\phi = const(t) \cdot \phi$. The term proportional to the velocity on the right hand side of the Euler equation (eq. 9) describes the drag force due to the uniform expansion of the universe.

Equations (8 – 10) make a set of *nonlinear* partial differential equations that should be solved for random initial conditions given at the decoupling epoch. An analytic solution is obviously impossible. Therefore, various approaches have been developed. The first and the easiest one is the linear analysis. Let us begin with an idealized problem: the universe filled with homogeneous collisionless medium is perturbed by the small smooth density and velocity fluctuations corresponding to the growing mode: $\delta_0 = -\nabla \mathbf{v}_0$.

3. Linear Regime

While the amplitude of the density fluctuations is small, their growth is adequately described by the linear theory. Usually it is formulated in the

Eulerian space by linearizing the continuity and Euler equations that is by dropping the nonlinear terms in both equations.

3.1 EULERIAN SPACE

Dropping the nonlinear terms $\delta \frac{\partial v_i}{\partial x_i}$, $v_i \frac{\partial \delta}{\partial x_i}$ in the continuity equation and $v_k \frac{\partial v_i}{\partial x_k}$ in the Euler equation one obtains

$$\frac{\partial \delta}{\partial D} = -\frac{\partial v_i}{\partial x_i}, \tag{11}$$

$$\frac{\partial v_i}{\partial D} = -\frac{3}{2} \frac{\Omega_0}{D \cdot f^2} \left(\frac{\partial \varphi}{\partial x_i} + v_i \right). \tag{12}$$

The Poisson equation is linear and therefore does not change. It is easy to show that the general growing solution in these variables has the following form

$$\delta(\mathbf{x}, D) = D \cdot \delta_0(\mathbf{x}), \quad \mathbf{v}(\mathbf{x}, D) = \mathbf{v}_0(\mathbf{x}), \quad \varphi(\mathbf{x}, D) = \varphi_0(\mathbf{x}), \tag{13}$$

where $\delta_0(\mathbf{x})$, $v_0(\mathbf{x})$, and $\varphi_0(\mathbf{x})$ are the initial perturbations in the growing mode related to each other by simple relations

$$\delta_0(\mathbf{x}) = -\nabla \mathbf{v}_0 = \nabla^2 \varphi_0. \tag{14}$$

To linear order the right hand side term in the Euler equation (12) vanishes and it thus becomes much simpler

$$\frac{\partial v_i}{\partial D} = 0. \tag{15}$$

Rigorously speaking solution (13) holds only while $|\delta| \ll 1$ but it is often used as a rough estimate until $\sigma_\delta \simeq 1$. It is worth recalling that this solution predicts negative densities in the underdense regions when $\delta < -1$[3]. This is obviously unphysical. It is also not always recognized that the positions of the density peaks can be qualitatively predicted only until $\sigma_\delta =< [(\rho - \bar{\rho})/\bar{\rho}]^2 >^{1/2} \approx l_\delta/l_v$ (Shandarin 1994), where l_δ and l_v are the scales of the initial density and velocity fields respectively

$$l_\delta = \sqrt{3} \frac{\sigma_\delta}{\sigma_{\delta'}}, \quad l_v = \sqrt{3} \frac{\sigma_v}{\sigma_\delta}. \tag{16}$$

In the realistic models the initial density, velocity, and gravitational potential fields have very different scales $l_\delta \ll l_v \ll l_\varphi$ (here $l_\varphi = \sqrt{3} \sigma_\varphi / \sigma_v$) (see Fig.1). Thus, the last restriction is important. Fortunately, this problem can be easily fixed by incorporating the motion into the linear theory.

---

[3] According to the linear theory at $\sigma_\delta = 1$ about 16% of space has negative densities

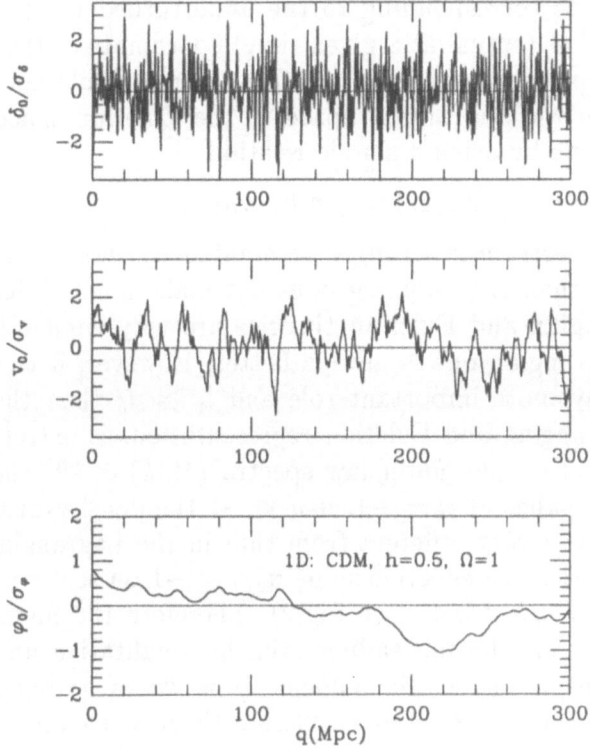

Figure 1. A one-dimensional analog of the initial density contrast, velocity, and gravitational potential perturbations of the standard CDM model. The one-dimensional analog of the initial spectrum was obtained from the three-dimensional spectrum: $P^{(1D)}(k) = k^2 P^{(3D)}(k)$. All fields are smoothed roughly on the galaxy scale ($1\,h^{-1}\,Mpc$).

3.2 LAGRANGIAN SPACE

Linearizing the dynamic equations in the Lagrangian space allows to keep two of three nonlinear terms dropped in the last section by incorporating them in the total derivatives

$$\frac{d\delta}{dD} \equiv \frac{\partial\delta}{\partial D} + v_i \frac{\partial\delta}{\partial x_i} = -\frac{\partial v_i}{\partial x_i}, \tag{17}$$

$$\frac{dv_i}{dD} \equiv \frac{\partial v_i}{\partial D} + v_k \frac{\partial v_i}{\partial x_k} = 0. \tag{18}$$

From the mathematical point of view the linear solution in the Lagrangian coordinates looks identical to the Eulerian one:

$$\delta(\mathbf{q}, D) = D \cdot \delta_0(\mathbf{q}), \quad \mathbf{v}(\mathbf{q}, D) = \mathbf{v}_0(\mathbf{q}), \quad \varphi(\mathbf{q}, D) = \varphi_0(\mathbf{q}), \tag{19}$$

The difference is in the physical interpretation: $\mathbf{q}$ in eq.(19) is the Lagrangian coordinate corresponding to the unperturbed position and $\mathbf{x}$ in eq.13 is the Eulerian (actual at a given time) coordinate of the fluid particle. In order to represent the solution (19) in the real (Eulerian) space one must map it from the Lagrangian space to the Eulerian space. To linear order it can be done by using a simple relation

$$\mathbf{x}(\mathbf{q}, D) = \mathbf{q} + D \cdot \mathbf{v}_0(\mathbf{q}). \qquad (20)$$

If the initial spectrum is strongly dominated by short wave perturbations and $l_v \simeq l_\delta$ than the mapping does not make much difference. As a result the Lagrangian and Eulerian theories are very similar in this case although the Lagrangian one is always better. However, if the long wave perturbations play more important role and $l_v \gg l_\delta$ then the difference between the Lagrangian and Eulerian representations is extremely significant. In the case of simple power law spectra ($P_0(k) \propto k^n$) the boundary between these cases lies at $n = -1$: if $n < -1$ the density distribution in the Eulerian space is very different from that in the Lagrangian space. In the realistic scenarios the effective slope $n_{eff} < -1$ on scales smaller than roughly $l \simeq k^{-1} < 5 h^{-1} Mpc$ (see Fig.2). Therefore the linear theory in the Eulerian space can be misleading even for qualitative understanding the present and especially earlier epochs. It is worth noting that in the literature the linear Lagrangian and Eulerian theories are often confused.

4. Quasilinear or Weakly Nonlinear Regime

In the quasilinear regime the nonlinear effects are weak but they are taken into account. Usually formal high order expansions are used to address this issue (Buchert 1993), (Bouchet et al 1995), (Catelan 1995) and references therein. For convergence the nonlinear corrections must be small. The topic is discussed in detail by (Juszkiewicz this volume) hence I make just a few comments. It is important to remember that the higher order expansions do not guarantee that all the outcomes of such a theory are at least qualitatively correct even some of the results are proven to be correct. This feature of nonlinear gravitational dynamics occurs already in the quasilinear regime especially when the theory is pushed to its limits. For instance, the comparison with the N-body simulations shows that the relation for the skewness and kurtosis of the probability distribution function predicted by the Eulerian peturbative theory holds for $\sigma_\delta \geq 1$ (Juszkiewicz this volume) but it is not clear whether this theory also predicts everywhere positive densities $\delta \geq -1$ or not.

At $\sigma_\delta \approx 1$ all orders in the perturbative series become comparable and a moderate gain in accuracy in a generic case is not worth the mathematical complexity of the higher orders. In particular, the high orders do not

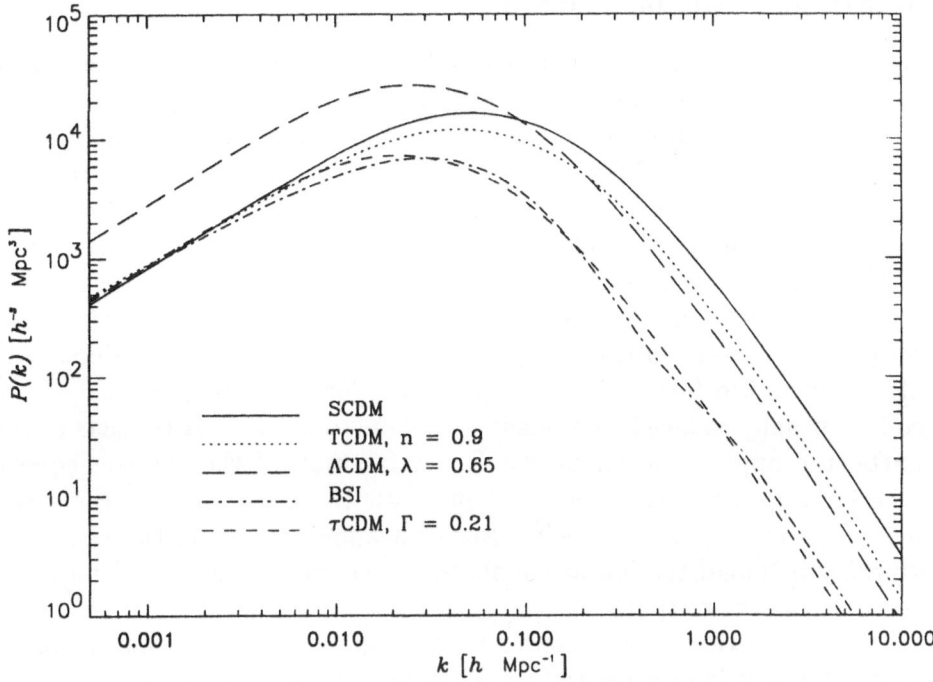

Figure 2. The COBE normalized spectra for some currently popular models: SCDM is the standard cold dark matter model (shown for the reference), TCDM is the tilted CDM model, ΛCDM is the model with non-zero Λ-term, BSI is the broken scale invariant model, τCDM is the decaying τ-neutrino model (see (Gottlöber, 1996) and the references to the original papers therein). Courtesy of S.Gottlöber.

improve the geometry of the density distribution (Buchert et al 1994). The perturbative expansions are asymptotic series, and their conversion ranges are poorly known. The higher order corrections usually improve only some aspects of the solution and worsen the others. Thus, the results of the high order expansions at $\sigma_\delta \approx 1$ should be taken with great caution (Sahni & Shandarin 1996). As often happens with the asymptotic series the first order theory provides the best overall approximation especially in the limit of applicability $\sigma_\delta \approx 1$.

The major achievement of the high order theories consist in the ability to calculate the high order moments (skewness, kurtosis, etc) of the distribution function in a weakly nonlinear regime. It is very useful for it may help to measure Ω and the biasing factor b.

5. Strongly Nonlinear Regime

Both linear and quasilinear theories break at a relatively low amplitude $\sigma_\delta \sim 0.5$. In order to study gravitational clustering in deeper nonlinear regime one has to use either the N-body simulations or approximations.

The advantages of the N-body simulation approach are related to its universality (e.g. the realistic initial spectra can be easily incorporated) and its effectiveness in modeling the observational environment (the effects of the redshift space, selection function, noise, etc). The advantages of the analytic and semianalytic models are different. First of all, successful models provide the insight into the physics and therefore provides understanding of the structure formation. The principle goal of a cosmological N-body code is usually reduced to modeling the "microphysics". It consists of three parts: the approximation of the Green function of the Poisson equation (the force between two point-like masses), the calculation of the density field, and moving the particles. Such an approach treats the code like a "black box": load the initial condition as the input and get the large-scale structure as the output.

Approximations try to understand what is inside the "black box". In particular, if it is possible to formulate some "macroscopic" principles that are supposed to catch the major feature of the process.

I begin with the discussion of some approximations. The basis of all the approximations I discuss here is the Zel'dovich approximation (Zel'dovich 1970), (Shandarin & Zel'dovich 1989).

5.1 ZEL'DOVICH APPROXIMATION

The right hand side of eq.(9) vanishes in the linear theory since $v_i = -\partial\varphi/\partial x_i$ to linear order. As we saw this greatly simplifies the Euler equation making the velocity independent of other variables. Zel'dovich suggested to extrapolate eq.20 until or even slightly after the orbit crossing. This is equivalent to the assumption that the relation $v_i = -\partial\varphi/\partial x_i$ also holds until the orbit crossing.

Incidentally, the form of equation (20) resulted in two major misconceptions. First, the Zel'dovich approximation is often characterized as kinematic, that is the one that does not take into account gravity. Second, the Zel'dovich approximation is said to be equivalent to the first order Lagrangian theory.

Actually, the kinematic form of eq.(20) results from the remarkable fact that the gravitational force generated by the density *inhomogeneities* exactly (in the linear approximation) balances the drag force resulted from the *uniform* expansion of the universe: $-\partial\varphi/\partial x_i = v_i$! Thus, the Zel'dovich approximation does take into account gravity although it usually underestimates the gravitational force in the overdense regions. The second statement also is not quite true. The Zel'dovich approximation does contain the correct first order term of the perturbation series but it also contains the higher order terms although they are not quite right. It uses the exact continuity equation (the conservation of mass) that results in the occurrence of singularities in the density distribution (Zel'dovich 1970) – definitely a nonlinear feature.

In principle, one can construct many approximations that are accurate to linear order (e.g. the Frozen Flow (Matarrese et al 1992), Frozen Potential approximations (Brainerd et al 1993), (Bagla & Padmanabhan 1994)). As a matter of fact they have different higher order terms that results in very different density and velocity distributions in the nonlinear regime (Sathyaprakash et al 1995). So far the Zel'dovich approximation is the simplest and the most accurate of all suggested in this class (Munshi et al 1994), (Bernardeau et al 1994), (Sahni & Coles 1995).

It is remarkable that the Zel'dovich approximation (eq.20) is a one step mapping of a homogeneous (Lagrangian) space at the initial time ($D \to 0$) into the Eulerian space at a 'time' D. Using this mapping one can calculate the density at the time D from the conservation of mass

$$\eta(\mathbf{x}, D) d^3 x = \bar{\eta} d^3 q. \tag{21}$$

It is straightforward to express the density $\eta(\mathbf{x}, D)$ in terms of the eigen values $\lambda_1(\mathbf{q})$, $\lambda_2(\mathbf{q})$, $\lambda_3(\mathbf{q})$ of the initial deformation tensor field $d_{jk}(\mathbf{q}) = -\partial v_{0j}/\partial q_k = \partial^2 \varphi_0/\partial q_j \partial q_k$

$$\eta(\mathbf{q}, D) = \frac{\bar{\eta}}{(1 - D\lambda_1)(1 - D\lambda_2)(1 - D\lambda_3)}. \tag{22}$$

Equation 22 is obviously singular. It predicts the origin of caustics having complex geometrical structures especially in the Eulerian space (Arnold, Shandarin, & Zel'dovich 1982). The simplest of all are pancakes ((Zel'dovich 1970). The three-dimensional N-body simulations (Shandarin et al 1995) have confirmed that the pancakes are the first singularities to form.

The velocity field remains constant in the Lagrangian space

$$\mathbf{v}(\mathbf{q}, D) = \frac{d\mathbf{x}}{dD} = \mathbf{v}_0(\mathbf{q}), \tag{23}$$

in other words, each particle moves with the constant velocity. In order to find the density distribution in the Eulerian space one has to make use of equation 20.

The Zel'dovich approximation describes *qualitatively* correctly the motion of particles until orbit crossing but breaks down within the multistream flow regions soon after the formation of the first caustics (pancakes). Should be extrapolated to later stages, the Zel'dovich approximation would predict that the thickness of pancakes grows unbounded that runs counter to the findings of N-body simulations. They show that the actual thickness of pancakes remains considerably smaller than both the pancake diameters an the mean distance between pancakes (Doroshkevich et al 1980).

The Zel'dovich approximation worked well in the Hot Dark Matter scenario but it cannot be applied in the original form to more realistic cosmological scenarios. In all the CDM based models the clumps with masses around $10^6\ M_\odot$ form first, then they cluster into galaxies ($\sim 10^{11} M_\odot$) in a hierarchical process, and finally clusters and superclusters of galaxies ($\sim 10^{14} M_\odot$) are assembled from galaxies and the DM halos. The process is often referred to as the bottom-up scenario.

One way to deal with this problem consists in introducing dynamical smoothing (the scale of smoothing grows with time) of the initial conditions similar to that used in the standard peak theory (see e.g. (Bardeen et al 1986), (Peebles 1980)).

5.2   TRUNCATED ZEL'DOVICH APPROXIMATION

The adhesion approximation (Kofman et al 1992) (see bellow) as well as the N-body simulations ((Beacom et al 1991), (Little et al 1991), (Melott & Shandarin 1993)) showed that the presence of small-scale initial perturbations do not change much the overall large-scale distribution of mass although the small-scale perturbations make the appearance of the large structures very clumpy. Thus, smoothing out the small-scale perturbations in the initial spectrum removes the small-scale clumpiness but retain most of the structures due to the long wave part of the spectrum.

For every epoch the border separating small and large scales can be roughly defined as the scale of nonlinearity $l_{n\ell} \equiv k_{n\ell}^{-1}$ that can be found from the following equation

$$\sigma_\delta(l_{n\ell}) = 4\pi \cdot D^2 \int\limits_0^{k_{n\ell}} P_0(k)k^2 dk = 1, \qquad (24)$$

here $P_0(k)$ is the initial spectrum of density fluctuations: $P_0(k) = <|\delta_k|^2>$.

The scale of nonlinearity $l_{n\ell}$ separates the perturbations in the linear regime

$$\sigma_\delta(l) < 1 \quad \text{if} \quad l > l_{n\ell} = k_{n\ell}^{-1} \qquad (25)$$

from the perturbations in the nonlinear regime

$$\sigma_\delta(l) > 1 \quad \text{if} \quad l < l_{n\ell} = k_{n\ell}^{-1}. \qquad (26)$$

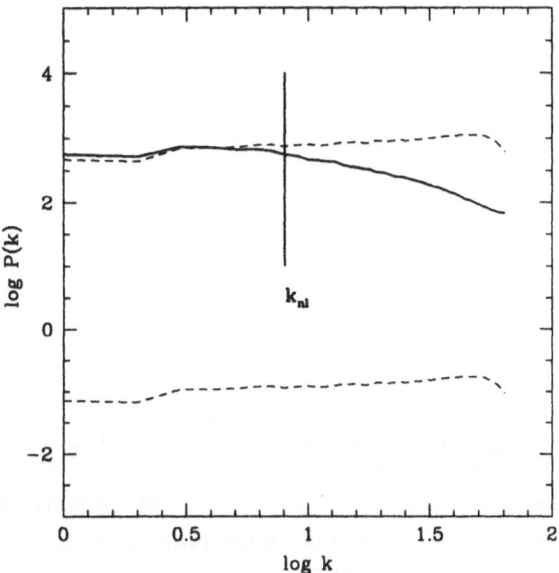

Figure 3. The lower dashed line shows the initial spectrum $P_0(k) \propto k^0$, the upper dashed line shows its extrapolation according to the linear theory. The solid line shows the actual spectrum at the same stage calculated from the three-dimensional N-body simulation. The short vertical line marks the nonlinear scale defined by eq.42.

This simple idea is illustrated in Fig.3. The standard Zel'dovich approximation applied to the initial conditions dynamically smoothed with the scale of nonlinearity $k_{n\ell}(D)$ has been labeled as the truncated Zel'dovich approximation (hereafter TZA) (Coles, Melott, & Shandarin 1993). The time dependence $k_{n\ell}(D)$ is determined by the initial spectrum and the cosmological model [4]. The truncation of small-scale perturbations allow to use the Zel'dovich approximation for arbitrary initial spectra. In terms of the visual resemblance, the probability distribution function, and the nonlinear spectrum, the truncated Zel'dovich approximation is much better than other analytic and semianalytic models except the adhesion approximation. It is important to stress that this remains true in the domain in which the slope of the initial spectrum $n > -3$, where the performance of the Zel'dovich approximation has often been assumed to be very poor (Peebles 1980).

The formation of gravitationally bound clumps implys that the mass is transported from one place to another. An interesting characteristics of this process is the average distance traveled by a fluid particle. It can be

---

[4](Melott, Pelman & Shandarin 1994) showed that smoothing with a Gaussian filter and small adjusting the smoothing scale for different spectra slightly improves the approximation.

estimated analytically from the truncated Zel'dovich approximation (Shandarin 1993)

$$d_{rms}^2 = \frac{\int_0^{k_{n\ell}} P_0(k)dk}{\int_0^{k_{n\ell}} P_0(k)k^2 dk}. \tag{27}$$

In the case of the power law initial spectra

$$P_0(k) \propto k^n, \quad k_1 \leq k \leq k_2 \tag{28}$$

one easily finds

$$d_{rms} \sim k_{n\ell}^{-1}, \quad \text{if } n > -1 \tag{29}$$

and

$$d_{rms} \sim \left(\frac{k_1}{k_{n\ell}}\right)^{\frac{n+1}{2}} k_{n\ell}^{-1}, \quad \text{if } -3 < n < -1 . \tag{30}$$

In the former case the characteristic displacement of mass roughly corresponds to the characteristic mass of nonlinear clumps $d_{rms} \sim k_{n\ell}^{-1}$, $M \sim \bar{\rho} k_{n\ell}^{-3}$ which intuitively is very clear. In the latter case d_{rms} can be much greater than $k_{n\ell}^{-1} : d_{rms} >> k_{n\ell}^{-1}$, if $k_1 << k_{n\ell}$. If the characteristic displacement d_{rms} is greater than the typical distance between clumps it obviously means that the clumps themselves are displaced coherently by the distance d_{rms}. This simple fact essentially explains why the standard linear theory in the Eulerian space failed so miserably when it was used for explaining the structure in the $n < -1$ case.

It is interesting but not surprising that the scale on which the phases of the Fourier components can depart significantly from their original (linear) values is close to d_{rms} ((Ryden & Gramann 1991)).

In the Cold Dark Matter model at present time $d_{rms} \approx 9 \ h^{-1} Mpc$ that is d_{rms} is about two times greater than $k_{n\ell}^{-1}$. The difference is not huge but significant. At earlier epochs when the effective slope of the spectrum at the scale of nonlinearity $k_{n\ell}(D)$ was smaller the effect of the mass motion was more substantial.

The major drawback of the truncated Zel'dovich approximation is that it artificially smoothes out all perturbation bellow the nonlinear scale $l_{n\ell} = k_{n\ell}^{-1}$. We know from the simulations that small clumps dissolve after merging but there are others residing in less dense environment that do not merge and therefore they remain to exist as separate units. The model that correctly describes this feature of gravitational clustering is outlined in the following section.

5.3 ADHESION APPROXIMATION

Another way of stabilizing the thickness of pancakes consists in using artificial viscosity to mimic some effects of nonlinear gravitational dynamics in the multistream flow regions. This is the basic idea of the adhesion approximation (hereafter AA) introduced by (Gurbatov, Saichev, & Shandarin 1989). Similarly to the Zel'dovich approximation it assumes that before entering the multistream flow regions particles move along straight lines with constant velocities. However, as we know from the N-body simulations, the trajectories of particles become quasi-chaotic after they enter the multistream flow regions. This is a quite complicated process. In pancakes only one component of the velocity – orthogonal to the pancake – is affected and the other two remain roughly as they were. In filaments two orthogonal components of the velocity are randomized and in clumps all three components become random. The adhesion approximation very roughly simulates this process. An important feature of AA is that it allows the pancakes, filaments, and clumps to move as a whole and the mass to flow inside the structure. This makes AA useful for the description of hierarchical clustering scenarios.

Constructing the adhesion approximation an artificial (non-physical) viscosity term has been added to the right hand side of the Euler equation

$$\frac{\partial v_i}{\partial D} + v_k \frac{\partial v_i}{\partial x_k} = \nu \frac{\partial^2 v_k}{\partial x_k^2}. \tag{31}$$

The form of the viscosity term was chosen to obtain the three-dimensional Burgers equation that had an exact analytic solution (Burgers 1974). In the original model (Gurbatov, Saichev, & Shandarin 1989) the viscosity was introduced by hand. A more physical derivation of Burgers' equation based on the notion of self-gravitating ideal gas rather than on sticky particles was recently suggested by (Buchert & Domingues 1996).

For potential flows Burgers' equation (31) has the exact solution solution for the velocity field

$$\mathbf{v}(\mathbf{x}, D) = \frac{\int d^3q \, \frac{(\mathbf{x}-\mathbf{q})}{D} \, \exp\left(\frac{1}{2\nu} S(\mathbf{q}, \mathbf{x}, D)\right)}{\int d^3q \, \exp\left(\frac{1}{2\nu} S(\mathbf{q}, \mathbf{x}, D)\right)}, \tag{32}$$

where

$$S(\mathbf{q}, \mathbf{x}, D) = \varphi_0(\mathbf{q}) - \frac{(\mathbf{x} - \mathbf{q})^2}{2D}. \tag{33}$$

Equation (32) has an elegant geometrical interpretation which can be summarized as follows: Whether or not a particle originally located at $\mathbf{q}$ is stuck within a pancake, filament or clump can be found by descending a paraboloid

$$P(\mathbf{q}, \mathbf{x}, D) = \frac{(\mathbf{x} - \mathbf{q})^2}{2D} + P_0 \tag{34}$$

with radius of curvature $2D$ and height P_0 onto the initial velocity potential φ_0, in such a manner so as to be tangential to the potential at $\mathbf{q}$. The height of the paraboloid P_0 is a free parameter uniquely determined by the condition that $P(\mathbf{q}, \mathbf{x}, D)$ osculates $\varphi_0(\mathbf{q})$ at $\mathbf{q}$. If the paraboloid so constructed touches or intersects the potential at any other point $\mathbf{q}_1 \neq \mathbf{q}$, then we say that the particle in question has already entered a caustic, otherwise it has not. Thus, one can divide the Lagrangian space at any given time into *stuck* and *free* regions. Stuck regions correspond to particles already in caustics, whereas free regions are still expanding via the Zel'dovich relations and correspond to voids. The apex of the paraboloid $\mathbf{x}$, which osculates the potential in two points (without intersecting elsewhere) describes the location of the pancakes, in three points - the filaments, and in four points - the compact clumps of matter in the Eulerian space. The geometrical technique gives the skeleton of the large-scale structure – a geometrical construction showing the positions of clumps, filaments and pancakes.

Another method of solving Burgers' equation developed by (Weinberg & Gunn 1990b) allows to calculate the particle distributions similar to the N-body simulations. But since the adhesion approximation does not calculate the gravitational force it is much more efficient computationally .

The adhesion ·approximation was thoroughly compared with N-body simulations and proved to be quite accurate in general description of the nonlinear stage ((Kofman, Pogosyan, & Shandarin 1990), (Weinberg & Gunn 1990b), (Nusser & Dekel 1990), (Kofman et al 1992), (Melott, Shandarin, & Weinberg 1994)) The density distribution predicted by AA look much clumpier than that obtained in TZA but it is still less clumpy than that obtained in the N-body simulation. The power spectrum and the mass distribution function behave correspondingly: for example, there is more scale power in the nonlinear spectra in AA than in TZA but less than in the N-body simulations.

Recently a new variant (COMA = COnserving Momentum Approximation) of the adhesion approximation has been suggested (Shandarin & Sathyaprakash 1996). In contrast with the original adhesion approximation COMA assumes physical viscosity described by the equation similar to the Navier-Stokes equation rather than by Burgers' equation.

The qualitative change of the type of the particle trajectory after entering a multistream region (from a straight line to a random walk) is mainly caused by the gravitational interaction with the other particles within the same region of multistream flow. Therefore this interaction must conserve the momentum on the scale of the multistream flow region. Thus, the major difference of the COMA with the standard adhesion approximation consists in that it conserves the momentum locally while the standard adhesion approximation does not.

The major features of the COMA code are:

- it is *universal* in terms of the initial and boundary conditions as well as the shape of the computational box;
- *extremely simple* both conceptually and practically;
- *very economical* in terms of memory – just four arrays, each as big as the size of the box, are sufficient to implement the algorithm;
- *very fast*;
- local and therefore *100% parallelizable*;
- it does not have an analytic solution.

In the system of coordinates employed, particles evolve simply by conserving the mass and momentum; as a consequence one circumvents the problem of having to compute the gravitational force after each time step. Generalization of the algorithm to deal with particles, rather than densities, is straightforward but would require additional memory.

The adhesion approximation has played an important role in establishing the universality of the cellular and network structures in cosmological gravitational clustering (Gurbatov, Saichev, & Shandarin 1989), (Nusser & Dekel 1990), (Kofman et al 1992). The statistical analysis of the nonlinear density distributions in an ensemble of the N-body simulations of the power law models (the initial spectrum is assumed to be a simple power law $P_{in} \propto k^n$) with $n = 1, 0, -1, -2$ has confirmed the inferences from AA (Yess & Shandarin 1996). In early 1990s the AA simulations (Weinberg & Gunn 1990a), (Weinberg & Gunn 1990b) were the largest simulations of the large-scale galaxy distributions. In particular, they revealed spectacular "spherical shells" in the galaxy distribution in redshift space. Later they were observed in the deep redshift surveys: first probably detected by (Broadhurst et al 1990), and then observed in the Las Campanas Redshift Survey (Landy et al 1996). Recently the adhesion approximation has been used as a dynamical model for the analytic calculations of the mass distribution function (Vergassola et al 1994), (Cavaliere et al 1996). (Sahni, Sathyaprakash, & Shandarin 1994) used AA for the study of the statistics of voids in the Cold Dark Matter model. In particular, they predict that voids can be populated by substructure such as mini-pancakes and mini-filaments.

The AA has been designed to model the overall mass distribution. But in the present forms it cannot model correctly the interior of the regions where densities are extremely high. This has been accomplished in the following semianalytic model.

5.4 PEAK PATCH PICTURE

A model labeled as the Peak Patch Picture is an attempt to unify the Zel'dovich approximation with the standard peak theory (Bond & Myers 1996). It splits the initial displacement field (proportional to the initial velocity field) into a background field smoothed over a large scale and a fluctuating small-scale field. The background field represents the initial condition for the Zel'dovich approximation similar but not identical to the truncated Zel'dovich approximation. The major difference is that TZA assumes that the scale of smoothing is universal throughout the space and close to the scale of nonlinearity $k_{n\ell}^{-1}$ (Coles, Melott, & Shandarin 1993), (Melott, Pelman & Shandarin 1994); in contrast PPP uses a quite elaborate scheme employing the locally adaptive filter that selects the patches – progenitors of the mass clumps. The small-scale dynamics (within the patches) is described by the homogeneous ellipsoidal collapse model that has been shown to be superior of the Zel'dovich approximation. In contrast to the previously discussed approximations PPP addresses the issue of the internal dynamics of the clumps which is certainly a step forward.

The model has been tested against the cosmological N-body simulations and showed a satisfyingly close agreement.

The peak-patch method has been used for the studies of clusters of galaxies. However, the better treatment of the small-scale dynamics has allowed to study the X-ray luminosity of the clusters. The model has been used for making deep field X-ray and Sunyaev-Zel'dovich maps. Combined with X-ray temperature data the model puts constraints on the amplitude and the slope of the primordial spectrum of perturbations.

Another interesting application of PPP is the study of superclusters and the large-scale structure as a whole (Bond, Kofman, & Pogosyan 1996). The model introduces a new probabilistic interpretation of filaments as bridges of the Zel'dovich-mapped correlation functions constrained by shear at two high density peaks – progenitors of the clusters of galaxies.

6. Cosmological N-body Simulations

Many principle aspects of cosmological gravitational clustering can be understood from analytic and semianalytic approximations outlined above. However, like any other method they have serious limitations. A very effective method to deal with the problems of the nonlinear dynamics that remain outside the reach of analytic methods is the N-body simulation technique. In the N-body simulations the gravitational force induced by the density distribution is calculated at each time step. The trajectory of every particle is integrated in a self-consistently varying gravitational field. Knowing the positions and velocities of particles at a given time one can easily compute the mass density smoothed on a specified scale. Then applying a chosen set of criteria (the simplest is the density threshold) one can identify objects: "galaxies" and/or "clusters of galaxies" and compute their velocities, masses, and other parameters. Comparing the statistical properties of the galaxy distribution in the N-body simulation with observations one can reject or approve the model.

From the theoretical point of view N-body simulations *themselves* do not bring satisfactory understanding of the nonlinear gravitational dynamics because they operate on a *microscopic* level. Ideally, a cosmological N-body code is a black box that accurately calculates the force between particles and integrates their trajectories. However, in cosmology N-body simulations play a role similar to experiments in physics. Thus, the numerical simulations have led to considerable increase of our understanding of the nonlinear gravitational dynamics in cosmology.

There are many good papers discussing the numerical techniques for cosmology (see e.g. (Bertschinger & Gelb 1991), (Klypin 1996) and references therein). Many papers discuss in detail the computational aspects of the simulations: the computational achievements are quite remarkable. However, some physical questions have not been adequately resolved yet. Unfortunately, they are not often discussed. Here I am going to discuss mainly physical problems not computational techniques.

One of the problems is the particle paradigm widely used in the cosmological simulations. The vast majority of current cosmological N-body codes operate with particles having fixed sizes. The collisionless medium in most cosmological scenarios is assumed to be made of tiny particles like axions or neutrinos. The gravitational interaction between particles of this medium is negligible. On the other hand the particles used to approximate the continuous medium of dark matter are usually huge and the volume of the simulation often is not sufficiently large. As a result the gravitational influence of closest few neighbors is not negligible in the N-body simulations. Although there is no doubt that increasing the number of particles in the simulations one will eventually reach the right limit it is not clear yet how far the present codes are from that goal. The masses of the particles in the simulations more than 70 orders of magnitude larger than the DM particles.

I begin with a brief description of the major features of the cosmological N-body simulations.

6.1 INITIAL AND BOUNDARY CONDITIONS

The initial condition is usually generated as a realization of a Gaussian random process with a given initial spectrum $P_0(k)$. The method that is often referred to as a quiet start was originally introduced to cosmology by (Doroshkevich, Ryabenkii, & Shandarin 1973) who simulated three-dimensional non-random fields, (Doroshkevich et al 1980) simulated two-dimensional random fields, and (Klypin & Shandarin 1983) simulated random fields in three dimensions [5]. The quiet start has many advantages compared to other techniques (Sellwood 1997). In early cosmological simulations the initial conditions were set up simply by the Poisson process (see e.g. (Aarseth et al 1979), (Gott et al 1979), (Efstathiou & Eastwood 1981)).

In the quiet start the homogeneous (unperturbed) state is modeled by placing particles on a regular Cartesian grid. Then the particles are slightly displaced according to the Zel'dovich approximation (eq.20). The initial vector field is assumed to be of the potential type. It is generated by the Fourier transform, assuming the amplitudes and phases statistically independent random numbers.

Cosmological simulations usually employ periodic boundary conditions that are designed to approximate unbound infinite space. In the non-periodic system the perturbations close to the border evolve in the absence of the perturbation forces induced by inhomogeneities outside the simulation volume. This is obviously unrealistic environment therefore the results in the most of the volume except the very center cannot be trusted.

6.2 POISSON EQUATION SOLVERS

The major difference that distinguishes the types of the codes is the method of solving the Poisson equation. In principle, the Poison equation can be solved in the real space by summation of forces from every particle (direct N-body, e.g. (Aarseth et al 1979)) or groups of particles of various sizes (tree codes, e.g. (Hernquist 1987), (Suginohara et al 1991)). It also can be solved in the space of the wave vectors by using the Fourier transform (PM, particle-mesh codes, e.g. (Klypin & Shandarin 1983)). Finally, these two methods can be combined in one code (PPPM = P3M , particle-particle particle-mesh codes, e.g. (Efstathiou & Eastwood 1981), (Couchman 1991)): the large-scale gravitational field is calculated by the PM techniques (Fourier transform) and the small-scale filed by the direct summation over the closest particles.

In the PM codes the Green function is smoothed roughly on the mesh scale [6]. The tree and P3M methods have an additional parameter – a softening length ϵ – which characterizes the size of the particles and determines

---

[5]The technique probably was used in plasma physics even earlier.

[6]Obviously, this is a qualitative statement; quantitatively the accuracy of the Green function is a function of the scale.

the scale bellow which the Green function is smoothed out. In a typical cosmological simulation ϵ is about 10 times smaller than the average interparticle separation. It is often claimed that these codes have about 10 times better dynamical range assuming the other parameters are same.

In the codes using periodic boundary conditions the Green function is also distorted on large scales by the artificial images of the particles. On the scale greater than about a half of the size of the simulation cube the accuracy in the gravitational potential becomes worse than about 70%.

6.3 MASS AND SPACE RESOLUTION

In particle-mesh (hereafter PM) codes with the number of particles to be equal or greater than the number of the mesh cites $(N \geq M)$[7] it is assumed that in the unperturbed state the particles having the shape of a cube fill the whole space without gaps. In the so called high resolution methods (tree and P3M codes) the particles have sizes much smaller than their initial separations. Therefore, they start from highly inhomogeneous state determined by the shot noise bellow the scale of the mean separation of the particles. Usually, it is argued that in the hierarchical clustering scenarios the initial stage actually represents a highly inhomogeneous distribution of mass that resulted from the previous stages. It is certainly true that the initial stage should be inhomogeneous but the properties of the inhomogeneities are very different from the shot noise modeled by the simulations. In realistic scenarios the slope of the initial spectrum on the scale corresponding to the mass of the particle in a typical cosmological simulation is certainly less than $-1.5 - 2$. This means that clustering on the corresponding scale is highly anisotropic and is very poorly represented by a discrete random set of particles. One can only guess how it affects the results of the simulations.

Having smaller particles allows to approximate the Green function of the Poisson equation much better on small scales: down to the size of the particles. This is the major achievement of all high resolution methods in cosmology. However, it is important to remember that having the Green function with high resolution does not necessarily mean that the dynamics of the gravitational clustering has also same high resolution. The overall accuracy of an N-body simulation depends also on the mass resolution. It is pretty obvious that if the source term in the Poisson equation (10) is erroneous then regardless of the accuracy of the Green function its solution is also erroneous. Both the Green function and the density distribution

---

[7]Actually, the number of particles greater than the number of mesh points is used relatively rarely, only in the simulations of the HCDM (Hot plus Cold Dark Matter) model (Klypin et al 1996).

should have similar accuracy. Otherwise, the accuracy is determined by the worst part.

It is also important to remember that the clumps made of DM do not behave like particles at all: they have internal degrees of freedom and are quite fragile. When they approach closely to each other they merge forming a single clump of a larger mass not a binary system. The numerical studies of the behavior of the *fluid particles* in the hierarchical clustering process showed that they constantly change their shapes and sizes; only a small fraction of fluid particles collapse to *a half* of their initial size in all three directions [8], most of them acquire very oblate or prolate shapes (Kuhlman, Melott, & Shandarin 1996).

The lack of mass resolution also results in the discreteness effects that should be completely absent in the cosmological scenarios that are supposed to be modeled. It is assumed that the most of mass of the universe comprises in the weakly interacting particles like axions and massive neutrinos. Both components make almost perfect continuous collisionless media.

The discreteness effects of the N-body simulations manifest themselves in various ways, e.g. by two-body scattering ((Efstathiou & Eastwood 1981), (?), (Melott etal 1996)). But testing for the discreteness effects is not an easy task. Unfortunately there is no exact solution that would be similar to real systems in the dynamical sense. A usual check of the the energy (the Irvine-Dmitriev theorem) is not affected by the discreteness effects at all, the scaling properties and the spherically symmetric solution (Bertschinger 1985) probably are not very sensitive to them.

Here I would like to report some results of one recent test that uses the symmetry of the solution. If the initial perturbation is one-dimensional then it should remain one-dimensional all the time. All possible violations of this symmetry are due to discreteness of the code. Actually, this type of test was used in (Klypin & Shandarin 1983) and later in other simulations but the direction of the plane collapse was always chosen along one of the axes of the simulation cube. To make the test a bit tougher we studied a simple sinusoidal wave perturbation along the tilted (with respect to the grid) direction (for the details see (Melott etal 1996)). Using the identical initial conditions we have tested two variants of the PM code (64^3 particles on equal mesh and on 128^3 mesh, in both cases the particles had the size of the mesh); the P3M code (Couchman 1991) with three different softening lengths 1, 0.5, and 0.1 (in units of the mean interparticle separation); the tree code with two different softening lengths 1, 0.1; and the nested grid particle-mesh code (Splinter 1996). Qualitatively, all codes showed the break of the symmetry but quantitatively it was very differ-

---

[8]Compare with one tenth used in the high resolution tree and P3M codes

ent. In short, the smaller the softening length the stronger the departure from one-dimensional symmetry regardless of the type of the code. The minor violation of the symmetry was observed in the PM (64^3 particles on equal mesh), the nested grid particle-mesh code, P3M and tree codes with the softening length being equal to the mean interparticle separation ($\epsilon = 1$). The both P3M and tree codes with small softening lengths ($\epsilon = 0.1$) showed the worst results. Figure 4 shows the scatter plot of the components of the velocity parallel and normal to the plane of the collapse. If the one-dimensional symmetry was preserved exactly the components parallel to the plane were exact zero. Similar distortions are present in the coordinate space.

It looks like that improving the resolution by making small particles may result in unwanted discreteness effects regardless of the type of the Poison equation solver. Another – in my opinion, more promising – way of improving the resolution is to use several grids of different sizes. This multigrid approach has been known in cosmology for about ten years already (Villumsen 1988) but has got enough attention only recently (Anninos et al 1994), (Splinter 1996), (Kravtsov et al 1996).

Summarizing this brief discussion, I would like to remind a cautious remark from the book that is referred to almost in every paper on the cosmological N-body simulations. "We must finish this section with a cautionary remark that it is by no means established that the representation of the universe as a set of discrete masses – as is done in N-body simulations – is a good model of the real universe. ... In addition, if (as is suggested by a wide body of data) much of the mass in the universe comprises an invisible component (the missing mass) there is no guarantee that the galaxies have ever acted as point particles. If this were the case, the results from N-body experiments would not apply to the real universe." (Hockney & Eastwood 1988).

7. Some Cosmological Issues

7.1 HIERARCHICAL CLUSTERING AND CONTINUOUS PANCAKING MODELS

The idea that the structure in the universe formed in a hierarchical clustering process is very old (see the discussion in (Peebles 1980), sec.26). The models from the currently popular CDM family definitely belong to this class. The structure grows from roughly $M \sim 10^6 M_\odot$ to galaxies $M \sim 10^{11} M_\odot$ and then to clusters $M \sim 10^{14} M_\odot$ and superclusters of galaxies. This model is also referred to as the bottom-up model.

The opposite model – a top-down scenario – was originally suggested for the adiabatic scenario of the structure formation in the baryon dom-

56

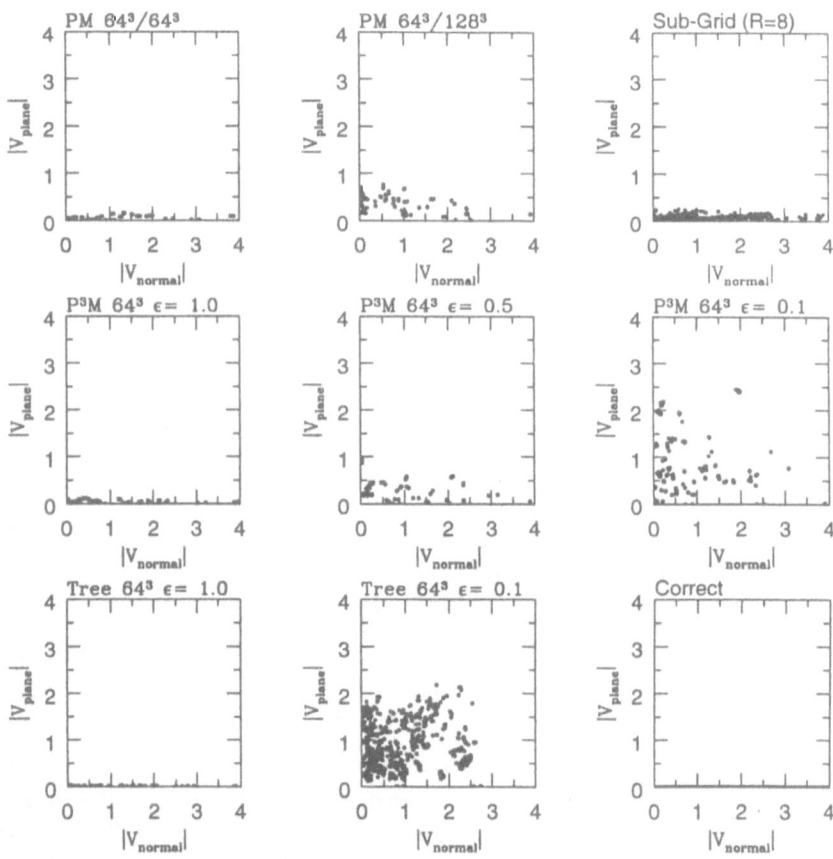

Figure 4. Scatter plot of the the absolute value of velocity components for 1000 randomly selected particles from each simulation. The velocity perpendicular to the plane of the collapse is on the horizontal and the velocity in the plane of the collapse is on the vertical. Velocity units are Hubble velocity across one cell. In the ideal case all points should lie on the horizontal as shown in the right bottom panel.

inated universe (Zel'dovich & Novikov 1983). In 1980 it was adopted for the structure formation in the Hot Dark Matter scenario assuming that the dark matter was in the form of neutrinos of some sort with masses about $30eV$ (for review see (Shandarin, Doroshkevich, & Zel'dovich 1983)). In both models the initial spectrum has a sharp cutoff on the scale of superclusters ($\sim 10^{14} M_\odot$). As a result the first objects to form are the Zel'dovich pancakes of the corresponding sizes. Galaxies ($\sim 10^{11} M_\odot$) were supposed to form in the process of fragmentation of the pancakes.

The Zel'dovich approximation was associated with the top-down scenarios and when the HDM model was rejected the approximation was rejected as well. As I showed in the previous section the latter was too premature. A simple modification of the Zel'dovich approximation – the dynamical truncation of the initial spectrum – allows to use it as a universal theoretical instrument for understanding cosmological clustering.

The clustering process is undoubtly hierarchical since large clumps formed by merging of small ones. However, this process is not completely random and structureless as it was usually portrayed in the standard hierarchical clustering model (Peebles 1980). The process has a significant coherent component on every scale when it reaches nonlinearity $l \approx k_{n\ell}^{-1}$. The smaller the effective index n_{eff} of the initial power spectrum the stronger the coherency of the process. The coherent part of the process is described by the truncated Zel'dovich approximation. Since the Zel'dovich approximation is associated with the pancakes this process is sometimes called continuous pancaking.

In TZA the hierarchical component is actually smeared out by the truncation of the small-scale part ($k > k_{n\ell}$) of the initial spectrum however, the adhesion approximation preserves most of it.

Roughly speaking if $n_{eff} < -3$ the coherency is dominant, if $-3 < n_{eff} < -1$ it is still quite strong, and if $-1 < n_{eff}$ it becomes weaker with the growth of n_{eff}. This has been measured by the crosscorrelation of the density fields obtained from identical initial conditions in TZA, AA, and the N-body simulations (Coles, Melott, & Shandarin 1993), (Melott, Pelman & Shandarin 1994), (Melott, Shandarin, & Weinberg 1994). Also the percolation statistics has measured the significant coherence in the density distributions in all power law models with $n = -2, -1, 0, +1$ (Yess & Shandarin 1996).

The standard hierarchical clustering model has failed to predict or explain the appearance of the large-scale structure after it has been discovered in the redshift surveys. The Zel'dovich approximation predicted pancakes in 1970, it also qualitatively predicted the connectedness of the structure in 1975 (see the discussion and Figure 12 in (Shandarin & Zel'dovich 1989)) that was accepted as a working hypothesis by some of the observers (Chincarini & Rood 1980), (Oort 1983).

7.2 GEOMETRY OF COSMIC STRUCTURES: PANCAKES AND FILAMENTS

The Zel'dovich approximation predicts that the first caustics forming from smooth initial conditions at the nonlinear stage have the shape of pancakes (Zel'dovich 1970). This conclusion has been confirmed by three-dimensional N-body simulations (Shandarin et al 1995). Recently (Bond, Kofman, & Pogosyan 1996) have challenged this result arguing that in *the hierarchical clustering process* the clumps are concentrated to filaments rather than to pancakes. There are several aspects of the this controversy:

- Zel'dovich has made a rigorous mathematical statement: the ratio of the thickness of the region bounded by the caustic surface to its diameter is proportional to the time elapsed from the origin of the singularity δt, thus at $\delta t \to 0$ the region is infinitesimally thin. (Bond, Kofman, & Pogosyan 1996) dealt with finite structures and based their judgment on the visual impression without an exact definition of what they labeled pancakes, filaments, and ribbons (see Fig. 1 in (Bond, Kofman, & Pogosyan 1996)).
- (Bond, Kofman, & Pogosyan 1996) used an arbitrary amplitude for the smooth part of the initial perturbations $\sigma_\delta \approx 0.5 - 0.7$; (Coles, Melott, & Shandarin 1993) showed that the best correlation between the density distributions obtained in TZA and the N-body simulations is reached at $\sigma_\delta \approx 1$. The density is controlled by three parameters λ_1, λ_2 and λ_3 (eq.22) and has varying rate of growth (it may be even non-monotonic) in the Zel'dovich approximation (eq.22) therefore a part of the disagreement may be related to this difference in the amplitude.
- Even in the HDM simulations (Klypin & Shandarin 1983) that are perfectly described by the Zel'dovich approximation the most prominent features were clumps and then filaments not pancakes. (Arnold, Shandarin, & Zel'dovich 1982) explained that by stressing that the high order singularities (one-dimensional and especially point-like ones) are always stronger than pancakes. Coarse graining that always is present in numerical calculations of the density fields affects stronger the weaker concentrations of mass. Thus pancakes can be smeared out considerably.

In short, there is no principle contradiction between two approaches since both use the same equations (the Zel'dovich approximation); the only difference is in attaching labels (pancakes, filaments, ribbons, etc.) to the objects having finite sizes. A simple inspection of Fig. 1 in (Bond, Kofman, & Pogosyan 1996) shows that all types of objects can be found.

8. Conclusions

The linear theory in the Eulerian space is very misleading when it is used for predicting the appearance of the large-scale structure.

The truncated Zel'dovich approximation is the simplest nonlinear model that gives rough but correct ideas about the overall appearance of LSS. The advantage of the approximation is its simplicity and the local character. Its major drawback consists in that it artificially smears out all the clumps except the largest ones. The density distribution calculated from TZA with the current scale of nonlinearity may be roughly interpreted as the probability density of finding a galaxy. However, TZA predicts correctly the locations of the largest clumps and can be successfully used for studies of the distribution of the clusters of galaxies, which does not require a resolution better than $5 - 10 \ h^{-1} Mpc$ (Borgani et al 1994),(Plionis et al 1995). Another drawback is that it cannot be used for the accurate calculation of higher moments of the galaxy distribution function (skewness, kurtosis, etc) in weakly nonlinear regime because the true nonlinear corrections are different from those in TZA[9] For this type studies one has to use the perturbative expansions (Juszkiewicz this volume).

The adhesion approximation predicts much clumpier distributions than the truncated Zel'dovich approximation. This is reflected in better correspondence of the power spectrum and density distribution function to the N-body simulations. However, in the particle realization of the adhesion model the density distribution is still smoother than in the N-body simulations (Weinberg & Gunn 1990b), (Melott, Shandarin, & Weinberg 1994).

The adhesion approximation stresses the importance on the initial gravitational potential as a primary quantity determining the properties of the large-scale structure (see (Kofman & Shandarin 1988), (Kofman 1988)). (Sahni, Sathyaprakash, & Shandarin 1994) showed that the size of voids strongly correlates with the value of the primordial gravitational potential at void centers: the higher the potential the larger the voids.

The peak-patch picture method combines the Zel'dovich approximation and the peak theory. Its advantage is a better description of the gravitational dynamics within the clumps but at the cost of considerably greater complexity.

All above approximation can be also viewed as the smoothing procedures in the *Lagrangian* space followed by the mapping into the Eulerian space. The truncated Zel'dovich approximation represents smoothing with a spatially uniform filter that roughly equals the scale of nonlinearity (24). AA and PPP use adaptive filters of different types. After filtering the dy-

---

[9]However, the overall shape of the distribution function look quite good (Kofman et al 1994), (Bernardeau & Kofman 1995).

60

namical mapping into the Eulerian space is made by means of the Zel'dovich approximation (20).

N-body simulations remain the most universal practical tool for studying LSS. However, a healthy amount of skepticism should be applied to some conclusions derived from the simulations because of small dynamical range. In my opinion, the most promising approach that will reliably increase the dynamical range is the multigrid PM techniques (Villumsen 1988), (Anninos et al 1994), (Suisalu & Saar 1995), (Splinter 1996), (Kravtsov et al 1996).

The analytic and semi-analytic models of nonlinear gravitational clustering based on the Zel'dovich approximation have recently reached the level when they can explain qualitatively (in some cases quantitatively) major features of the large-scale galaxy distribution. In particular, these models predict that highly anisotropic shapes of superclusters (filaments and sheets) as well as their connectedness are natural generic consequences of cosmological gravitational clustering resulted from primordial density fluctuations of Gaussian type. The difference between different scenarios is mainly quantitative. Deriving the observational predictions from the models is a challenge for theorists. Ironically enough, history shows that theorists did not make many interesting observational predictions from simple Zel'dovich's theory even the theory itself (properly applied) allowed to do so. Measuring the statistical parameters of the large-scale structure in existing and especially upcoming large redshift surveys (e.g. Sloan Digital Sky Survey (Gunn & Weinberg 1995)), comparing them with the theoretical predictions will be an exciting field of cosmological research in coming years.

Acknowledgments
I thank A. Melott, R. Splinter and Ya. Suto for allowing me to use some of common but unpublished results. I thank S. Gottlöber for providing Fig.2. I am grateful for financial support from NASA Grant NAGW-3832, NSF Grants AST-9021414.

References

AARSETH, S.J., GOTT, J.R., & TURNER, E. 1979 *Astrophys. J.* **288**, 664.
ANNINOS, P., NORMAN, M., CLARKE, D.A. 1994 *Astrophys. J.* **436**, 11.
ARNOLD, V.I., SHANDARIN, S.F., & ZEL'DOVICH, YA.B. 1982 *Geophys. Astrophys. Fluid Dyn.* **20**, 111.
BAGLA, J.S., & PADMANABHAN, T. 1994 *MNRAS*, **266**, 227.
BARDEEN, J.M., BOND, J.R., KAISER, N., & SZALAY, A.S. 1986 *Astrophys. J.* **304**, 15.
BEACOM, J., DOMINIK, K., MELOTT, A.L., PERKINS, S., & SHANDARIN, S.F. 1991 *Astrophys. J.* **372**, 351.

BERNARDEAU, F., & KOFMAN, L. 1995 *Astrophys. J.* **443**, 479.

BERNARDEAU, F., SINGH, T.P., BANERJEE, B., & CHITRE, S.M. 1994 *MNRAS*, **269**, 947.

BERTSCHINGER, E. 1985 *Astrophys. J. Suppl.* **58**, 39.

BERTSCHINGER, E. 1994 *Physica D* **77**, 354.

BERTSCHINGER, E., & GELB, J. 1991 *Comp. Phys.* **5**, 164.

BLUMENTHAL, G.R., FABER, S.M., PRIMACK, J.R., & REES, M.J. 1984 *Nature* **311**, 517.

BOND, R. 1996 *this volume.*

BOND, R., & MYERS, S. 1996 *Astrophys. J. Suppl.* **103**, 1.

BOND, R., KOFMAN, L., & POGOSYAN, D. 1996 *Nature* **380**, 603.

BORGANI, S., COLES, P., & MOSCARDINI, L. 1994 *MNRAS* **271**, 223.

BOUCHET, F.R., COLOMBI, S., HIVON, E., & JUSZKIEWICZ, R. 1995 *Astron. Astrophys.* **296**, 575.

BRAINERD, T., SCHERRER, R.J., & VILLUMSEN, J.V. 1993 *Astrophys. J.*, **418**, 570.

BROADHURST, T.J., ELLIS, R., KOO, D., & SZALAY, A. 1990 *Nature* **343**, 726.

BUCHERT, T. 1993 *Astron. Astrophys.* **267**, L51.

BUCHERT, T., MELOTT, A.L., & WEISS, A.G. 1994 *Astron. Astrophys.* **288**, 349.

BUCHERT, T. & DOMINGUES, A. 1996 preprint

BURGERS, J.M. 1974 *The Non-Linear Diffusion Equation*, Reidel, Dordrecht.

CATELAN, P. 1995 *MNRAS* **276**, 115.

CHINCARINI, G., & ROOD, H.J. 1980 *Sky and Telescope*, **59**, 364.

CAVALIERE, A., MENCI, N., & TOZZI, P. 1996 *Astrophys. J.* in press.

COLES, P., & LUCHIN, F. 1995 *Cosmology: The Origin and Evolution of Cosmic Structures* John Wiley & Sons, Chichester.

COLES, P., MELOTT, A., & SHANDARIN, S.F. 1993 *MNRAS* **260**, 765.

COUCHMAN, H. 1991 *Astrophys. J. Lett.*, **368**, L23.

DEKEL, A. 1994 *Annu. Rev. Astron. Astrophys.*, **32**, 371.

DOROSHKEVICH, A.G., RYABENKII, V.S., & SHANDARIN, S.F. 1973 *Astrophysica* **9**, 144.

DOROSHKEVICH, A.G., KOTOK, E., NOVIKOV, I.D., POLYUDOV, A., SHANDARIN, S.F., & SIGOV, YU. 1980 *MNRAS* **192**, 321.

EFSTATHIOU, G. & EASTWOOD, J.W. 1981 *MNRAS* **194**, 503.

GIOVANELLI, R 1996 *this volume.*

GOTT, J.R., TURNER, E., & AARSETH, S. 1979 *Astrophys. J.* **234**, 13.

GOTTLÖBER, S. 1996 *Proc. of the Varenna summer school* .

GUNN, J., & WEINBERG, D. 1995 in *Wide-Field Spectroscopy and the Distant Universe*, eds. S.J.Maddox ans A. Aragón – Salamanca, World Scientific.

GURBATOV, SAICHEV, & SHANDARIN 1989 *MNRAS* **236**, 385.

HERNQUIST, L. 1987 *Astrophys. J. Suppl.*, **64**, 715.

HOCKNEY, R.W. & EASTWOOD, J.W. 1988 *Computer Simulations Using Particles*, Institute of Physics Publishing, Bristol.

JUSZKIEWICZ, R. 1996 *this volume.*

KLYPIN, A. 1996 *astro-ph/9605183*

KLYPIN, A., & SHANDARIN, S.F. 1983 *MNRAS* **204**, 891.

KLYPIN, A., HOLTZMAN, J., PRIMACK, J., & REGOS, E. 1993 *Astrophys. J.*, **416**, 1.

KOFMAN, L. 1988 in *Morphological Cosmology* Proc. of the XIth Cracow Cosmological School, eds. P.Flin and H.W. Duerbeck.

KOFMAN, L., & STAROBINSKY, A. 1985 *Sov. Astron. Lett.* **11**, 271.

KOFMAN, L., & SHANDARIN, S.F. 1988 *Nature* **334**, 129.

KOFMAN, L., POGOSYAN, D., & SHANDARIN, S.F. 1990 *MNRAS* **242**, 200.

KOFMAN, L., POGOSYAN, D., SHANDARIN, S.F., & MELOTT, A.L. 1992 *Astrophys. J.* **393**, 450.

KOFMAN, L., BERTSCHINGER, E., GELB, J.M., NUSSER, A., & DEKEL, A. 1994 *Astrophys. J.* **420**, 44.

KRAVTSOV, A.V., KLYPIN, A.A., & KHOKHLOV, A.M. 1996 *preprint.*

62

KUHLMAN, B., MELOTT, A.L.,& SHANDARIN, S.F. 1996 *Astrophys. J. Lett.* **470**, L41.
LANDY, S.D., SHECTMAN, S.A., LIN, H., KIRSHNER, R.P., OEMLER, A.A., & TUCKER, D. 1996 *Astrophys. J. Lett.* **456**, L1.
LIFSHITZ, E.M. 1946 *ZhETF* **16**, 587.
LITTLE, B., WEINBERG, D.H., & PARK, C. 1991 *MNRAS* **253**, 295.
MATARRESE, S., LUCCHIN, F., MOSCARDINI, L., & SAEZ, D. 1992 *MNRAS*, **259**, 437.
MELOTT, A.L., & SHANDARIN, S.F. 1993 *Astrophys. J.* **410**, 469.
MELOTT, A.L., PELMAN, T., & SHANDARIN, S.F. 1994 *MNRAS* **269**, 626.
MELOTT, A.L., SHANDARIN, S.F., & WEINBERG, D.H. 1994 *Astrophys. J.* **428**, 28.
MELOTT, A.L., SHANDARIN, S.F., SPLINTER , R.J. ,& SUTO, Y. 1997 *Astro-ph.*
MUNSHI, D., SAHNI, V., & STAROBINSKY, A.A. 1994 *Astrophys. J.*, **436**, 517.
NOVIKOV, I.D. 1964 *ZhETF* **46**, 686.
NUSSER, A., & DEKEL, A. 1990 *Astrophys. J.* **362**, 14.
OORT, J.H. 1983 *Annu. Rev. Astron. Astrophys.*, **21**, 373.
PAGE, L. 1996 *this volume.*
PEEBLES, P.J.E. 1980 *The Large-Scale Structure of the Universe*, Princeton University Press, Princeton.
PEEBLES, P.J.E. 1993 *Principles of Physical Cosmology*, Princeton University Press, Princeton.
PLIONIS, M., BORGANI, S., MOSCARDINI, L., COLES, P. 1995 *Astrophys. J.* **441**, L57.
RYDEN, B., & GRAMANN, M. 1991 *Astrophys. J. Lett.* **383**, L33.
SAHNI, V., & COLES, P. 1995 *Physics Reports* **262**, 1.
SAHNI, V., & SHANDARIN, S.F. 1996 *MNRAS* **282**, 641.
SAHNI, V., SATHYAPRAKASH, B., & SHANDARIN, S.F. 1994 *Astrophys. J.* **431**, 20.
SATHYAPRAKASH, B.S, SAHNI, V., MUNSHI, D., POGOSYAN, D., & MELOTT,A.L. 1995 *MNRAS* **275**, 463.
SCHRAMM, D. 1996 *this volume.*
SELLWOOD, J.A. 1997 to appear in *Computational Astrophysics*, eds. D.A. Clarke & M.J. West, ASP Conference Series.
SHANDARIN, S.F. 1993 *in Cosmic Velocity Fields* eds. F.R.Bouchet and M. Lachièze-Rey, Edition Frontieres, p.383.
SHANDARIN, S.F. 1994 *Physica D* **77**, 342.
SHANDARIN, S.F. & SATHYAPRAKASH, B. 1996 *Astrophys.J. Lett.* , **467**, L25.
SHANDARIN, S.F., & ZEL'DOVICH, YA.B. 1989 *Rev. Mod. Phys.* **61**, 185.
SHANDARIN, S.F., DOROSHKEVICH, A.G., & ZEL'DOVICH, YA.B. 1983 *Sov. Phys. Usp.* **26**, 46.
SHANDARIN, S.F., MELOTT, A.L., McDAVITT, K., PAULS, J.L., & TINKER, J. 1995 *Phys. Rev. Lett.* **75**, 7.
SPLINTER, R. 1996 *MNRAS* **281**, 281.
STRAUSS, M.A., & WILLICK, J.A. 1995 *Phys. Rep.*, **261**, 271
SUISALU, I. & SAAR, E. 1995 *MNRAS*, **274**, 287.
SUGINOHARA, T., SUTO, Y., BOUCHET, F.R., & HERNQUIST, L. *Astrophys.J.Suppl.* **75**, 631.
SZALAY, A. 1996 *this volume.*
TURNER, M. 1996 *this volume.*
TUROK, N. 1996 *this volume.*
VERGASSOLA, M., DUBRULLE, B., FRISCH, U., & NOULLEZ, A. 1994 *Astron. Astrophys.* **289**, 325.
VILLUMSEN, J.V. 1988 *Astrophys.J.Suppl.* **71**, 407.
WEINBERG, D.H., & GUNN, J. 1990a *Astrophys.J. Lett.* **352**, L25.
WEINBERG, D.H., & GUNN, J. 1990b *MNRAS* **247**, 260.
YESS, C., & SHANDARIN, S.F. 1996 *Astrophys.J.* **465**, 2.
ZEL'DOVICH, YA. B. 1970 *Astron. Astrophys.* **5**, 84.
ZEL'DOVICH, YA.B. & NOVIKOV, I.D. 1983 *The Structure and Evolution of the Universe* University of Chicago Press, Chicago, London.

THE LARGE SCALE STRUCTURE OF THE UNIVERSE

A.S. SZALAY

Department of Physics and Astronomy
The Johns Hopkins University, Baltimore

1. Introduction

The study of large scale structure is one of the most dynamically evolving areas of astrophysics today. Cosmology and large scale structure is growing into an accurate science and requires correspondingly more sophisticated methods of analysis. Twenty years ago the estimates of the fluctuation amplitude were about 10^{-3}, almost a factor of 100 off of today's measurements. Ten years ago we could only hope for high precision measurements of large scale structure, there were less than 5000 redshifts measured, and only a handful of normal galaxies with $z > 1$ were known. Computer models of structure formation had just begun to consider non-power-law spectra based on physical models like hot/cold dark matter. As a consequence there was considerable freedom in adjusting parameters in the various galaxy formation scenarios. In contrast, many of today's debates are about factors of 2 and soon we will be arguing about 10% differences. The shape of the primordial fluctuation spectrum, first derived from philosophical arguments [2, 1], can now be quantified from detections of fluctuations in the CBR made by COBE[3]. The number of available redshifts is beyond 50,000, and soon we will have redshift surveys surpassing 1 million galaxies. N-body simulations are becoming more sophisticated, of higher resolution, and incorporating complex gas dynamics. The unprecedented number of new observations currently under way give us hope that over the next decade we will gain a clear understanding of the shape and evolution of the primordial fluctuation spectrum, understand from first principles how galaxies were formed, and make quantitative comparisons and tests to differentiate among the various galaxy formation scenarios.

D. N. Schramm and P. Galeotti (eds.), Generation of Cosmological Large-Scale Structure, 63–74.
© 1997 *Kluwer Academic Publishers.*

2. Quantifying Large Scale Structure

Structure in the universe evolves from the initially small primordial fluctuations. These fluctuations can arise during an inflationary expansion or
come from topological defects later. They grow in amplitude, due to gravitational instability, and the shape of the fluctuation spectrum is altered by
different physical processes. The nature of the dark matter, whether hot or
cold, believed to dominate the mass density of the universe, determines the
shape of the power spectrum on small (< 100 Mpc) scales. On the other
hand, the shape of the large scale part of the fluctuations (> 200 Mpc) remains remarkably unchanged, because no scale in the evolutionary process
becomes this large.

The COBE measurements constrain both the amplitude and the initial
spectrum of the fluctuations in this regime, and demonstrate extremely
good agreement ($n = 1.1 \pm 0.4$) [4] with the Harrison- Zeldovich predictions
of $P(k) = k^n$, with $n = 1$. These fluctuations are due to differences in
the gravitational potential at the surface of last scattering[5], reflecting the
state of the universe at a redshift of ≈ 1000. Galaxy surveys (at $z < 0.3$) are
rapidly increasing in size, thus providing increasingly better measurements
of the fluctuations on small scales (CfA slices[6], IRAS[7], APM[8], APM
redshift surveys[9], LasCampanas[10]). One can use theoretical scenarios to
evolve and extrapolate the large scale CBR measurements into the structure
of the local universe, but the two regimes do not yet overlap directly.

The currently most popular scenario is the Cold Dark Matter dominated universe, where most of the mass is dark, interacting only via gravity, consisting of particles of such a large mass that their thermal motion
is negligible. To match the observed clustering of galaxies without producing too large a velocity dispersion, the concept of 'biasing' has been
invoked[11, 12]: mass is converted into light only at the densest regions
in the universe, creating a luminous component more clustered than the
mass. This scenario, modulo a properly chosen initial normalization, has
been remarkably successful over the last fifteen years.

The COBE measurements create a conflict with the minimal biased
CDM model: if a Harrison-Zeldovich spectrum is assumed and the normalization is locked to COBE, then the biasing parameter must be unity to
match the small scale part of the fluctuation spectrum, leading to very
large small scale velocities. Several alternative models have been rapidly
suggested. Gravity waves, which decay with time, may contribute to the
largest scale modes observed by COBE and produce a 'tilt' of the spectrum[13]. Alternative scenarios invoke a large cosmological constant[14]. A
mixture of cold and hot dark matter would also help [15].

What are the most important measurements we can make in order to differentiate between proposed models? Overlap between scales probed by CBR experiments and redshift surveys in the 'local' universe would place strong constraints on the power spectrum. Measurements of galaxy clustering on scales of 200-500 Mpc from well-sampled redshift surveys would tell us whether the gravity wave/tilted model is relevant, measure the bias factor, and determine the shape of the spectrum on scales where most of today's models differ but which are too small for COBE and beyond the scale of current galaxy measurements. For the same reason, many CBR experiments are probing 1-2 degree scales, corresponding to a comoving scale of about ≈ 120 Mpc. These experiments together with the redshift surveys will soon yield an unambigous answer.

3. Observing Walls

Several surveys have now found evidence for sharp, wall-like structures in the universe. The existence of such features is by no means unexpected, Zeldovich[18] predicted, that the generic features in a pressure-free gravitational collapse will be highly flattened 'pancakes'. Observational confirmation took a few years, Chincarini & Rood[16], and Gregory & Thompson[17] identified the excess of galaxies between Coma and A1367 with a supercluster, resembling a 'pancake'.

In 1980, Kirschner etal[19] identified the Bootes void, showing the first big 'void' in the galaxy distribution. A major breakthrough in our understanding of large scale structure came from the CfA 'slice' by deLapparent, Geller and Huchra[20], 6 degrees wide in declination but over 100 degrees in right ascension. In the region where the slice maps the universe, at a radial distance of $70h^{-1}$ Mpc, a distinct pattern appears: a 'Great Wall' containing hundreds of galaxies, connecting several of the known Abell clusters. Its tranverse spatial extent exceeds 100 by $50h^{-1}$ Mpc. The general trend has been profoundly summarized by Geller and Huchra [6]: 'all surveys have detected structures as large as they could ...'

If the universe were full of 'Great Walls', i.e. if they are typical of the very large scale structure, already from the surface density of galaxies one can get an estimate what would a 'fair sample' consist of. If we assume (in the extreme), that all bright galaxies are on these surfaces, with the surface density of $\mu = 0.4$ galaxies Mpc^{-2}, we can estimate the characteristic 'cell' size by requiring that the corresponding 'local' volume density of bright galaxies, $n = 0.01$ galaxies h^{-1} Mpc^{-3}, be approximately reproduced. Assuming spherical bubbles, and counting only half of the surface area, since the walls separate two volumes, the typical size of the voids is

$$\lambda = 2R = \frac{3\mu}{n} = 120h^{-1} \text{ Mpc.} \qquad (1)$$

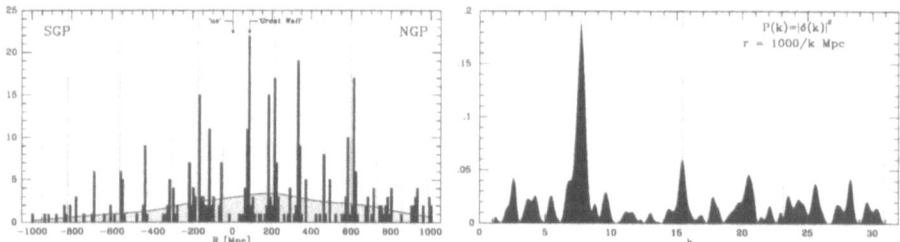

Figure 1. The redshift distribution of galaxies in the BEKS survey, comprised of two narrow pencilbeams towards the Galactic poles. The left hand figure shows the histogram of all galaxies, while the right hand side shows the 1D power spectrum. The big spike corresponds to the comoving scale of 128 h^{-1} Mpc.

This gives us some idea what cell sizes can one expect in a universe dominated by 'Great Walls', derived solely from the observations.

In 1990 Broadhurst, Ellis, Koo & Szalay[21] (BEKS) published results from a redshift survey in two opposite pencilbeams. The angular diameter of the survey is 30', and the depth is about 0.5 in redshift, both at the North and South Galactic Poles. The combined surveys have a joint length in excess of $2000h^{-1}$ Mpc, considerably deeper than any other survey before. To compensate for the small physical size of the survey at low redshifts, data from two bright surveys in almost the same directions were used, resulting in a combined selection well approximated by a cylinder of constant comoving radius.

The Northern pencilbeam is in the CfA slice, and one can find the 'Great Wall' without much difficulty. Surprisingly, however, at very large radial distances one still cannot see a homogeneous distribution, rather most galaxies are in a few large 'spikes' along the line of sight, separated typically by more than $100h^{-1}$ Mpc. The simplest explanation was that further 'walls' were found, meaning that the 'Great Wall' is by no means unique, and that these structures contain a large percentage ($\approx 50\%$) of the galaxies.

4. Observing Bumps

Even more suprising was the fact, that in the one-dimensional Fourier transform of the redshift distribution, a highly significant peak was found at the wavelength of 128 h^{-1} Mpc, with a probability of $P < 3 \times 10^{-4}$. This observation prompted many debates, and even more exotic theories. The main question was, of course, whether the peak in the Fourier spectrum is just a random accident, or does this scale arise as a result of a physical process? Extending the BEKS survey to 9 pencilbeams, randomly distributed

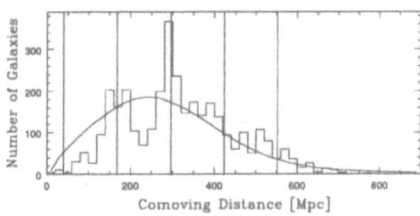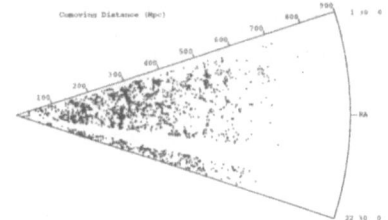

Figure 2. The distribution of galaxies in the ESP survey, a slice near the South Galactic pole. The left hand figure shows the redshift histogram. The vertical lines denote the corresponding spikes in BEKS. Note the excellent agreement. The right hand figure shows the cone diagram of the galaxy redshift agains right ascension. Figure courtesy of G. Chincarini and E. Zucca.

over a 6x6 degree region at both galactic poles, it was shown, that the cross-correlation signal stays strong up to about 60 h^{-1} Mpc transverse separation[22], indicating that the redshift spikes are indeed 'Great-Wall' like structures, also that the power spectrum peak was not due to a random alignment of small groups.

Several years later bigger redshift surveys became available. The Las-Campanas survey[10], consisting of 6 slices of 450 h^{-1} Mpc depth, found evidence for statistically significant excess power on 100 h^{-1} Mpc scales[23]. A similar slice near the South Galactic Pole, the ESP project[24] confirms the BEKS spikes in the overlap region. Deeper surveys on the Keck telescope[25], and the CFRS[26] found evidence for the existence of sharp walls at $z = 1$. Excess power on 100^+h^{-1} Mpc scales is present at even higher redshifts, in QSO absorption systems[27] and in galaxies[28]. These high redshift observations are extremely important — if the bumps appear on the same comoving scale at much earlier Hubble times, then there is a built-in feature in the power spectrum!

5. Sampling Strategies

If fluctuations in the universe are strictly Gaussian, their full statistical description is contained in the two-point correlation function or in its Fourier transform, the power spectrum. The phases of the individual Fourier components are random for such a process, and all high-order measures of clustering vanish. In this case, Kaiser[29] shows that measuring the redshifts of a small fraction of the galaxies is the most efficient way to measure the power spectrum on large scales. However, if the universe contains sharp large scale features like the "Great Wall," a sparsely sampled survey may fail to identify them because it is less sensitive to the higher-order correlations (equivalently, the phase correlations) that characterize such struc-

Voronoi foam, smoothed original   Voronoi foam, random phases

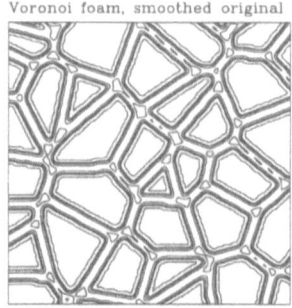 

Figure 3. Two simple realizations of a two-dimensional universe, with identical second order statistical properties. The left hand figure is a two dimensional Voronoi foam, generated by the median surfaces between Poisson 'seeds' at the mean separation of 100 Mpc. In this simple toy model galaxies reside only on the walls of the foam, smoothed, so the walls have a finite thickness. The structure has a well defined second order statistic, but also has well correlated phases. This picture has been Fourier transformed, all the phases randomized, then transformed back again. The result is shown on the right hand side, with the same second order properties, but with a Gaussian distribution. It is easy to see, that placing well-sampled pencilbeams across both surveys will easily distinguish between the two, whereas a sparse sample drawn from the two realizations cannot differentiate.

tures. A distribution with high-order clustering can be very different from a homogeneous, isotropic, Gaussian random field with the same second-order statistical properties. A sparse survey optimized to measure the two-point correlation function would miss the very real differences between these distributions.

Thus we can understand now how sparse sampling affects power spectrum estimators: in the ensemble average, the power spectrum as a second order statistic is invariant with respect to sampling. Averaging over an infinite number of finite size realizations, the correct power spectrum is recovered, even for a very low sampling rate. On the other hand, if there is a network of sharp 'walls' present, they are manifested as a set of sharp 'spikes' in Fourier space. These sharp spikes will vary from realization to realization, and in an ensemble average they will converge to the underlying power spectrum. Even though both scenarios converge to the true power spectrum in the infinite limit, it is much harder to detect the sharp Fourier spikes with a low sampling rate in a single local realization, like our nearby Universe. This is why the well-sampled pencilbeams may yield seemingly quite different results for the statistics of power spectrum amplitudes than wide angle sparsely sampled surveys.

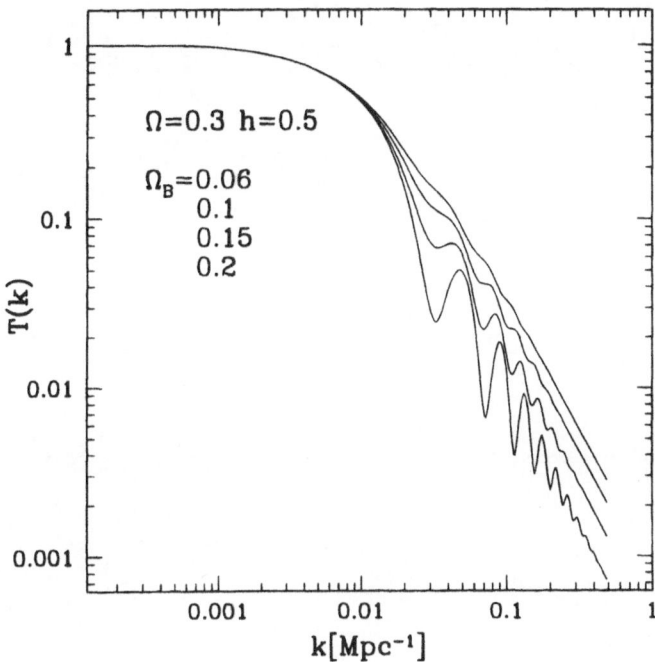

Figure 4. The transfer function for the fluctuations in a high baryon content universe. In order to get the dimensionless contribution to the variance, $k^3 P(k)$, this curve needs to be multiplied by k^4, for a Zeldovich spectrum. Note, that the scale of the first peak is at $k = 0.05$, corresponding to $125\ h^{-1}$ Mpc. This figure was kindly provided by Wayne Hu.

6. Origin of the 100 Mpc Bumps

Here I would like to discuss, how can such $100\ h^{-1}$ Mpc bumps arise in the power spectrum. It has been understood for a long time [31, 32] that around recombination due to the high pressure in the photon-baryon plasma, fluctuations oscillate like sound waves. On smaller wavelengths, these oscillations damp, but on larger scales, near the horizon scale at recombinations, they may survive longer. The motion of baryons due to these sound waves gives rise to the Doppler-peaks in the CMB fluctuations. At the same time it was understood early on, that after recombination, as the sound speed approaches zero, an interference pattern may emerge, the so-called Sakharov-oscillations[33]. Sound waves which exactly fit inside the horizon will amplify, others with opposite phases will cancel. Since the horizon scale at recombination is very close to the regime of interest (between 100-200 h^{-1} Mpc, depending on $\Omega_0 h$), it is worth to consider what does it take for these sound-waves to have an appreciable effect on not only the CMB but on the galaxy distribution!

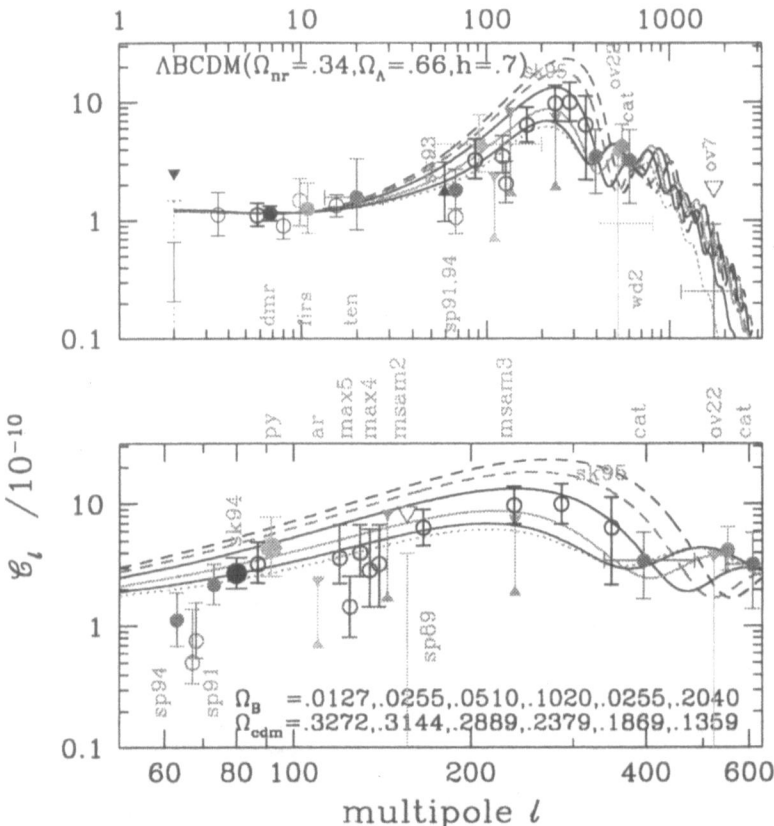

Figure 5.   The angular multipole amplitudes of the CMB fluctuations for a family of models with high baryon density plotted with the results of recent small-scale anisotropy measurements. This figure is from Bond and Szalay (1997).

Since galaxy clustering is only affected by gravity, the fluctuations in the baryons due to the sound waves need to leave an imprint in the gravitational potential. This requires as high a baryon fraction as possible. How high can this number be? From observations of the primordial deuterium[34], the 1σ limit $\Omega_B h^2 \leq 0.025$, can be combined with reasonably low estimates of the Hubble constant, and values of $\Omega_B \approx 0.1$ are not unimaginable. At the same time, in order for a large imprint, Ω_t has to be low, in the range of $\Omega_t \approx 0.4$. Given the faint number counts of galaxies, this is again not an outrageous idea any longer. Calculations of W. Hu [35] indicate, that in these scenarios there will be several bumps in the power spectrum, the first at $kh = 0.05$. The amplitude for $\Omega_B = 0.1$, and $\Omega_t = 0.3$ is quite substantial, although nowhere as high as e.g. the BEKS or the LasCampanas detections.

On the other hand, there are several other amplification mechanisms at work. The surveys measure the distribution of galaxies, while the above calculations refer to the linear fluctuations in all the mass. First of all, the formation of the pancakes is a highly non-linear process, which will amplify fluctuations, if there is a distinct scale associated with them. Second, the galaxy surveys are analysed in redshift space, thus infall on to the walls will enhance these structures, and will result in a further amplification. This effect will depend on the survey geometry: it is very important for pencil-beams, less so for slices, and spherical volumes. Even in the Las Campanas survey one can notice, that some of the walls curve to stay perpendicular to the line of sight – a consequence of redshift space enhancements. The high baryon content has another effect: the increased viscous damping will decrease the fluctuations on small scales. If this effect is too large, it cannot be compensated for by the bias factor. This provides a practical upper limit on how high Ω_B/Ω_t can become. The relevant parameter range ($h = 0.5 - 0.65, \Omega_B = 0.06 - 0.12, \Omega_t = 0.2 - 0.4$) has not been particularly well studied. We are currently undertaking such a linear analysis, combining the results with the possible nonlinear amplifications mechanisms, to sharpen these constraints further[36] (Bond and Szalay 1997).

7. The Future: The Sloan Digital Sky Survey

The Sloan Digital Sky Survey (SDSS) is a collaboration between the University of Chicago, Princeton University, the Johns Hopkins University, the University of Washington, Fermi National Accelerator Laboratory, the Japanese Promotion Group, the United States Naval Observatory, and the Institute for Advanced Study, Princeton, with additional funding provided by the Alfred P. Sloan Foundation and the National Science Foundation. In order to perform the observations, a dedicated 2.5 meter Ritchey-Chretien telescope was constructed at Apache Point, New Mexico, USA. This telescope is designed to have a large, flat focal plane which provides a 3 field of view. This design results from an attempt to balance the areal coverage of the instrument against the detector's pixel resolution.

The survey has two main components: a photometric survey, and a spectroscopic survey. The photometric survey is produced by drift scan imaging of 10,000 square degrees centered on the North Galactic Cap using five broad-band filters that range from the ultra-violet to the infra-red. The photometric imaging will use a CCD array that consists of 30 2K x 2K imaging CCDs, 22 2K x 400 astrometric CCDs, and 2 2K x 400 Focus CCDs. The data rate from this camera will exceed 8 Megabytes per second, and the total amount of raw data will exceed 40TB.

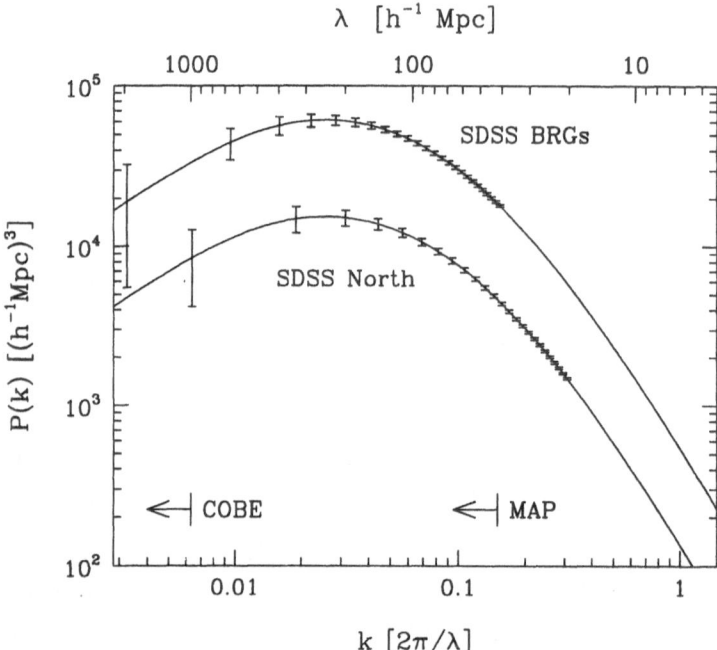

Figure 6. Model power spectra. The power spectra for the SDSS Northern galaxy survey and for the luminous red galaxies (BRGs) are shown. The latter are assumed to be biased by a factor of two with respect to the galaxies. The individual points and their error bars are statistically independent. On small scales, the errors are smaller than the smallest bars. Also shown are the smallest comoving wavelength scales accessible to COBE and to the upcoming Microwave Anisotropy Probe (MAP).

The spectroscopic survey will target over a million objects chosen from the photometric survey in an attempt to produce a statistically uniform sample. This survey will utilize two multi-fiber medium resolution spectrographs, with a total of 640 optical fibers, of 3 seconds of arc in diameter, that provide spectral coverage from 3900 - 9200 Å. The telescope will gather about 5000 galaxy spectra in one night. The total number of spectra known to astronomers today is about 60,000 - only 12 days of SDSS data! Whenever the Northern Galactic cap is not accessible from the telescope site, a complementary survey will repeatedly image several areas in the Southern Galactic cap to study fainter objects and identify any variable sources.

We are in the middle of designing and constructing an extremely ambitious project, aiming to provide a useful tool for almost all astronomers in the world. This endavour would not have been possible ten years ago, and even now we are pushing the limits of technology. We hope that our efforts will be successful, and the result will substantially change the way scientists do astronomy today. Having 200 million objects at our fingertips will undoubtedly lead to new major discoveries, and the spectroscopic followup of even a fraction of our fainter objects occupy astronomers for decades. The day when we have a "Digital Sky" at our desktop may be nearer than most astronomers think. Given the enormous public interest in astronomy, we hope that the resulting archive will also provide a challenge and inspiration to thousands of interested high-school students, and a lot of fun for the web-surfing public.

8. Summary

In summary, there are several observations pointing to excess power on 100-130 h^{-1} Mpc scales, which manifests itself in a small number of sharp spikes in Fourier space. These reflect the presence of walls and voids on similar scales. The emergence of this comoving scale at high redshift implies that this is imprinted on the fluctuations. Such a scale occurs naturally at recombination. The Sakharov oscillations, remnants of the sound waves at that epoch may provide an intriguing explanation. In such scenarios the baryon content of the Universe must be high, the Hubble constrant low, and the Universe open. This family of models deserves further investigation, and it is just barely possible that the 100 Mpc bumps may be the first preview of the elusive Doppler peaks — a fascinating preview of further connections between the galaxy distribution and the Cosmic Microwave Background. The next generation of redshift surveys, such as the SDSS and 2dF will carry also the galaxy clustering studies into the high precision era of cosmology.

Acknowledgments

The author would like to acknowledge useful discussions with Dick Bond, Joe Silk, Lyman Page, Wayne Hu and Istvan Szapudi. The author is supported by NASA LTSA and a grant from the Seaver Institute.

References

1. Zel'dovich, Ya.B. 1972, *M.N.R.A.S.*, **160**, 1P.
2. Harrison, E.R. 1970, *Phys.Rev.D*, **1**, 2726.
3. Smoot, G.F. et al. 1992, *Ap.J. Lett.*, **396**, L1.
4. Gorski, K.M., Hinshaw, G., Banday, A.J., Bennett, C.L., Wright, E.L., Kogut, A., Smooth, G.F. & Lubin, P. 1994, *Ap.J. Lett.*, **430**, L89.
5. Sachs, R.K., & Wolfe, A.M. 1967, *Ap.J.*, **147**, 73.
6. Geller, M.J. & Huchra, J.P. 1989, *Science*, **246**, 897.

7. Saunders, W., Frenk, C., Rowan-Robinson, M., Lawrence, A. & Efstathiou, G. 1991, *Nature*, **349**, 32.

8. Maddox, S.J., Efstathiou, G., Sutherland, W.J. & Loveday, J. 1990, *M.N.R.A.S.*, **242**, 43P.

9. Loveday, J., Efstathiou, G, Peterson B.A. & Maddox, S.J. 1992, *Ap.J.*, **400**, L43.

10. Shectman, S.A., Landy, S.A., Oemler, A., Tucker, D., Lin, H., Kirschner, R.L. & Schechter, P.L. 1996, *Ap.J. Supp.*, **470**, 172.

11. Kaiser, N. 1984, *Ap.J. Lett.*, **284**, L9.

12. Bardeen, J.M., Bond, J.R., Kaiser, N., & Szalay, A.S. 1986, *Ap.J.*, **304**, 15.

13. Davis, R.L., Hodges, H.M., Smoot, G.F., Steinhardt, P.J., & Turner, M.S. 1992, *Phys. Rev. Lett.*, **69**, 1856.

14. Kofman, L., Gnedin, N., & Bahcall, N.A. 1993, *Ap.J.*, **413**, 1.

15. Klypin, A., Holtzman, J., Primack, J., & Regös, E. 1993, *Ap.J.*, **416**, 1.

16. Chincarini,G. & Rood, H.J. 30, *Ap.J.*, **206**, 1976.

17. Gregory,S.A. & Thompson, L.A. 784, *Ap.J.*, **222**, 1978.

18. Zeldovich,Ya.B. 1970, *Astron. Astrophys.*, **5**, 84.

19. Kirshner, R.P., Oemler, A.J., Schechter, P.L. & Shectman, S.A. 1983, *Astron.J.*, **88**, 1285.

20. de Lapparent, V., Geller, M.J., & Huchra, J.P. 1986, *Ap.J. Lett.*, **301**, L1.

21. Broadhurst, T.J., Ellis, R.S., Koo, D.C., & Szalay, A.S. 1990, *Nature*, **343**, 726.

22. Broadhurst,T.J., Ellis,R.S., Ellman,N.E., Koo,D.C. & Szalay,A.S. eds: H. McGillivray and C. Collins, *Proc. of ROE Meeting on Digital Sky Surveys*, **1992**, p.397.

23. Landy, S.D., Shectman, S.A., Lin H., Kirschner, R.P., Oemler, A.A. & Tucker, D. 1996, *Ap.J. Lett.*, **456**, 1L.

24. Vettolani, G., Zucca, E. etal 1997, *Astron. Astrophys.*, **in press**, .

25. Cohen, J.G., Cowie, L.L., Hogg D.W., Songalia, A., Blandford, R., Hu, E.M. & Shopbell, P. 1996, *Ap.J.*, **471**, 5.

26. Lilly, S.J., Tresse, L., Hammer, F., Crampton, D. & Le Fevre, O. 1996, *Astron.J.*, , in press.

27. Quashnock, J.M.,VanDen Berk, D.E., York, D.G 1996, *Ap.J.*, **472**, 69.

28. Steidel, C. 1997, this volume.

29. Kaiser, N. 1986, *M.N.R.A.S.*, **219**, 785.

30. Szapudi,I. and Szalay, A.S. 1996, *Ap.J.*, **459**, 504.

31. Peebles, P.J.E. 1968, *Ap.J.*, **153**, 1.

32. Sunyaev, R.A. & Zeldovich, Ya.B. 1970, *Astrophys. Space Sci.*, **7**, 3.

33. Sakharov, A. 1966, *JETP(english)*, **22**, 241.

34. Tytler, D. 1997, this volume.

35. Hu, W. *Ph.D. Thesis*, (The University of California, Berkeley, 1996).

36. Bond, J.R. & Szalay, A.S. 1997, *Nature*, in preparation.

ON OBSERVING THE COSMIC MICROWAVE BACKGROUND

L.A. PAGE

Princeton University
Dept. of Physics
Princeton, NJ

Abstract.

The cosmic microwave background (CMB) comprises the oldest photons in the universe and is arguably our most direct cosmological observable. All precise and accurate measurements of its attributes serve to distinguish between cosmological models. Detector technology and observing techniques have advanced to the point where fluctuations in the CMB of order a few microkelvin are measured almost routinely. In these lecture notes, we review recent measurements of both the absolute temperature and the anisotropy of the CMB and discuss the relation between the data and the general theoretical framework. Future directions are indicated and the upcoming satellite experiments are discussed.

1. Introduction

The CMB is a powerful probe of cosmology because essentially no steps separate what is measured from what is of cosmological import; what you see is what you get. The CMB photons have free-streamed through the cosmos since last scattering off electrons some 100,000 years after the big bang. The spectrum of the CMB is indistinguishable from a Planck function to roughly 0.01% accuracy. This tells us that there were not any highly energetic cosmic processes, that coupled to photons, before $z \approx 10^3$. The near perfect shape of the spectrum is the strongest evidence to date that the hot big-bang model is correct. The pattern of minute spatial variations or anisotropy in the CMB, which are of order $\delta T/T \approx 10^{-5}$, is a fossil of the early universe. Furthermore, most models that give rise to cosmic structure affect the CMB, leaving an imprint of a small temperature fluctuation.

D. N. Schramm and P. Galeotti (eds.), Generation of Cosmological Large-Scale Structure, 75–104.
© 1997 *Kluwer Academic Publishers.*

There are many review articles on both the spectrum of and anisotropy in the CMB. For the anisotropy see [1]-[6]; for a recent review of both the spectrum and anisotropy results see [7] and [8]; and for reviews of the theory and results on the spectrum see [9], [10], or [11]. In addition, Partridge has written a new book [12] devoted to the subject.

The outline for these notes is as follows. We discuss the microwave/far-infrared sky in Section 2. Next, in Section 3, we discuss the recent results of absolute temperature measurements of the CMB. After a model of the anisotropy is developed in Section 4, Section 5 provides a discussion of the measurement technologies and an overview of the canonical formalism for describing the anisotropy. In Section 6, we review the state of the field. We end with a discussion of the upcoming satellite missions in Section 7.

2. The Microwave/Far-Infrared Sky

The CMB is the brightest broad-band diffuse emitter in the sky between about 1 and 500 GHz, completely dominating the Galactic foreground emission. At the low frequency end of this range, Galactic synchrotron emission exceeds the CMB and at the high frequency end, interstellar dust emission dominates. Galactic bremsstrahlung, or free-free emission, is the largest foreground near 90 GHz. Figure 1 shows the frequency spectrum of the diffuse Galactic emission between 3 and 3000 GHz near a Galactic latitude of $20°$. There are two sources not shown on this plot. Near 3000 GHz, thermal emission from the interplanetary dust (Zodiacal light) is roughly ten times smaller than the interstellar dust and its brightness scales with frequency as ν^4. Throughout the plotted range, interstellar molecular line emission is also observed [13].

In addition to the diffuse foreground, galactic and extragalactic point-like objects such as quasars, blazars, gigahertz-peaked sources, and loud radio sources also emit microwave and far-infrared radiation. The spectrum of most sources falls with increasing frequency, though not of every source.[1] Relatively little is know about sources near 90 GHz. There are no deep unbiased surveys from which to ascertain the ensemble properties with confidence. Also, we know that many of the extragalactic high-frequency sources are variable.

In Figure 1 it is evident that to probe either the spectrum or the anisotropy to a part in 10^3 or 10^5 the foreground emission must be confronted. Before COBE[14], the best full-sky maps were the "Haslam *et al.*" map at 408 MHz[15] and the IRAS dust map, with Zodiacal light subtracted, at 3000 GHz (100 μm)[16]. Extrapolation of these maps to fre-

---

[1]For sources, "flat spectrum" means that $S(\nu)$ is independent of ν, similar to free-free emission.

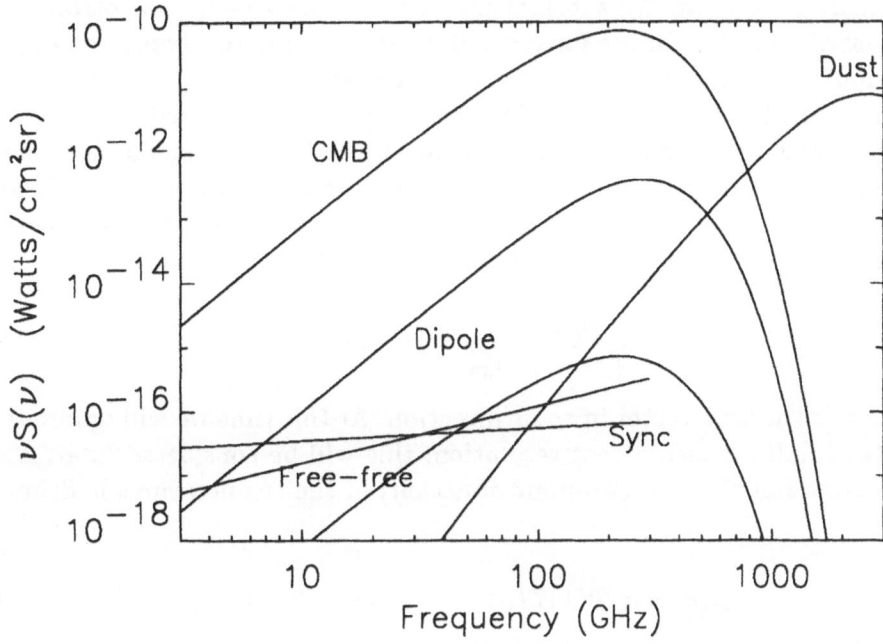

Figure 1. The microwave sky from 3 to 3000 GHz near a Galactic latitude of $b = 20°$. The ordinate is the brightness of the sky times the
frequency. With this convention, the plot indicates the distribution of
power. For synchrotron emission $S(\nu) \propto \nu^{-0.7}$; for free-free emission
$S(\nu) \propto \nu^{-0.1}$; and for dust emission near 100 GHz $S(\nu) \propto \nu^{3.7}$. The
scaling in effective temperature is $T(\nu) \propto \nu^{-2} S(\nu)$. The lowest Planck-
like curve is for $T = 27\ \mu K$.

quencies of interest is problematic because the spectral index varies from
place to place on the sky and there are components, such as free-free and
cold dust emission, that are missing from these maps. A believable CMB
spectrum or anisotropy experiment *must* measure the CMB and foreground
emission to similar precision.

In addition to the frequency dependence of the foreground emission,
there is also a spatial dependence. At the largest angular scales, Galactic emission falls off with Galactic latitude roughly as $1/\sin(b)$. In other
words, its intensity distribution, to first order, resembles a quadrupole: hot
around a circle, cold near the poles. Indeed this quadrupole confounded
some of the early measurements of the anisotropy and its removal from the
COBE/DMR data requires ingenuity [17].

The spatial distribution of celestial sources is commonly quantified with the angular spectrum. This formalism is used to describe the anisotropy, radio sources, and the diffuse foreground. It allows a direct comparison of the contribution from each. To illustrate it, the relatively simple case of foreground emission by extra-galactic radio sources is considered. We assume radio sources are randomly placed on the sky and so the angular spectrum should be that of "white noise." If we measure the total source emission temperature in each sky pixel, then that distribution may be represented as an expansion in spherical harmonics,

$$T(\hat{x}) = \sum_{l,m} a_{lm} Y_{lm}(\hat{x}), \tag{1}$$

where $\hat{x}$ is the unit vector in some direction. At this time we will ignore the effects of finite measurement resolution; this will be considered later. The correlation function (or two-point function) of the temperatures is defined as

$$C(\theta) = < T(\hat{x})T(\hat{y}) >, \quad \text{where} \quad \theta = \hat{x} \cdot \hat{y}, \tag{2}$$

and where the diagonal bracket indicates an ensemble average over many universes. For small angular scales, the ensemble average can be replaced by an average over positions in our universe. For a Gaussian random field, all the information is contained in the two-point correlation function (eq. 2). We do not know if the CMB is a Gaussian field, and we know that Galactic emission is certainly not. However, we still use the two-point function; a more complete description would use the higher-point correlation functions. Because $C(\theta)$ depends only on the angular separation between two directions, it may be expanded in Legendre polynomials as

$$C(\theta) = \sum_l \frac{2l+1}{4\pi} C_l P_l(\cos(\theta)) \quad \text{where} \quad C_l = < |a_{lm}|^2 > . \tag{3}$$

Here the brackets denote averaging over the $2l + 1$ values of m. The variance of the pixel temperatures is given by $C(0)$. Because $P_l(0) = 1$, the variance in each "mode" l is just $(2l + 1)C_l/4\pi$. For "white noise," $C_l = const$. We may see this by considering that the magnitude squared of the Fourier transform of a uniform distribution of sources will also be a uniform distribution in 2-D Fourier space; there is no preferred scale. Call the uniform value U. For small angles, the conjugate variable to angular separation is $l+1/2$, where $l+1/2 \approx l = 1/\delta\theta$ where $\delta\theta$ is the angular scale. Now, the variance for a band of modes of width δl at a radius l is given by the area in Fourier space or $U2\pi l\delta l$. In other words, the $C_l \propto U = const$.

To visualize the angular spectrum ("power spectrum" is usually reserved for the C_l), one generally plots $(l(2l + 1)C_l/4\pi)^{1/2} = \delta T_l$ versus l. The

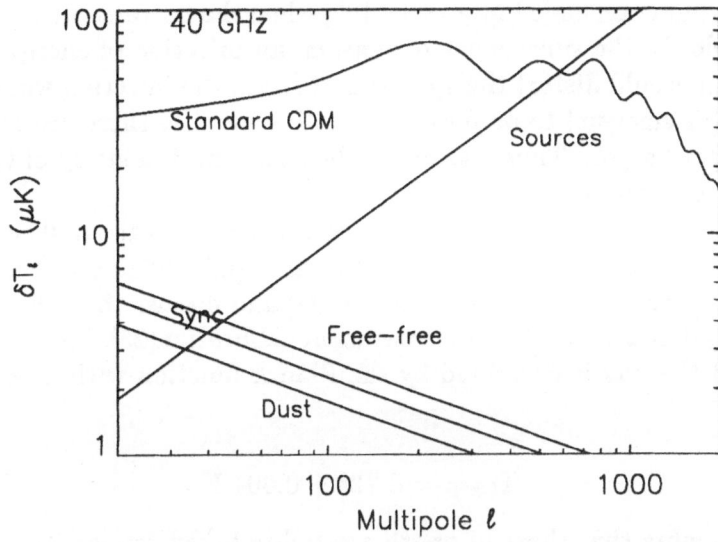

Figure 2. Spatial spectrum of the foregrounds from Netterfield *et al.* [18] and references therein. This is for the region near the North Celestial Pole at a frequency of 40 GHz. At higher frequencies, the flux from dust increases and that from synchrotron and radio sources decreases. The curved line is for standard CDM (from [19]).

extra factor of l means that δT_l is the root mean squared fluctuation per logarithmic interval, $\delta l/l$. The results from such an analysis for radio sources and Galactic emission at 40 GHz are shown in Figure 2.

From Figures 1 and 2, it is clear that the foreground emission must be well understood before the absolute CMB spectrum is known to 0.01% near 3 GHz or before the angular spectrum is known to the few percent level. One mitigating factor for the anisotropy is that the foreground fluctuations add to the signal in quadrature in the approximation that they are random fields. The absolute measurements do not enjoy this benefit. We are a long way from understanding the Galactic emission and radio sources. For a thorough up-to-date assessment, see Tegmark and Efstathiou [20].

3. Measurements of the Spectrum

The spectrum of the CMB is as close to that of a blackbody as can be measured; no distortions have been detected. This is our best evidence that the universe went through a hot dense phase when everything that interacts with photons was in thermal equilibrium. In current models, the epoch of photon production ended at $z \approx 3 \times 10^6$, when the universe was roughly two months old. In the subsequent expansion, an injection of energy mediated by baryons would distort the spectrum, though the injection would have to be large (or efficient) to be detectable today because there are roughly 10^9 photons per baryon. Thus a study of the spectrum is a study of the history of cosmic energetics.

The FIRAS experiment [21] aboard the *COBE* [14] satellite measured the flux from the sky between 2 and 96 cm^{-1} (60 – 2880 GHz). Fixsen and colleagues give the most recent results [22] and discuss the exhaustive program of systematic checks and instrument calibration[23]. The FIRAS team finds that the flux is described by the Planck function with a temperature of

$$T_{\text{CMB}} = 2.728 \pm 0.004 \text{ K}. \tag{4}$$

One doubts that short of another satellite-based experiment this result will be matched or bettered at frequencies above 100 GHz. It is comforting that the UBC rocket experiment [24] gives a consistent result.

The error on the FIRAS result, 4 mK, is due entirely to systematic effects and Fixsen *et al.* interpret it as a 95% confidence limit. (The statistical error is 7 μK.) The error exists to tell the readers how confident the authors are in the results. It should not be interpreted in the sense that "if one hundred experiments were performed only ten would lie outside the error bounds."

When interpreting this result, one must bear in mind that it comes from a *model* of the data. If the model is not correct, the results must be re-interpreted. To be more specific, the analyzed data come from maps of the sky in multiple frequency bands between 2 and 21 cm^{-1}. A four parameter fit is made to the maps at $|b| > 5°$ at each frequency,

$$\text{FIRAS}(\nu) = \alpha_0(\nu)\text{Uniform} + \alpha_1(\nu)\text{Dipole} + \alpha_2(\nu)\text{D9} + \alpha_3(\nu)\text{D10}. \tag{5}$$

The fit parameters scale the spatial distributions of a uniform background, the CMB dipole, channel 9 from the DIRBE experiment (D9, 72 cm^{-1}) and channel 10 from the DIRBE experiment (D10, 42 cm^{-1}). The last two maps are found to be good measures of the interstellar dust distribution. Two maps are needed because there may be multiple dust components or, in an explanation that Fixsen [25] prefers, the dust temperature may be a function of position.

The set of coefficients of the uniform component, $\alpha_0(\nu)$, is then fit to a combination of four frequency distributions. They are a) a blackbody, b) the derivative of a blackbody (to fit an error to the temperature scale), c) the

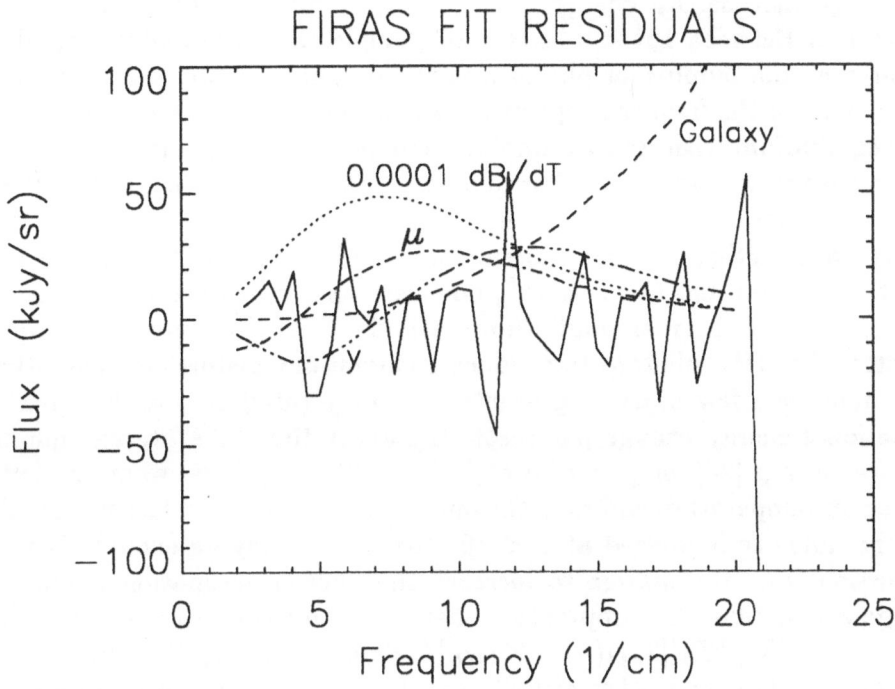

Figure 3. Residuals of the fit to $\alpha_0(\nu)$ in eq. 5 from [22]. If one starts
with the $\alpha_0(\nu)$ in eq. 5 and subtracts a blackbody (curve not shown),
a small calibration correction term (0.0001 dB/dT, where B(ν) is the
Planck function), and a model of the Galactic spectrum, the solid line
is obtained. The Galactic spectrum is shown scaled to 1/4 its value
at the galactic poles. In other words, if a Galactic-type spectrum were
not subtracted, the spectrum of the uniform sky map would rise with
frequency. The peak of the CMB is near $\nu = 5.5$ cm^{-1} = 165 GHz.
The intensity measured there is roughly 385 MJy/sr. The μ and y
distortions are shown at the 95% CL values in eq. 6. The units on the
abscissa are converted to GHz by multiplying by 30.

spectrum of the Galaxy (that accounts for residual Galactic signal in the
monopole!), and d) a spectral distortion. The distortion is parameterized by
either a chemical potential μ, or a Compton y−factor. Only one distortion
is fit at a time because the spectral signatures of the two are anti-correlated.
The residuals to this fit and some of the basis functions are shown in Figure
3.

Before $z \approx 3 \times 10^6$, double Compton scattering and free-free emission
maintain the thermal equilibrium of the CMB with the surroundings by

creating photons. An energy input simply results in a hotter CMB temperature. Between $10^5 < z < 3 \times 10^6$, single Compton scattering, which conserves the number of photons, is the dominant scattering mechanism over most of the frequency spectrum. In this epoch, the CMB is in *statistical* equilibrium with its surroundings and the distribution is characterized by a chemical potential μ. (The quoted numbers are for the unitless chemical potential; the flux is $S_\nu(T,\mu) = 2h\nu^3/[\exp(h\nu/kT_{CMB} + \mu) - 1]$). At long wavelengths, free-free emission is still effective and "fills in the tail" of the distribution. For $z < 10^5$, hot electrons, which are neither in statistical nor in thermal equilibrium with the CMB, can inverse Compton scatter the CMB photons to produce a Compton y distortion. When there are relatively few scattering events, one may think of y as the average fractional energy change per scattering event times the average number of scatterings [26], or $y = 1/m_e c^2 \int [k(T_e - T_{CMB})]d\tau_e$ [9] where T_e is the electron temperature and τ_e is the optical depth due to scattering. Finally, if the universe is ionized at $z < 10^3$ then there may be enough free-free emission from the plasma to increase the photon occupation number at large λ and thus the temperature there. This distortion is parameterized by $Y_{ff} = (h\nu/kT)^2[T_{eff}(\nu) - T_{CMB}]/T_{CMB}$, with T_{eff} the plasma temperature. These and other distortions, along with their interpretation, are discussed in [9], [11], [27], [4].

The best limits on y and μ come from FIRAS [22]. From these, Wright *et al.* [27] constrain energy injection in the early universe as shown in Figure 4. The limit on Y_{ff} [28] comes from a fit of the low frequency data. The limits are:

$$
\begin{aligned}
|y| &< 1.5 \times 10^{-5} \quad 95\% \ CL, \\
|\mu| &< 9 \times 10^{-5} \quad 95\% \ CL, \\
Y_{ff} &< 1.9 \times 10^{-5} \quad 95\% \ CL.
\end{aligned}
\tag{6}
$$

The y distortion is manifest at high frequencies and it will be a long time before the FIRAS limit is improved. The limit on y also strongly constrains alternative models of the origin of the CMB. One may try to mimic a Planck spectrum with a superposition of multiple grey bodies. At long wavelengths, the Rayleigh-Jeans region, the results cannot be distinguished. However, near the peak of the spectrum, such a superposition will result in a y-distortion. We also note that if the universe is inhomogenous on the largest scales, and we are not at a preferred center, a distortion will result [29].

The signatures of any μ and Y_{ff} distortions are evident at low frequencies. While the current generation of experiments will just barely, if at all, improve on the FIRAS limits, they are paving the way for the next genera-

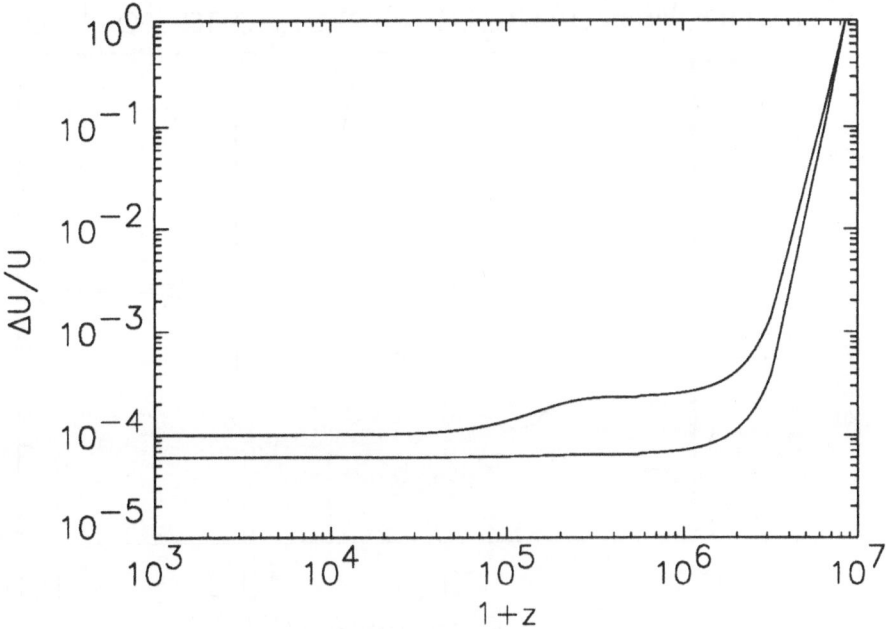

Figure 4. Limits on energy injection into the CMB prior to decoupling. The top curve uses $y = 2.5 \times 10^{-5}$ and $\mu = 3.3 \times 10^{-4}$. These were the limits from FIRAS in 1994 (Wright *et al.* [27]). The different epochs are clearly evident. The bottom curve is based on the FIRAS limits in eq. 6. This plot was produced with a program written by Ned Wright and is based on models in Burigana *et al.* [32], [33].

tion which may detect a distortion. Figure 5 shows a plot of the spectrum along with the μ and Y_{ff} distortion limits.

Two groups [30], [31] are pursuing long-wavelength measurements of the spectrum. Outside of the precise instrumentation necessary to perform absolute measurements between 0.1% and 1%, one must contend with Galactic and atmospheric emission. At 600 MHz, the temperature of the Galaxy is roughly 5.8 K [34] and falls as $\nu^{-2.7}$. In Figure 1, this is roughly where the "synchrotron" and "CMB" lines cross. Observing from the ground, the atmosphere emits at roughly 1.5 K between 0.6 GHz and 1.4 GHz [34], [35]. Ground-based experiments thus require precise modeling and sky dips to subtract the atmospheric signal. Compact experiments may be flown from balloons to rise above it. Finally, in the not-too-distant future, narrow-band measurements will be limited by the FIRAS error when characterizing distortions.

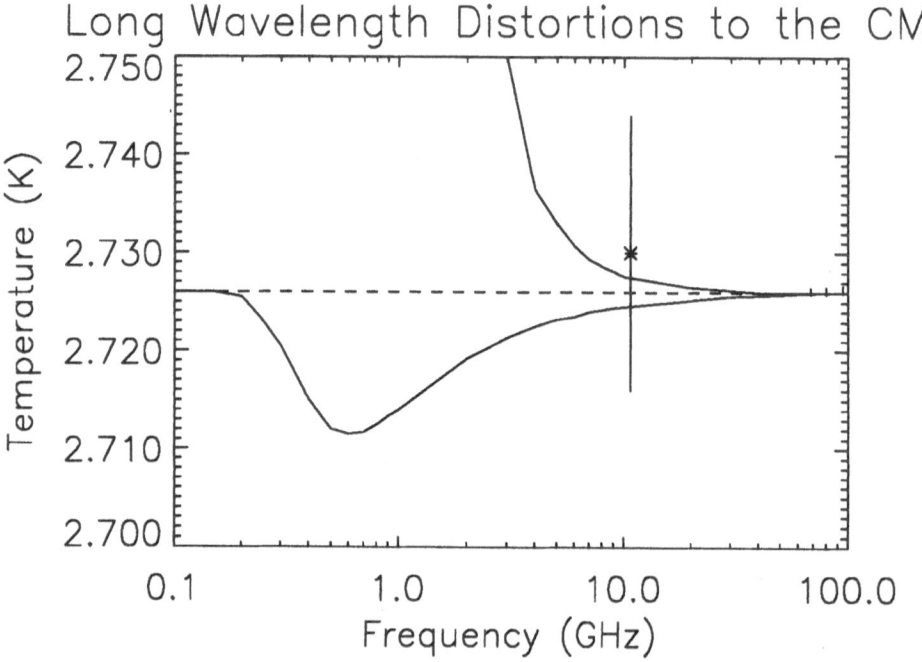

Figure 5. Plot of the long wavelength (low frequency) distortions to the CMB. The flat line is for a perfect blackbody. The curve with a pronounced minimum near 700 MHz shows a μ distortion with $\mu = 9 \times 10^{-5}$, the *COBE* limit. The curve that begins to rise near 10 GHz is for a Y_{ff} distortion with $Y_{ff} = 1.5 \times 10^{-5}$. All data with error bars small enough to fit on this plot are shown. The measurement at 10.7 GHz comes from Staggs *et al.* [30]. Note the three FIRAS data points near 100 GHz. This plot was adapted from a similar plot made by Al Kogut.

4. The Anisotropy

The photons that end their lives in our detectors were last scattered off electrons when the universe had a temperature of 5000 K and was evolving from a plasma to a state of neutral hydrogen and helium plus a thermal background[2]. This era is called the epoch of decoupling. The plasma was responding to the gravitational potential wells that would eventually foster galaxies and clusters of galaxies. The photons, now decoupled, bring to us an imprint of the potential wells and a signature of the dynamics of the

---

[2]We assume the standard inflationary model in this discussion, as well as a nearly complete transition from plasma to neutral matter.

plasma's response to the wells. From the angular spectrum of the fluctuations, one can distinguish among various possible mechanisms of structure formation. In recent years, it has become evident that for a certain class of models (eg. standard CDM[3]), a measurement of the detailed shape of the power spectrum will yield values for many of the cosmological parameters [4] such as Ω_0, Ω_B, H_0, and Λ [36].

Figure 6 shows the angular spectra for a few of the many models of structure formation. The point of the plot is to indicate that the model predictions are rather different and that given measurements with uncertainties of order the width of the plot line, the best model could be identified.

We divide the spectrum into three regions. At $l < 80$, or large angular scales ($\delta\theta > 2°$, $l \approx 180/\delta\theta$ with $\delta\theta$ in degrees), separate regions of the sky are not causally connected at decoupling. Indeed, the relative isotropy at these scales was one of the motivations for the inflation model. If inflation is correct, these potential wells and hills are the manifestation of quantum fluctuations that were superluminally expanded beyond the Hubble radius, grew with the expansion of the universe, and then re-entered the Hubble radius at a later time. The largest scales entered the Hubble radius most recently. In 10 billion years, a new CMB quadrupole will come into view as the universe expands and what we now call the quadrupole will be distributed among the higher moments. At these large scales, the anisotropy is produced by photons climbing out of the gravitational potential wells or sliding down the hills, just after their last scattering event, as discovered by Sachs and Wolfe. The anisotropy in this region reflects the primordial power spectrum of fluctuations, P(k). One of the strengths of inflation is that it predicts the shape of this spectrum[5]. At large angular scales, $C_l \propto 1/l(l+1)$, in other words, a nearly flat line in Figure 6.

At smaller angular scales, greater than $l \approx 80$, there was time for the primordial plasma to communicate. Hu and colleagues [6], [5] have presented an intuitive physical picture of the mechanisms behind the anisotropy although models date back to Silk [39], Sunyaev and Zel'dovich [40], Peebles & Yu [41] and others. To first order, we may think of the plasma as a photon-baryon fluid that acoustically oscillates in response to fluctuations in the gravitational potential produced by the dark matter. As the fluid flows into a potential well, it heats up (the phase is opposite to that of

[3] Standard CDM has $\Omega_0 = 1$, $h = 0.5$, $\Omega_B = 0.05$, and $n = 1$. Ratra points out that "fiducial" is a better description because many observations disagree with standard CDM.

[4] Ω is the fraction of the critical density. For standard Cold Dark Matter models, $\Omega_0 = \Omega_{CDM} + \Omega_B$; for Lambda models, $\Omega_0 + \Omega_\Lambda = 1$; while for open models, $\Omega_0 + \Omega_\Lambda + \Omega_{curv} = 1$.

[5] The shape of the spectrum was surmised independently by Harrison, Peebles & Yu, and Zel'dovich long before inflation.

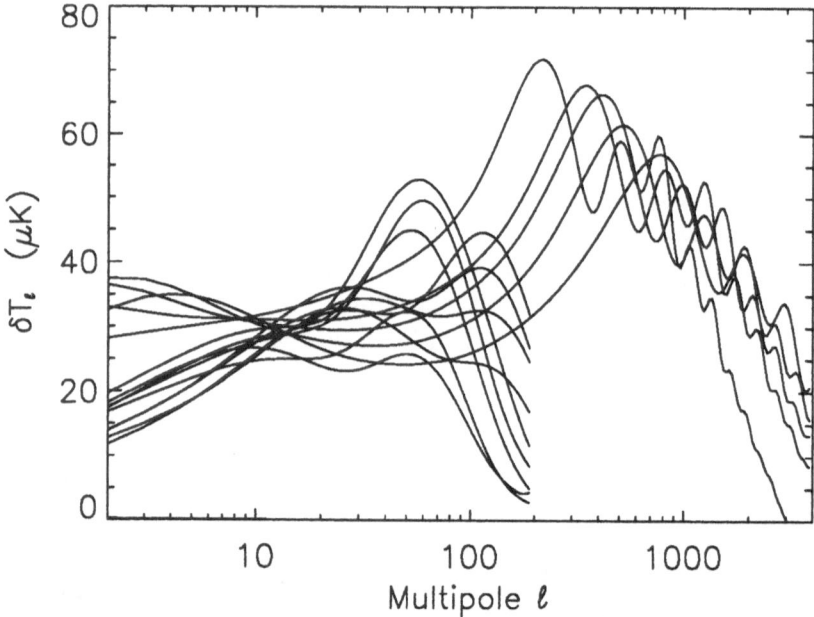

Figure 6. Angular spectrum for a sample of anisotropy models. The curves
grouped on the left hand side are the predictions of isocurvature models from Peebles [37]. The curves that peak on the right are for the
standard and open bubble inflation models (Ratra and Sugiyama [38],
[19]) with $\Omega_0 = 1.0, 0.4, 0.3, 0.2$ and 0.1 for peaks going from left to
right.

the Sachs-Wolfe effect). This is responsible for the large peak in the power
spectrum which occurs at roughly the angular scale of the largest potential
that can support plasma oscillations. Oscillations of the fluid in response to
fluctuations at smaller scales give rise to the other peaks. A full description
accounts for the doppler effect, the self gravity of the photon-baryon fluid,
the inertia of the baryons, and the evolution of the fluid with time.

Because decoupling happened so fast, $\delta z/z \approx 0.1$, the thickness of the
surface of last scattering is less than the horizon size at $z \approx 1400$. The scale
of the fluctuations near the first peak is of order the horizon size, about 100
Mpc in comoving units[6], and we observe the full signature of the potential
wells there. On smaller angular scales, corresponding to smaller physical
scales, there are more fluctuations contained within the horizon and their

---

[6]Alex Szalay spoke of a possible connection between this scale and the scale of the
largest structure in the galaxy surveys.

effects average out. In addition, diffusion of photons out of the potential wells diminishes the temperature fluctuations. The combination of these effects leads to the suppressed anisotropy near $l = 2000$ as can be seen in Figure 6.

Cosmological parameters are extracted from the shape of the power spectrum. For standard CDM type models this is done, for instance, by noting that the distance between the peaks depends on the sound speed at decoupling. This in turn depends on $h^2\Omega_B$. Also, as Ω_B increases, the photon-baryon fluid has more inertia (the sound speed decreases) and the compressional peaks (the odd ones starting at $l = 200$) get bigger. On the other hand, if the universe goes through a period of reionization, all the peaks in the anisotropy can be wiped out. Extracting generic cosmological information, regardless of model, is an active area of research [42].

One particularly nice demonstration of what the anisotropy can tell us was noted by Kamionkowski *et al.* [43]. The location in l of the first peak is a good indicator of $\Omega_0 + \Omega_{Lambda}$. This happens because a universe of any geometry, in its early stages, evolves as though $\Omega_0 = 1$. Because the physical size of a fluctuation depends on the sound horizon at decoupling, it is a "standard yardstick." The angular size of the standard yardstick, as viewed today, depends on the overall geometry of the universe. If the universe is flat, it will appear at $2°$, if it is open it will appear at a smaller angle; this last is because there is "more space" far away. This effect is seen in the CDM models in Figure 6. The lower Ω_0, the further to the right the peak moves. The scaling is roughly $l_{peak} \approx 200/\Omega_0^{1/2}$.

In models of the formation of large scale structure, the same potential fluctuations that produce the anisotropy also produce the large scale clustering of galaxies. We still do not know what type of mass comprises galaxies, or how it couples to gravitational fluctuations, or even what produced the fluctuations. In the past few years, a number of large galaxy surveys have been undertaken and older surveys have been re-analyzed. Over a region of l-space between $l = 30$ and $l = 600$, there is overlap between the two probes of the fluctuations: galactic surveys and CMB anisotropy data. There is a nice plot of this in White, Scott, & Silk's review [3]. Unfortunately, the data are not sufficiently good, from either probe, to draw firm conclusions. However, the indications are that the simplest models for the formation of structure are incorrect.

5. Anisotropy Measurements: Technologies & Techniques

5.1 TECHNOLOGIES

The desire to characterize the anisotropy has led to improvements in detectors and instrument technology. For the anisotropy, three classes of detector are now in use: HEMT based amplifiers (20-90 GHz), superconductor-insulator-superconductor (SIS) based mixers (90-250 GHz), and a variety of types of bolometers (90-1000 GHz). Anisotropy measurements require very stable observing conditions. This is especially true for "configuration-space" measurements, as opposed to interferometric measurements. To overcome atmospheric fluctuations, the primary culprit, experiments are performed at stable or high sites (Mauna Kea, South Pole, Owens Valley, Saskatoon, & Tenerife) and from balloon platforms. Plans are underway for long duration balloon flights that circumnavigate the Antarctic and for balloons that can stay aloft for 100 days. Interferometers, which currently operate mostly below 30 GHz, are intrinsically less sensitive to the atmosphere.

The HEMT amplifiers for most of the HEMT-based experiments were designed by Marian Pospieszalski at the National Radio Astronomy Observatory (NRAO) electronic research lab. One couples celestial radiation to the amplifiers with waveguide. The incident electric field is simply amplified to a reasonable level and then detected with a diode. The amplifiers are special because they are rugged, require only modest cooling (≈ 20 K), and have a 30% bandwidth. A good wide-band (10 GHz) sensitivity for an optimized system near 40 GHz is roughly 500 μKs$^{1/2}$. In other words, one can detect 1/2 mK signal with a signal-to-noise of one with one second of integration. A number of semi-conductor groups are pushing to make better high-frequency transistors (for instance TRW and Hughes) and to make entire radiometers on a single chip (TRW and Lockheed-Martin).

SIS-based systems have been used for a number of years (Timbie [44], Meinhold [45], Robertson [46]) though the anisotropy has not yet been detected with them. The currently favored designs, and devices, come from Anthony Kerr and S-K Pan at NRAO. The SIS is a mixer. It is simultaneously illuminated with celestial radiation and with a "local oscillator." The output of the SIS is a signal containing the sum and difference frequencies of the LO and sky. The signal is low-pass filtered, amplified with HEMTs, and detected. For a system with a LO at 144 GHz and a 4 GHz IF bandwidth, a reasonable sensitivity is 400 μKs$^{1/2}$. These devices must operate below the superconducting transition of niobium, or below roughly 4.2 K. They are more difficult than HEMTs to operate but are still straightforward.

Both HEMT and SIS systems are coherent; in other words, one works with the electric field right up until the final detection. This allows possibilities for phase sensitive techniques such as correlation receivers and interferometers. In addition they both have a very large audio bandwidth. Signals at many megahertz are easily detected.

The current bolometers are essentially thermistors held at 0.3 K or below. When they are illuminated, they heat up and change resistance, and the change in resistance is electronically read out. This is called incoherent detection because the phase information is lost. The great advantage to bolometers is that they are very sensitive. A decade ago, they were near 400 $\mu Ks^{1/2}$ [47], and some current devices (the "spider" bolometers, Bock et al. [48]) achieve better than 100 $\mu Ks^{1/2}$. The disadvantage is that they are more difficult to use than HEMTs and SISs and their intrinsic time constants are longer, but both of these problems are actively being worked on. There are different types of bolometers in various stages of development. They include frequency sensitive bolometers[49], hot electron bolometers[50], monolithic silicon bolometers[51], spider web composite bolometers[48], and transition edge bolometers[52].

The bolometer's advantage is not its intrinsic sensitivity per photon, which is comparable to that of HEMTs or SIS, but rather the fact that one can detect in multiple electromagnetic modes (in conventional use, waveguide supports just one mode) and almost arbitrary bandwidth. In other words, bolometers detect more photons than single mode waveguide systems. The power on a device is

$$P = \int_\nu \int_A \int_\Omega S_\nu(T) d\nu dA d\Omega \rightarrow \int_\nu kT(\nu) d\nu \quad \text{Watts,} \qquad (7)$$

where $S_\nu(T)$ is the flux from some source, ν is the RF frequency, A is the area of the detector (or antenna), and Ω is the solid angle incident on that detector (or antenna). The quantity $\int dA d\Omega$ is called the throughput or étendue. Generally, $A\Omega = n\lambda^2$ where λ is the wavelength at the passband center and n is the number of modes. For a single mode system, $n = 1$, and we get the quantity on the right of the arrow, where $T(\nu)$ is the effective temperature of the source.

5.2 TECHNIQUES

The experimental challenge is to measure a variance of a random field, which is of order 30 μK, with a noisy detector and a background signal of 300 K. As much effort has gone into determining robust ways to do this as has gone into understanding the detector systems. The configuration-space techniques are the better developed so we will focus on those. The interferometric techniques are rapidly maturing and have a lot of promise; the first detection of the anisotropy with an interferometer was just reported in Scott et al. [53].

A typical telescope has a beam described by $P(\theta) \propto \exp(-\theta^2/2\sigma^2)$. If the telescope observes the sky which has temperature $T(\hat{x})$, it measures $t(\hat{x}) = \int P(\theta)T(\hat{x})dx$. If two measurements made near each other, $t(\hat{x}_1)$ and $t(\hat{x}_2)$ are subtracted, the common atmospheric signal drops out. One is then sensitive to only the gradient in the atmosphere and to the CMB temperature difference. If two single differences, with one position in common, are performed and then subtracted the "double difference" is $s_i = 2t(\hat{x}_1) - t(\hat{x}_2) - t(\hat{x}_3)$. The double difference is still sensitive to the CMB but is insensitive to atmospheric gradients as well as the atmospheric temperature. This extra difference helps because the fluctuation spectrum of the atmosphere drops with both increasing spatial and spectral frequency. Let us call the effective beam profile for the double difference measurement $H_i(\hat{x})$; it will have one large positive central lobe and two negative lobes as shown in Figure 7. We can write

$$s_i = \int H_i(\hat{x})T(\hat{x})dx. \tag{8}$$

We will consider this a single measurement at a pixel $\hat{x}$. In practice, we make a set of N similar measurements over a patch of sky. For a first order estimate, we may find the intrinsic variance of the sky from

$$\sigma^2_{sky} = \sigma^2_{meas} - \sigma^2_{data}, \tag{9}$$

where σ^2_{data} is the square of the average statistical error per measurement and σ_{meas} is the variance of the N data points. This answer is usually only correct to 25% so it is used as a sanity check. Also, it does not give the correct error. Note too that this method ignores all intrinsic correlations in the data.

What variance does one expect from such a measurement? After working through the math, we find that the inclusion of finite beams modifies eq. 3 to the following:

$$C_T^{ij} = \sum_L \frac{2l+1}{4\pi} C_l W_l^{ij}, \text{ where } W_l^{ij} = \int dx_1 \int dx_2 H_i(\hat{x}_1)H_j(\hat{x}_2)P_l(\hat{x}_1 \cdot \hat{x}_2). \tag{10}$$

This is the full theoretical covariance matrix for the observing pattern for measurements i and j.[7] W_l is called the window function, it tells us the portion of l-space being examined. Generally, just the diagonal elements are plotted. For the beam in Figure 7, the window function is shown in Figure 8. When $i = j$, we get a prediction for σ^2_{sky}.

The most frequently used analyses follow Bond's work [4]. A complete analysis requires knowledge of the covariance matrix of the data; we call this C_D. The full theory-plus-data covariance matrix is given by $M = C_D + C_T$. Our goal is to determine the probability of a theory given the data, $P(T|D)$. To do this, we use Bayes's theorem with a uniform prior, $P(T) = 1$, and set the probability of getting the data, $P(D)$, equal to 1:

---

[7]From a theoretical perspective, our universe is one realization of a cosmological model that can only predict the ensemble average of C_l. Even if we knew the correct physics, the data would be scattered around the predicted C_l with a "cosmic variance."

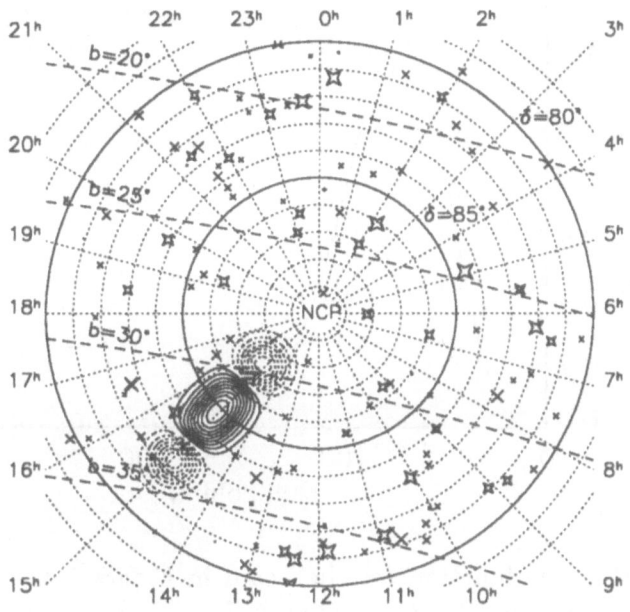

Figure 7. Contour plot of a typical double-difference beam superimposed on a source map of the north celestial polar region from Netterfield *et al.* [54]. This profile corresponds to $H_i(\hat{x})$ in eq. 8. The dashed lines are negative and the solid lines are positive. The sources come from the Kühr survey[55]. The stars mark the flat spectrum sources. The symbol size is proportional to the log of the flux. Lines of Galactic latitude are also shown.

$$P(T|D) = \frac{P(D|T)P(T)}{P(D)} = L(D|T) = \frac{\exp(-t^T M^{-1} t/2)}{(2\pi)^{N/2}|M|^{1/2}}, \qquad (11)$$

where L is the likelihood function and t is a vector of the data. The argument of the exponent is proportional to χ^2. This boils down to saying that the likelihood of the data plotted as a function of some parametrization, for instance σ_{sky}, is the probability of obtaining σ_{sky} with a given set of data. When the signal to noise is high, the likelihood is fairly Gaussian. To get an error, we find bounds symmetric around the maximum, that contain 68% of the area under the likelihood curve.

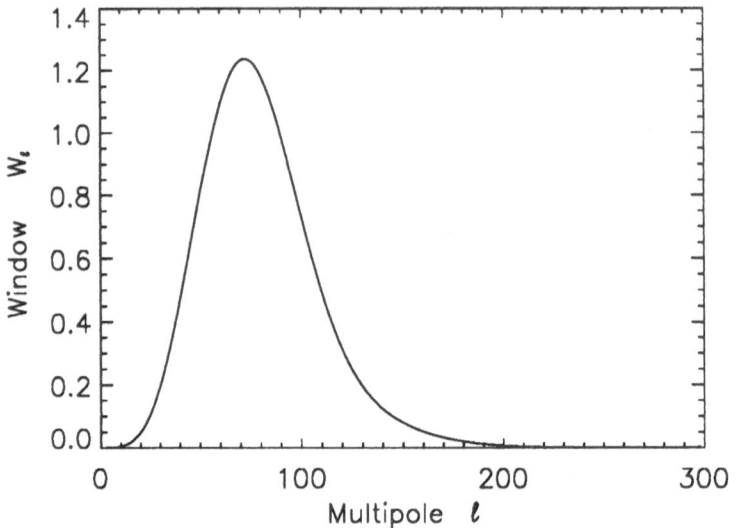

Figure 8. Window function for the beam in Figure 7 using eq. 10 with $i = j$. The amplitude of the window depends on the normalization of the beam. For this plot, $\int d\hat{x}|H(\hat{x})| = 2$ was used.

To estimate the angular spectrum, we take the most likely value of σ_{sky} with its error, convert it into a "band-temperature"[8] and plot it at the l-weighted center of the window function. A horizontal error bar is often plotted indicating the width of the window function. Any one experiment observes in multiple windows and so the spectrum may be mapped out. One example is given in Figure 9 from the Saskatoon experiment [18].

6. Anisotropy Measurements:
The Current Results and Immediate Future

The anisotropy of the CMB was first unambiguously measured by the DMR experiment aboard the *COBE* satellite [57] using 7° resolution full-sky maps at 30, 53, and 90 GHz. To date, these are still the cleanest and best checked data. All indications are that the fluctuations are thermal, though I am not aware of any formal limits on, say, the Compton y parameter of the anisotropy. When smoothed to a 10° resolution, the *rms* of the temperature fluctuations is about 30 μK. This is the canonically quoted value. However,

---

[8]See Bond [56] for details. One obtains a band power by dividing σ_{sky} by $\sqrt{\sum_l W_l/l}$. This makes sense because $\sigma_{sky} = \sqrt{\sum (\delta T_l)^2 W_l/l}$ can be written as $\sigma_{sky} = \delta T_l \sqrt{\sum W_l/l}$.

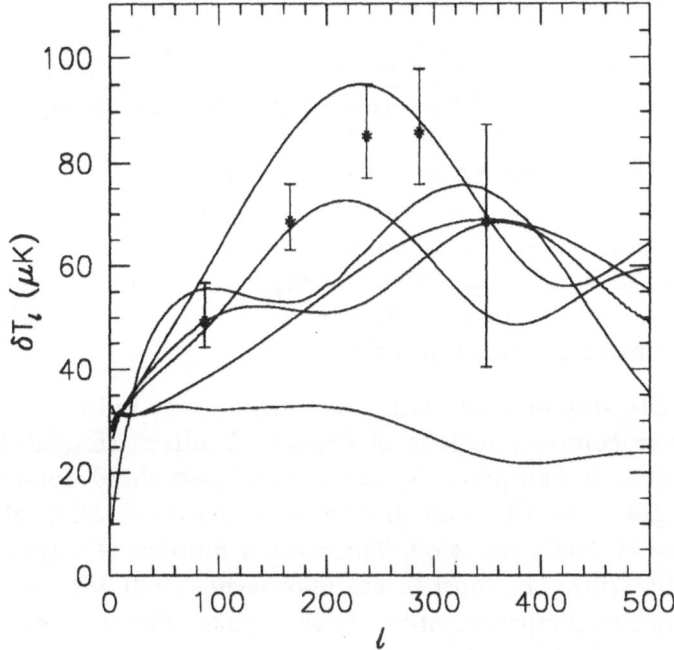

Figure 9. Results from the Saskatoon experiment [71], [18]. The spectra from six theories are also shown. From top to bottom at $l = 160$ they are a flat Λ+CDM model with $\Omega_\Lambda = 0.7$ [38], Standard CDM [19], a PPI model [62], an $\Omega_0 = 0.4$ open bubble model [19], a texture model [63], and a model with reionization [64]. There is a 14% overall calibration uncertainty that is not included in the error bars. This affects the normalization of the spectrum, but not the shape.

with a $1/2°$ resolution map, the *rms* is closer to 90 μK. The final DMR results (the satellite is now turned off) are published in Ap.J Vol 464, 1996 [58], [17], [59], [60], & [61]. The lasting contribution will be the maps of the sky at 30, 53, and 90 GHz. A combination of these maps, optimized to give the anisotropy, has a signal-to-noise of two per $10° \times 10°$ pixel. The two primary results derived from these maps are:

1. From a fit of the data to a power spectrum parameterized by the spatial index and the quadrupole amplitude, the DMR team finds $n_{DMR} = 1.21 \pm 0.3$ and $Q_{rms-PS} = 15.3^{+3.8}_{-2.8}$ μK. The quadrupole of the raw maps is slightly below Q_{rms-PS}, but not by a statistically significant amount. Note that for "standard CDM" one expects $n_{DMR} = 1.1$ because DMR probes the low-l tail of the acoustic peak.

In this notation, the C_l (eq. 10) are given by [66]:

$$C_l = \frac{4\pi}{5} Q^2_{rms-PS} \frac{\Gamma[l + (n_{DMR} - 1)/2]\Gamma[(9 - n_{DMR})/2]}{\Gamma[l + (5 - n_{DMR})/2]\Gamma[(3 + n_{DMR})/2}$$ (12)

with $n_{DMR} = 1$ this reduces to $C_l \propto 1/l(l+1)$.

2. The data appear best described by Gaussian statistics [65]. At these large angular scales this is not surprising because even non-Gaussian processes at small scales, when averaged over a large enough volume, appear Gaussian. However, it is reassuring that the statistics we all assume have some basis in reality.

My favorite way of quantifying the DMR maps is to show the power spectrum over compact regions of l-space. A direct computation of the power spectrum is hampered by the unevenly-weighted non-uniform sky coverage imposed by the scan pattern and the elimination of data contaminated by Galactic emission. There are a number of ways around this problem ([69], [70]). The most recent approaches ([68], [67]) work to minimize the width of a representative bin in l-space. The results are shown in Figure 10. Górski's method in particular shows the power within $\Delta l = 0$. Though one should expect to get exactly the same results from two different methods applied to the same data set, we can be pleased by the general concordance.

Many groups are working to measure the anisotropy. Though some are focussing on large angular scales and frequencies not observed with DMR, most concentrate on smaller angular scales. Table 1 contains a list for recent, current and planned experiments. It does not include the satellite experiments nor does it claim to be comprehensive. I apologize for any omissions or misrepresentations.

Unlike measurements of the absolute temperature of the CMB, where the final result is completely dominated by one's control of subtle systematic errors, anisotropy measurements require a combination of high sensitivity and immunity to systematic effects. The state-of-the-art in absolute measurements, excluding FIRAS, is 1% [30]; the anisotropy has yet to be measured to 15% accuracy.

To give a broad and almost un-biased sense of what the anisotropy data are telling us, we take the compilation from Ratra [97] (which I believe is the most comprehensive and thoroughly checked compilation to date) and bin the data according to the following:

- Select logarithmically spaced bins in l with four bins per decade.
- Ignore the widths of window functions and add data to a bin according to the value of the weighted mean of the window, l_e in [97]. Many of the data have $\Delta l/l_e < 1/4$ so this is not the sin it may at first appear.

TABLE 1. Recently Completed, Current and Planned Anisotropy Experiments

Experiment	Resolution	Frequency	Detectors	Type	Groups
ACE(c)[72]	0.2°	25-100 GHz	HEMT	C/B	UCSB
APACHE(c)[73]	0.33°	90-400 GHz	Bol	C/G	Bologna, Bartol Rome III
ARGO(f)[74]	0.9°	140-3000 GHz	Bol	C/B	Rome I
ATCA[75]	0.03°	8.7 GHz	HEMT	I/G	CSIRO
BAM(c)[76]	0.75°	90-300 GHz	Bol	C/B	UBC, CfA
Bartol(c)[77]	2.4°	90-270 GHz	Bol	C/G	Bartol
BEAST(p)[72]	0.2°	25-100 GHz	HEMT	C/B	UCSB
BOOMERanG(p)[78]	0.2°	90-400 GHz	Bol	C/G	Rome I, Caltech UCB, UCSB
CAT(c)[53]	0.17°	15 GHz	HEMT	I/G	Cambridge
CBI(p)[79]	0.0833°	26-36 GHz	HEMT	I/G	Caltech, Penn.
FIRS(f)[47]	3.8°	170-680 GHz	Bol	C/B	Chicago, MIT, Princeton, NASA/GSFC
HACME/SP(f)[80]	0.6°	30 GHz	HEMT	C/G	UCSB
IAB(f)[81]	0.83°	150 GHz	Bol	C/G	Bartol
MAT(p)[82]	0.2°	30-150 GHz	HEMT/SIS	C/G	Penn, Princeton
MAX(f)[83]	0.5°	90-420 GHz	Bol	C/B	UCB, UCSB
MAXIMA(p)[84]	0.2°	90-420 GHz	Bol	C/B	UCB, Caltech
MSAM(c)[85]	0.4°	40-680 GHz	Bol	C/B	Chicago, Brown, Princeton, NASA/GSFC
OVRO 40/5(c)[86]	0.033°, 0.12°	15-35 GHz	HEMT	C/G	Caltech, Penn
PYTHON(c)[87]	0.75°	35-90 GHz	Bol/HEMT	C/G	Carnegie Mellon Chicago, UCSB
QMAP(f)[88]	0.2°	20-150 GHz	HEMT/SIS	C/B	Princeton, Penn
SASK(f)[89]	0.5°	20-45 GHz	HEMT	C/G	Princeton
SuZIE(c)[90]	0.017°	150-300 GHz	Bol	C/G	Caltech
TopHat(p)[91]	0.33°	150-700 GHz	Bol	C/B	Bartol, Brown, DSRI,Chicago, NASA/GSFC
Tenerife(c)[92]	6.0°	10-33 GHz	HEMT	C/G	NRAL, Cambridge
VCA(p)[93]	0.33°	30 GHz	HEMT	I/G	Chicago
VLA(c)[94]	0.0028°	8.4 GHz	HEMT	I/G	Haverford, NRAO
VSA(p)[95]	–	30 GHz	HEMT	I/G	Cambridge
White Dish(f)[96]	0.2°	90 GHz	Bol	C/G	Carnegie Mellon

1. For "Type" the first letter distinguishes between configuration or interferometer, the second between ground or balloon.

2. An "f" after the experiment's name means it's finished; a "c" denotes current; a "p" denotes planned, building may be in progress but there is no data yet.

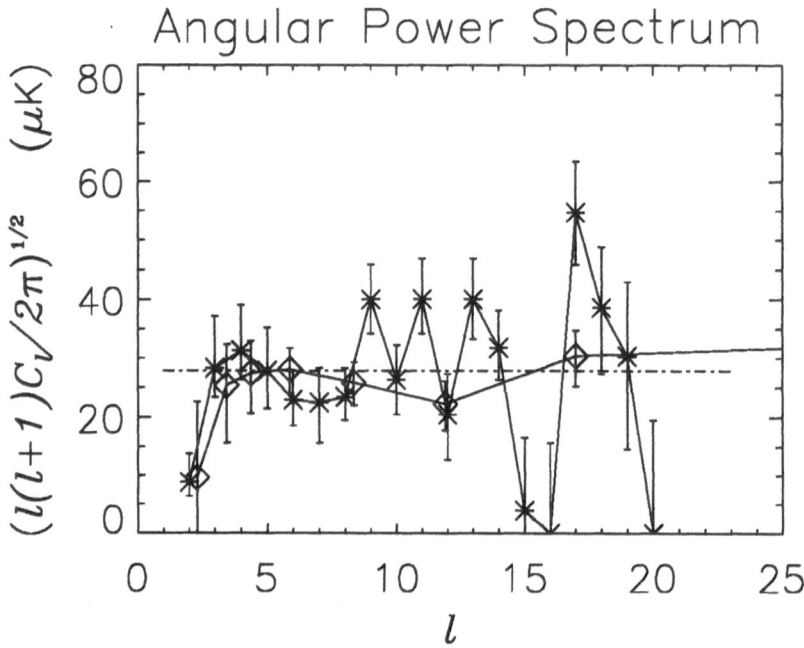

Figure 10. The angular power spectrum from DMR. Results from Górski
[67] (stars) and Tegmark [68] (diamonds) are shown. Both have analyzed the data in a manner to produce narrow window functions.
Górski's, in particular, have $\Delta l/l = 0$. One should bear in mind that
Górski's results are derived from non-Gaussian likelihood distributions. The flat dashed line is for a Harrison-Zel'dovich spectrum with
$Q_{rms-PS} = 18 \ \mu K$.

- Compute the weighted mean of the data that fall into each bin and
  call that the band-power δT. Use the inverse root of the total weight
  as the error bar. Use all the *unique* data in [97]. Be aware that many
  of the data points are unconfirmed!
- Compute the arithmetic mean value $\bar{l}$, of the l_e that fall in a given bin.
- Plot $\bar{l}$ versus δT and connect the ends of the error bars.
- Ignore the intrinsic calibration uncertainties and the relative calibra-
  tion uncertainties.
- Ignore upper limits for $l < 500$. For some cases, (eg SASK[89]), the
  data can be combined to give a detection [18].
- In addition to the Ratra compilation, add the Tegmark results [68],
  ATCA [75], and the new CAT [53] results.

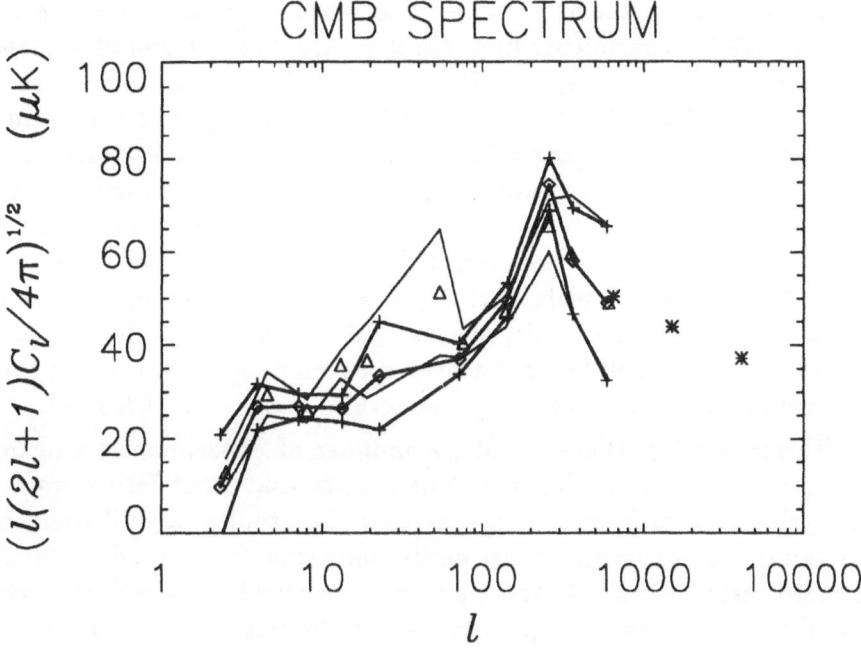

Figure 11. Angular spectrum of the anisotropy. The thick lines (diamonds) were obtained following the prescription given in the text. Bharat Ratra has used somewhat different criteria and obtained the results indicated by the thin lines. The asterisks near $l = 1000$ are upper limits. Clearly, the data indicate a rise in the spectrum from $l = 10$ to $l \approx 200$. For $l > 20$, the quantity on the y-axis is the same as in Figure 3.

The principle conclusion one should draw from Figure 11 is that there is a general rise in δT_l as one moves from the *COBE* scales to smaller angular scales. This is a stunning observation that was predicted long before the anisotropy was discovered. Has a peak to the spectrum been detected? Possibly, but it is still too early to say this with any confidence. All indications are that the power spectrum at $l \approx 10$ is lower than at $l \approx 200$ but we cannot say where the spectrum turns over above $l \approx 200$. The errors are simply too large and there are too many systematic effects hidden in the data. Also, there are plenty of examples where 95% upper limits have become detections at a higher level and examples where detections have become upper limits. However, it is somewhat reassuring that the $\chi^2/(\nu - 1)$ for most of the individual bins is not too different from one.

The measurements have come a long way in the past two years. In three cases multiple experiments have observed the same region of sky and seen the same thing. They are DMR and FIRS [98], DMR and Tenerife[92], and MSAM ([99], [100] & [101]) and SASK [18], [8]. The spectrum of the fluctuations for MSAM and SASK is thermal from 25 to 200 GHz. However, one should still view the data with some caution. In the analyses that give rise to Figure 11, an entire data set is reduced to give one measurement and one statistical error bar. When this is done, there is simply not enough signal-to-noise to quantify systematic effects that are lurking at the 1σ to 2σ level. On top of this, the analysis of these experiments is tricky; "new" effects are still being discovered by many groups. Finally, the inter-calibration of the experiments is uncertain to the 10%-15% level.

To improve on these results, a number of experimental and observational challenges must be met. The results that went into Figure 11 are primarily from difference measurements. Eventually we will want maps of the sky so that experiments are easily compared, foreground contamination is more easily identified, and powerful statistical tests can be performed. Interferometers offer one proven way to do this at low frequencies (and eventually at higher frequencies) but other techniques and strategies are needed. The calibration of the experiments must be better than 10% in order to distinguish between the various models. Currently, the best reference is the intrinsic dipole in the CMB. Finally, to distinguish features near $l \approx 1000$, high resolution will be needed. One desires at least $\delta l/l \approx 1/10$. Lower resolutions smear the features in the power spectrum.

A measurement of the polarization in the CMB is now within grasp. In the standard CDM models, the signal is predicted to be at 1% to 5% of the anisotropy[66]. John Ruhl at Santa Barbara, Suzanne Staggs at Princeton, and Peter Timbie at Wisconsin are actively working on these measurements. The polarization is caused by Thompson scattering. At angular scales of order ten degrees, the polarization may be used to identify any primordial gravity waves (tensor modes) though the signal is expected to be largest at degree scales. Crittenden and Turok [102] point out that there is a correlation between the polarization and the anisotropy that is different for the scalar and tensor modes. This correlation should also aid in separating the CMB polarization from polarized foreground emission, about which very little is known.

7. The New Satellite Experiments

Two satellite missions are planned that will endeavor to map the CMB anisotropy over the entire sky. The ESA mission is called *PLANCK* (originally *COBRAS/SAMBA*). The NASA mission, which is in the final design and definition phase, is called *MAP*. Because I am part of the *MAP* team, I will focus on it.

PLANCK[9] uses both bolometers and HEMT- based amplifiers. It will carry cryogens so that the bolometers may be operated at 0.1 K. At this stage in the design, the instrument is planned to span the frequencies between roughly 30 and 900 GHz with an angular resolution between 30′ and 4.4′. The lower frequency channels will be polarization-sensitive. The planned launch date is 2005.

MAP is based on the HEMT amplifiers developed by Marian Pospieszalski. The radiometers are intrinsically polarization-sensitive and differential, similar in some regards to the successful *COBE*/DMR design. The instrument will span from 20 to 106 GHz in five frequency bands with an angular resolution ranging between 54′ and 15′. The sensitivity per $0.3° \times 0.3°$ pixel (of which there are roughly 400,000 in the sky) will be about 35 μK. Because the instrument is passively cooled, it can in principle observe longer than the 15 month design life.

The primary goal of *MAP* is to make multi-frequency, high-fidelity, high-sensitivity maps of the sky. This requires extreme control of systematic effects. We believe the best vantage for these observations is L2, the Earth-Sun Lagrange point. At L2, the Sun, Earth, and moon are $\approx$ 90° out of the beams and the environment is essentially isothermal. From the work on DMR and balloons, the team has found that successful map production requires reference of one pixel to another over many directions and over many time scales. *MAP* plans to do this with the scan strategy shown in Figure 12.

From a high-quality map, one may not only obtain the power spectrum, but may also compare the data to the results of other CMB experiments and to maps of the foreground emission at different frequencies. Also, a map gives the best data set for testing the underlying statistics of the fluctuations. For instance, we will be able to tell from the *MAP* data if the CMB is a Gaussian random field. Finally, with its high sensitivity and large scale coverage, the time-line data from *MAP* will be ideal for searching for transient radio emission.

The question of how well one can determine the parameters of cosmological models is still an active area of research. The most recent published work on parameter estimation for inflation-based models is in [36] but one must remember that there are other classes of promising models. At this school, Dick Bond discussed an approach where one works in an eigen-parameter space to circumvent the strong covariance between many of the standard parameters such as Ω_B, Λ, h, etc. At any rate, if the anisotropy is

---

[9] See web site http://astro.estec.esa.nl/sa-general/projects/cobras/cobras.html for additional information.

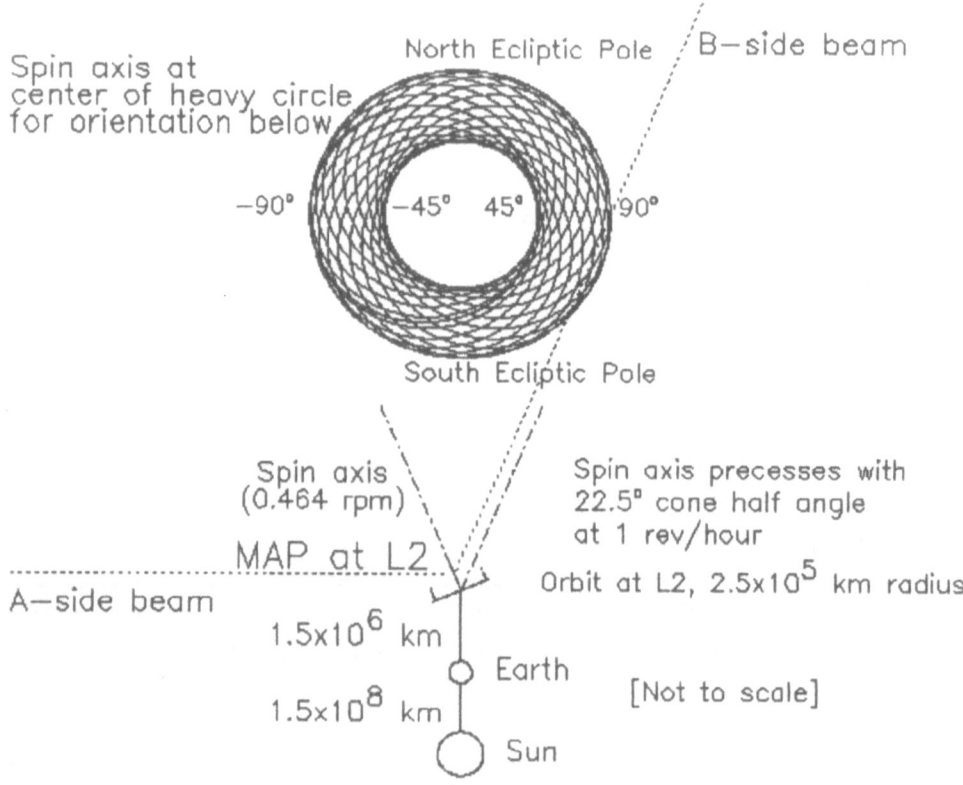

Figure 12. The *MAP* scan pattern for one hour of observation. The lines show the path for one side of a differential pair. The other pair member follows a similar path, only delayed by 1.1 min. There are four principal time scales for the observations. The phase of the difference signal is switched by 180° at 2.5 KHz. The spacecraft spins around its symmetry axis with a 2.2 min period (bold circle) with cone opening angle of roughly 135°. This pattern precesses about the Earth-Sun line with a period of 60 minutes. Thus, in about 1 hour, over 30% of the sky is covered. Every six months, the whole sky is observed. Note that any pixel is differenced to another pixel in many directions.

normally distributed, the *MAP* CMB data will be cosmic variance limited up to $l \approx 600$ (assuming the foreground/radio source emission is success-fully removed) and will probe multipoles up to $l \approx 1000$.

In the current schedule, the satellite design and definition will be complete by November 1997 and then the building will begin. *MAP* is scheduled for launch late in 2000. The *MAP* science team is comprised of Chuck Bennett (PI) at NASA/GSFC, Mark Halpern at UBC, Gary Hinshaw at NASA/GSFC, Norm Jarosik at Princeton, John Mather at NASA/GSFC, Steve Meyer at Chicago, Lyman Page at Princeton, Dave Spergel at Princeton, Dave Wilkinson at Princeton, and Ned Wright at UCLA. More information about *MAP*, the CMB, and other experiments may be obtained from http://map.gsfc.nasa.gov/.

I would like to thank Roberta Bernstein, Venya Berezinsky, Piero Gallotti, David Schramm and the Ettore Majorana Center staff for organizing a wonderful school. Marsala will never taste the same. Conversations with many colleagues were helpful in preparing these notes. I would especially like to thank Tom Herbig, Gary Hinshaw, Bharat Ratra, Suzanne Staggs, and Ned Wright. Ned gave me the computer code to produce Figure 4. This work was supported by the US National Science Foundation and the David and Lucile Packard Foundation.

References

1. Weiss, R., Ann. Rev, Astron. Astrophys, 1980, 18:489.
2. Readhead, A. C. S. and Lawrence, C. R., Ann. Rev, Astron. Astrophys, 1992, 30:653.
3. White, M. Scott, D. and Silk, J. Ann. Rev, Astron. Astrophys, 1994, 32:319.
4. Bond, J. R., in *Cosmology and Large Scale Structure*, ed R. Schaeffer, 1995, Elsevier Science Publishers, Netherlands
5. Tegmark, M. Proc. Enrico Fermi Course CXXXII, Varenna, 1995. (astro-ph/9511079)
6. Hu, W., Sugiyama, N. & Silk, J. Review for Nature, 1996, (astro-ph/9604166).
7. Smoot, G. & Scott, D. Current Summary of Results. in L. Montanet et al. Phy. Rev D50, 1173 (1994), off-year addition (astro-ph/9603157).
8. Page, L., Proceedings from the Critical Dialogues in Cosmology Conference, Princeton NY, June 1996.
9. Sunyaev, R. A. & Zel'dovich , Ya. B. Ann. Rev, Astron. Astrophys, 1980, 18:537.
10. Danese, L., Burigana, C. Toffolatti, L., De Zotti, G. & Franceschini, A. *The Cosmic Microwave Background: 25 Years Later*, 153, 1990, Kluwer Academic Publishers. Mandolesi & Vittorio (eds.)
11. Bartlett, J. G., Stebbins, A. 1991, ApJ, 371:8.
12. Partridge, B. "3K: The Cosmic Microwave Background Radiation", Cambridge University Press, 1995.
13. Wright, E. L. et al., 1991, Ap.J. 381:200.
14. *COBE* is the COsmic Background Explorer. The three experiments aboard the satellite are the Differential Microwave Radiometers (30-90 GHz, DMR), the Far-InfraRed Absolute Spectrophotometer (60-630 GHz, FIRAS), and the Diffuse InfraRed Background Experiment (1.2-240 μm, DIRBE) All the experiments produce maps of the sky.
15. Haslam et al., 1982 A&AS, 47, 1 1991b, ApJ, 379, 1
16. Available through the Infrared Processing and Analysis Center in Pasadena, CA.
17. Kogut, A. et al. 1996, Ap.J. 464:L5-L9.
18. Netterfield, C. B. et al., 1997, ApJ. accepted, (astro-ph/9601197).
19. Ratra, B.Banday, A.J., Górski, K.M., & Sugiyama, N. 1995, (astro-ph/9512145).
20. Tegmark, M. and Efstathiou, G. Submitted to MNRAS (astro-ph/9507009).

21. Mather, J. C., et al. 1994, Ap. J. ApJ, 420:439.
22. Fixsen, D. J., et al. 1996, ApJ, 473:576. (astro-ph/9605054).
23. Fixsen, D. J., et al. 1994, ApJ, 420:457
24. Gush, H. et al., 1990, PRL, 65, 537.
25. Private communication, June 1996.
26. Rybicki and Lightman. "Radiative Processes in Astrophysics" John Wiley& Sons, New York, 1979.
27. Wright, E. L., et al. 1994, Ap. J. ApJ, 420:450.
28. Bersanelli, M. et al., 1994, Ap.J. 424:517.
29. Peebles, P. J. E. et al. 1991, Nature, 352:769.
30. Staggs, S. T., et al, 1996, ApJ 473:L1 (astro-ph/9609128).
31. The team of R. Shafer, M. Mather, M. DiPirro, A. Kogut, D. Fixsen, M. Seiffert, P. Lubin, & S. Levin have designed a satellite mission called DIMES (Diffuse Microwave Emission Survey) to measure the CMB temperature between 2 and 100 GHz to 0.1 mK accuracy. A subset of this team is also building a balloon experiment for a multi-frequency absolute measurement. They hope to fly in early 1997. More information may be obtained from http://ceylon.gsfc.nasa.gov/DIMES/
32. Burigana, C., Danese, L.& De Zotti, G.F. 1991a, A&A, 246, 49
33. Burigana, C., Danese, L.& De Zotti, G.F. 1991b, ApJ, 379, 1
34. Smith, C., 1996, Ph.D. Thesis, Princeton University.
35. Staggs, S., 1993, Ph.D. Thesis, Princeton University.
36. Jungman, G., Kamionkowski, M., Kosowsky, A. and Spergel, D. Phys. Rev. D, 1996, 54 1332.
37. Peebles, P. J. E., 1994, ApJ 419:L49.
38. Ratra, B. & Sugiyama, N. 1995, (astro-ph/9512157)
39. Silk., J. 1968, ApJ, 151:459.
40. Sunayev, R.A. and Zel'dovich Y. B., 1970, Astrophysics and Space Science, 7:3-19, Reidel, Dordrecht-Holland.
41. Peebles P.J.E. and Yu ,J.T., 1970, AP.J 162:815.
42. Hu, W. and White, M. 1996, IAS preprint IASSNS-AST 96/47. Submitted to Ap.J. Available through http://www.sns.ias.edu/ whu.
43. Kanionkowski, M., Spergel, D. N., & Sugiyama, N. 1994, ApJ, 426:L57.
44. Timbie, P, 1985, PhD. Thesis, "A Novel Interferometer to Search for Anisotropy in the 2.7 K Background Radiation." Princeton.
45. Meinhold, P., 1989 PhD. Thesis, "Anisotropy Measurements of the Cosmic Microwave Background Radiation at 3 mm Wavelength and an Angular Scale of 30 Arcminutes." UCB
46. Robertson, T. 1996, Senior Thesis, Princeton University.
47. FIRS. The Far InfraRed Survey. This is an experiment that started at MIT but has since moved to Princeton, University of Chicago and NASA/GSFC. It is a bolometer-based balloon-borne radiometer. It confirmed the initial *COBE*/DMR discovery.
48. Bock, J. et al. Proceedings of "Submillimeter and Far-Infrared Space Instrumentation," 30th ESLAB Symposium, 24-26 Sept. 1996, ESTEC, Noordwijk, The Netherlands.
49. Kowitt, M. S. et al. 1996, Appl. Opt. 35:5630.
50. Nahum, M. & Martinis, J. M. 1993, "Novel Hot-electron Bolometer" Proceedings from 20th conference on Low-Temperature Physics, Eugene, OR.
51. Downey, P. M. et al. 1984, Appl. Opt. 239:10
52. Lee, A. T. et al. 1996, Preprint, To appear in Appl. Phys. Lett.
53. CAT. This is the Cambridge Anisotropy Telescope. It operates near 15 GHz and produces images of the microwave background. Early results are reported in Scott et al. 1996, Ap.J. 461:L1.
54. Netterfield, B et al., 1995 ApJ 445:L69
55. Kuhr, H. Pauliny-Toth, I.I.K., Witzel, A. &Schmidt, J. 1981, AJ, 86, 854.
56. Bond, J. R., Astro. Lett. & Comm. Vol 32, No. 1, 1995. Presented in 1994.

57. Smoot, G. F. et al. 1992, Ap.J. 396:L1.
58. Bennett, C. et al. 1996, Ap.J. 464:L1-L4.
59. Górski K. et al. 1996, Ap.J. 464:L11-L15.
60. Hinshaw G. et al. 1996, Ap.J. 464:L17-L20.
61. Wright E. et al. 1996 Ap.J. 464:L21-L24.
62. Peebles, P.J.E. 1995, Private Communication
63. Crittenden, R. G. & Turok, N. 1995, PRL, 75, 14
64. Sugiyama, N. 1995, Ap.J., 100, 281
65. Kogut, A. et al., 1996, ApJ 464:L29-L33.
66. Bond, J. R. & Efstathiou, G. 1987, MNRAS, 226, 655.
67. Górski, K. M. 1996, Private Communication
68. Tegmark M. 1996, Ap.J. 464:L35.
69. Hauser, M. G. & Peebles, P.J.E., 1973, Ap.J. 185:757.
70. Górski, K. M. 1994, Ap.J. 430:L85.
71. Wollack E. et al. 1997, ApJ, Accepted (astro-ph/9601196).
72. ACE and BEAST. These two new projects are aimed at using HEMTs between 26 and 100 GHz on both super-pressure and conventional long-duration balloon platforms. The finest angular resolution will be near 1/5°.
73. APACHE. This experiment will observe from Dome-C in the Antarctic. Web site http://tonno.tesre.bo.cnr.it/ valenzia/APACHE/apache.html contains more information.
74. ARGO. A balloon-borne bolometer based experiment. Results are reported in de Bernardis, et al. 1994, Ap.J. 422:L33.
75. ATCA: Australia Telescope Compact Array. An interferometer operating at 8.7 GHz with a 2' resolution produced a map that was analyzed for anisotropy. The results are reported in Subrahmanyan R., Ekers, R. D., Sinclair, M. & Silk, J. 1993, MNRAS 263:416.
76. BAM: Balloon Anisotropy Measurement. This uses a differential Fourier transform spectrometer to measure the spectrum of the anisotropy between 90 and 300 GHz. Recent results are reported in astro-ph/9609108. More information may be obtained from http://cmbr.physics.ubc.edu.
77. Bartol. This is a bolometer-based experiment designed to look at 2° angular scales. It observed from the Canary Islands. Results are reported in Piccirillo et al., astro-ph/9609186.
78. BOOMERanG is a collaboration between the Caltech, Berkeley, Santa Barbara (Ruhl) and Rome groups. It will use bolometers to measure the anisotropy in the CMB between 90 and 410 GHz. The ultimate goal is a circumpolar Antarctic flight.
79. CBI: Cosmic Background Imager. This is an interferometer that plans to produce maps of the microwave sky near 30 GHz.
80. HACME/SP. This uses HEMTs on the ACME telescope. Observations were made from the South Pole. Recent results are reported in Gundersen, J. et al. 1995, Ap.J. 443:L57.
81. IAB. A bolometer-based experiment carried out at the Italian Antarctic Base. Results are reported in Piccirillo, L. & Calisse, P. 1993, Ap.J. 413:529.
82. MAT. This is the Mobile Anisotropy Telescope. It is similar to QMAP but is designed to operate from the ground in Chile.
83. MAX was a collaboration between UCSB and Berkeley. It is a balloon-borne bolometer-based radiometer spanning roughly between 90 and 420 GHz. Recent results are reported in Lim et al. 1996, Ap. J. 469:L69. It flew on the ACME telescope.
84. MAXIMA is a collaboration between Caltech and Berkeley. It is the next generation of MAX. Web site http://physics7.berkeley.edu/group/cmb/gen.html contains more information.
85. MSAM. There are a number of versions of MSAM. All use bolometers of various sorts and fly on balloons. The MSAM collaboration includes NASA/GSFC, Bartol Research Institute, Brown University, and the University of Chicago.

86. OVRO. The Owen's Valley Radio Observatoty telescopes operate with various receivers between 15 and 30 GHz. The 40 m dish has a 2' beam, and the 5.5 m has a 7.3' beam. The experiments are aimed primarily at small angular scales.

87. PYTHON. A multi-pixel bolometer- and HEMT- based experiment operated from the ground at the South Pole. The experiment has run in a number of configurations. Recent results are reported in Ruhl, J., et al. 1995, Ap.J., 453:L1.

88. QMAP. This is a balloon-borne experiment that uses a combination of HEMTs and SIS detectors. The angular resolution is 1/5°. This experiment is designed to produce "true" maps of the sky.

89. SASK. These experiments are based on HEMT amplifiers operating between 26 and 46 GHz. They were performed in Saskatoon, Saskatchewan CA. Three years of observations have gone into the final data set.

90. SuZIE is a bolometer-based experiment that observes from the ground. It is primarily intended to measure the SZ effect at high frequencies though it will also give information on the anisotropy at small scales.

91. TopHat is a collaboration between Bartol Research Institute, Brown University, DSRI, NASA/GSFC, and the University of Chicago. The group plans to observe with an extremely light-weight bolometer-based payload mounted on top of a scientific balloon that circumnavigates the Antarctic. For more information see http://cobi.gsfc.nasa.gov/msam-tophat.html.

92. Tenerife. Ground-based differential radiometers with 10-33 GHz receivers. The resolution is about 6°. The experiment observes from the Observatorio del Teide in Tenerife, Spain. It has operated for many years. Recent results are discussed in Hancock et al. 1994, Nature, 367, 333.

93. VCA: Very Compact Array. This is an interferometer being developed at the University of Chicago. It will produce maps of the CMB at 30 GHz and be sensitive to larger angular scales than CBI.

94. VLA. This is work done near 5 GHz, on arcminute and smaller angular scales. It uses the Very Large Array. Recent results are reported in Fomalont et al. 1993, Ap.J. 404:8-20.

95. VSA: Very Small Array. This is a 30 GHz interferometer; the next generation of CAT. Web site http://www.mrao.com.ac.uk/telescopes/cat/vsa.html contains more information.

96. White Dish. This experiment uses and on-axis Cassegrain telescope and a 90 GHz single-mode bolometer. It observed at the South Pole and is sensitive to small angular scales. Results are reported in Tucker et al., 1993, Ap.J. 419:L45.

97. Ratra, B. 1996. The original compilation was reported in Ratra & Sugiyama, 1995. This is available through astro-ph/9512157. Ratra has kept the list up-to-date and kindly supplied his more recent results.

98. Ganga, K. M. et al. 1993, Ap.J. 432:L15-L18.

99. Cheng, E. S. et al., 1994, Ap.J. 422:L37-L40.

100. Cheng, E. S. et al., 1996, Ap.J. 456:L71-L74.

101. Inman, C. A. et al, Submitted to Ap.J. Letters; astro-ph/9603017.

102. Crittenden, R. G., Coulson D., & Turok, Phys. Rev. D, 1995, D52, 5402. See also astro-ph/9408001 and astro-ph/9406046.

INTRODUCTION TO CORRELATION PROPERTIES OF GALAXY DISTRIBUTION

L. PIETRONERO, M. MONTUORI AND F. SYLOS LABINI
Dipartimento di Fisica, Universitá di Roma "La Sapienza", Piazzale A. Moro 2, 00185 Roma, Italy and INFM unit of Roma 1

Abstract.
Statistical analysis of spatial galaxy distribution is usually performed through the two point function $\xi(r)$. This analysis allows one to determine a correlation length r_0 ($\xi(r_0) = 1$), which separates a correlated regime ($r < r_0$) from an uncorrelated one ($r > r_0$). Some years ago we criticized this approach and proposed a new one based on the *concepts and methods of modern Statistical Physics*. Here we present an introduction to these methods and report the results of the analysis to all the available three dimensional catalogues of galaxies and clusters, i.e. CfA, Perseus-Pisces, SSRS, IRAS, Stromlo-APM, LEDA, Las Campanas and ESP for galaxies and Abell and ACO for clusters. All the data analyzed are consistent with each other and show fractal correlations (with dimension $D \simeq 2$) up to the deepest scales probed until now ($1000h^{-1}Mpc$) and even more as indicated from the new interpretation of the number counts. The very first consequence of this result is that the usual statistical methods (as for example $\xi(r)$), based on the assumption of homogeneity, are therefore inconsistent for all the length scales probed until now. In the range of self-similarity theories should shift from "amplitudes" to "exponents". These facts lead to fascinating conceptual implications about our knowledge of the universe and to a new scenario for the theoretical challenge in this field.

1. Introduction

From the experimental point of view there are four main facts in Cosmology:
- *The space distribution of galaxies and clusters*: the recent availability of

D. N. Schramm and P. Galeotti (eds.), Generation of Cosmological Large-Scale Structure, 105–126.

several three dimensional samples of galaxies and clusters permits the direct characterization of their correlation properties. - *The cosmic microwave background radiation* (CMBR), that shows an extraordinary isotropy and an almost perfect black body spectrum. - *The linearity of the redshift-distance relation*, usually known as the Hubble law. This law has been established by measuring independently the redshift and the distance of galaxies. - *The abundance of light elements in the universe*. Each of these four points provides an independent experimental fact. The objective of a theory should be to provide a coherent explanation of all these facts together.

Our work refers mainly to the first point, *the space distribution of galaxies and clusters* which, however, is closely related to the interpretation of all the other points. In particular we claim that the usual methods of analysis are intrinsically inconsistent with respect to the properties of these samples. The correct statistical analysis of the experimental data, performed with the methods of modern Statistical Physics, shows that the distribution of galaxies is fractal up to the deepest observed scales (Sylos Labini *et al.*, 1997) (Coleman and Pietronero, 1992). These methods which are able to identify self-similar and non-analytical properties, allow us to test the usual homogeneity assumption of luminous matter distribution.

In section 2 and 3 we introduce the main statistical tools we use for the analysis, together with the basic properties of fractal distributions, while in section 4 we discuss the properties of standard autocorrelation function $\xi(r)$. In section 5 we discuss the results of our analysis on all the available three dimensional catalogues of galaxies and clusters, and in sec.6 we show the consistency of the various different catalogs. In Sec.7 we briefly discuss the limit of statistical valid of the three dimensional samples and we define the concept of a "statistically fair sample". Finally we present our conclusions in sec.8.

2. Statistical Methods and Correlation Properties

In this section we mention the essential properties of fractal structures because they will be necessary for the correct interpretation of the statistical analysis. However in no way these properties are assumed or used in the analysis itself. A fractal consists of a system in which more and more structures appear at smaller and smaller scales and the structures at small scales are similar to the ones at large scales. The self similarity of these structures is then incompatible with analyticity. Standard mathematical tools based on analytical functions can not characterize these distributions. The first quantitative description of these forms is the metric dimension. One way to determine it, is the mass-length method. Starting from an point occupied by an object, we count how many objects N ("mass") are present within a volume of linear size r ("length") (Mandelbrot, 1982):

$$N(r) = B \cdot r^D \tag{1}$$

D is the fractal dimension and characterizes in a quantitative way how the system fills the space. The prefactor B depends to the lower cut-offs of the distribution; these are related to the smallest scale above which the system is self-similar and below which the self similarity is no more satisfied. In general we can write:

$$B = \frac{N_*}{r_*^D} \tag{2}$$

where r_* is this smallest scale and N_* is the number of object up to r_*. For a deterministic fractal this relation is exact, while for a stochastic one it is satisfied in an average sense.

Eq.(1) corresponds to a average behavior of real $N(r)$, that is a very fluctuating function; a fractal is, in fact, characterized by large fluctuations and clustering at all scales. We stress that eq.(1) is completely general: for an homogeneous distribution, for example, one has $D = 3$.

From eq.(1), we can compute the average density $< n >$ for a sample of radius R_s which contains a portion of the structure with dimension D. The sample volume is assumed to be a sphere ($V(R_s) = (4/3)\pi R_s^3$) and therefore

$$< n >= \frac{N(R_s)}{V(R_s)} = \frac{3}{4\pi} B R_s^{-(3-D)} \tag{3}$$

If the distribution is homogeneous ($D = 3$) the average density is constant and independent from the sample volume; in the case of a fractal, the average density depends explicitly on the sample size R_s and it is not a meaningful quantity. In particular in this case the average density is a decreasing function of the sample size and $< n >\to 0$ for $R_s \to \infty$.

It is important to note that eq.(1) holds from every point of the system, when considered as the origin. This feature is related to the non-analyticity of the distribution. In a fractal distribution every observer is equivalent to any other one, i.e. it holds the property of local isotropy around any observer (Sylos Labini, 1994).

It is useful to define the conditional density from an occupied point i as:

$$\Gamma(r)_i = S(r)^{-1} \frac{dN(r)}{dr} = \frac{D}{4\pi} B r^{-(3-D)} \tag{4}$$

where $S(r)$ is the area of a spherical shell of radius r. $\Gamma(r)_i$ is then the density at distance r from the $i - th$ point in a shell of thickness dr. Such a function will be useful in the determination of the statistical properties of the real galaxy samples.

Usually the exponent that defines the decay of the conditional density $(3 - D)$ is called the codimension and it corresponds to the exponent γ of the galaxy distribution.

3. The Conditional and Conditional Average Density

The correlation function suitable to study homogeneous and inhomogeneous distribution is described by (Pietronero, 1987) (Coleman and Pietronero, 1992)

$$G(r) = < n(\vec{r} + \vec{r_i})n(\vec{r_i}) >_i \approx r^{-\gamma} \tag{5}$$

where the exponent $\gamma = 3 - D$ (in 3-dimensional space)(Eq. 4) and the index i of the average means that this is performed on all the occupied points r_i of the system. If the sample is homogeneous, $D = 3$, $G(r) \approx < n >^2$ and then $G(r)$ is constant; if the sample has correlations on all scales, it is fractal, $D < 3$, $\gamma > 0$ and $G(r)$ is a *power law*. For a more complete discussion we refer the reader to (Coleman and Pietronero, 1992). We can normalize the $G(r)$ to the size of the sample under analysis and define, following (Coleman and Pietronero, 1992):

$$\Gamma(r) = \frac{< n(\vec{r} + \vec{r_i})n(\vec{r_i}) >_i}{< n >} = \frac{G(r)}{< n >} \tag{6}$$

where $< n >$ is the average density of the sample. This normalization does not introduce any bias even if the average density is sample-depth dependent, as in the case of fractal distributions (Eq.3), as one can see from Eq. 7. The $\Gamma(r)$ (Eq. 6) can be computed by the following expression

$$\Gamma(r) = \frac{1}{N} \sum_{i=1}^{N} \frac{1}{4\pi r^2 \Delta r} \int_r^{r+\Delta r} n(\vec{r_i} + \vec{r'})d\vec{r'} =$$

$$= \frac{1}{N} \sum_{i=1}^{N} \Gamma(r)_i = \frac{D}{4\pi} B r^{3-D} \tag{7}$$

where N is the number of objects in the sample. $\Gamma(r)$ is the average of $\Gamma(r)_i$ and hence it is a smooth function away from the lower and upper cutoffs of the distribution (r_* and the dimension of the sample). From eq.(7), we can see that $\Gamma(r)$ is independent from the sample size, depending only by the intrinsic quantities of the distribution (B and D). In such a way the comparison between different samples, extracted from the same distribution, is straightforward. If the distributions under analysis come from the same fractal structure, the conditional density computed in the various samples has the same amplitude and scaling behaviour, independently of their sizes.

If the distribution is fractal up to a certain distance λ_0, and then it becomes homogeneous, we have that

$$\Gamma(r) = \frac{BD}{4\pi} r^{D-3} \quad r < \lambda_0$$

$$\Gamma(r) = \frac{BD}{4\pi} \lambda_0^{D-3} \quad r \geq \lambda_0 \tag{8}$$

It is also very useful to use the *conditional average density* defined as

$$\Gamma^*(r) = \frac{3}{4\pi r^3} \int_0^r 4\pi r'^2 \Gamma(r') dr' \tag{9}$$

This function produce an artificial smoothing of $\Gamma(r)$ function, but it correctly reproduces global properties (Coleman and Pietronero, 1992).

Given a certain sample of solid angle Ω and depth R_d, it is important to define which is the maximum distance up to which it is possible to compute the correlation function ($\Gamma(r)$ or $\xi(r)$). As discussed in (Coleman and Pietronero, 1992) (see also (Sylos Labini et al., 1996b) (Di Nella et al., 1996) (Sylos Labini et al., 1996a)), we have limited our analysis to an effective depth R_s that is of the order of the radius of the maximum sphere fully contained in the sample volume. For a catalog with the limits, for example, in right ascension ($\alpha_1 \leq \alpha \leq \alpha_2$) and declination ($\delta_1 \leq \delta \leq \delta_2$) we have that

$$R_s = \frac{R_d sin(\delta\theta/2)}{1 + sin(\delta\theta/2)} \tag{10}$$

where $\delta\theta = min(\alpha_2 - \alpha_1, \delta_2 - \delta_1)$. In such a way that we eliminate from the statistics the points for which a sphere of radius r is not fully included within the sample boundaries. Hence we do not make use of any weighting scheme with the advantage that we do not make any assumption in the treatment of the boundaries conditions. Of course in doing this, we have a smaller number of points and we stop our analysis at a smaller depth than that of other authors. In Tab.1 we report the values of Ω, R_s and R_d for the various catalogs. We can see that, although LCRS or ESP are very deep, the value of R_s is of order of the one of CfA1, and this is the reason why the value of r_0 is almost the same in these different surveys. On the other hand, in CfA2 the value of r_0 has been measured to be $r_0 \approx 11h^{-1}Mpc$ (Park et al., 1994) and in SSRS2 $r_0 \approx 15h^{-1}Mpc$ (Benoist et al., 1996), because their solid angle is quite large.

The reason why $\Gamma(r)$ (or $\xi(r)$) cannot be computed for $r > R_s$ is essentially the following. When one evaluates the correlation function (or power spectrum (Sylos Labini and Amendola, 1996)) beyond R_s, then one makes explicit assumptions on what lies beyond the sample's boundary. In fact, even in absence of corrections for selection effects, one is forced to consider incomplete shells calculating $\Gamma(r)$ for $r > R_s$, thereby implicitly assuming that what one does not see in the part of the shell not included in the sample is equal to what is inside (or other similar weighting schemes). In other words, the standard calculation introduces a spurious homogenization which we are trying to remove.

If one could reproduce via an analysis that uses weighting schemes, the correct properties of the distribution under analysis, it would be not necessary to produce wide angle survey, and from a single pencil beam deep survey it would be possible to study the entire matter distribution up to very deep scales. It is evident that this could not be the case.

By the way, we have done a test (Sylos Labini *et al.*, 1997) on the homogenization effects of weighting schemes on artificial distributions as well as on real catalogs, finding that the flattening of the conditional density is indeed introduced owing to the weighting, and does not correspond to any real feature in the galaxy distribution.

4. $\xi(r)$ **Analysis**

Before explain in detail the problems related to the standard correlation analysis, it is useful to recall briefly the basic properties of mathematical self-similarity. From a mathematical pont of view self-similarity implies that a rescaling of the length by a factor b

$$r \rightarrow r' = br \tag{11}$$

leaves the correlation function unchanged apart from a rescaling that depends on b but not on the variable r. This leads to the functional relation

$$\Gamma(r') = \Gamma(b \cdot r) = A(b) \cdot \Gamma(r) \tag{12}$$

which is clearly satisfied by a power law with any exponent. In fact for

$$\Gamma(r) = \Gamma_0 r^\alpha \tag{13}$$

we have

$$\Gamma(r') = \Gamma_0 (br)^\alpha = (b)^\alpha \Gamma(r) \tag{14}$$

The same does not hold, for example, for an exponential behavior

$$\Gamma(r) = \Gamma_0 e^{-r/r_0} \tag{15}$$

This reflects the fact that power laws do not possess a characteristic length while for the exponential decay r_0 is a characteristic length. Note that the characteristic length has nothing to do with the prefactor of the exponential and it is not defined by the condition $\Gamma(r_0) = 1$, but from the intrinsic behavior of the function. This brings us to a common misconception that sometimes occurs in the discussion of galaxy correlations. Even for a perfect power law as Eq.13 one might use the condition $\Gamma(r_0) = 1$ to derive a "characteristic length":

$$r_0 = \Gamma_0^{-1/\alpha} \tag{16}$$

This however is completely meaningless because the power law refers to a fractal structure constructed as self-similar and therefore without a characteristic length. In Eq.16 the value of r_0 is just related to the amplitude of the power law that, as we have discussed, has no physical meaning. The point is that the value 1 used in the relation $\Gamma(r_0) = 1$ is not particular in any way so one could have used as well the condition $\Gamma(r_0) = 10^{10}$ or $\Gamma(r_0) = 10^{-10}$ to obtain other lengths. This is the subtle point of self-similarity; there is no reference value (like the average density) with respect to which one can define what is big or small.

At this point it is instructive to consider the behaviour of the standard correlation function $\xi(r)$. Coleman & Pietronero (1992) clarify some crucial points of the such an analysis, and in particular they discuss the meaning of the so-called "correlation length" r_0 found with the standard approach (Peebles, 1993) (Davis and Peebles, 1983) and defined by the relation:

$$\xi(r_0) = 1 \tag{17}$$

where

$$\xi(r) = \frac{< n(\vec{r_i}) n(\vec{r_i} + \vec{r}) >_i}{< n >^2} - 1 \tag{18}$$

is the two point correlation function used in the standard analysis. If the average density is not a well defined intrinsic property of the system, the analysis with $\xi(r)$ gives spurious results. In particular if the system has fractal correlations, the average density is simply related to the sample size as shown by eq.(3).

Following (Coleman and Pietronero, 1992), the expression of the $\xi(r)$ in the case of fractal distribution, is:

$$\xi(r) = ((3 - \gamma)/3)(r/R_s)^{-\gamma} - 1 \tag{19}$$

where R_s (the effective sample radius) is the radius of the spherical volume where one computes the average density from Eq. (3). From Eq. (14) it follows that

i.) the so-called correlation length r_0 (defined as $\xi(r_0) = 1$) is a linear function of the sample size R_s

$$r_0 = ((3 - \gamma)/6)^{\frac{1}{\gamma}} R_s \tag{20}$$

and hence it is a quantity without any correlation meaning but it is simply related to the sample size.

ii.) the amplitude of the $\xi(r)$ is:

$$A(R_s) = ((3 - \gamma)/3)R_s^\gamma \qquad (21)$$

iii.) $\xi(r)$ is a power law only for

$$((3 - \gamma)/3)(r/R_s)^{-\gamma} >> 1 \qquad (22)$$

hence for $r \lesssim r_0$: for larger distances there is a clear deviation from a power law behavior due to the definition of $\xi(r)$. This deviation, however, is just due to the size of the observational sample and does not correspond to any real change of the correlation properties. It is clear that if one estimates the exponent of $\xi(r)$ at distances $r \lesssim r_0$, one systematically obtains a higher value of the correlation exponent due to the break of $\xi(r)$ in the log-log plot. This is actually the case for the analyses performed so far: in fact, usually, $\xi(r)$ is fitted with a power law in the range $0.5r_0 \lesssim r \lesssim 2r_0$. In this case one obtains a systematically higher value of the correlation exponent. In particular, the usual estimation of this exponent by the $\xi(r)$ function leads to is $\gamma \approx 1.7$, different from $\gamma \approx 1$ (corresponding to $D \approx 2$) that we found by means of the $\Gamma(r)$ analysis.

5. Analysis of the Galaxy Distributions

Here we discuss the correlation properties of the galaxy distributions in terms of volume limited catalogues (Coleman and Pietronero, 1992) arising from most of the 50.000 redshift measurements that have been made to date. In Table 1 we report the geometrical and statistical properties of the analyzed catalogues.

A first important result will be that *the samples are statistically rather good* and their properties are in agreement with each other. This gives a new perspective because, using the standard methods of analysis, the properties of different samples appear contradictory with each other and often this is considered to be a problem of the data (unfair samples) while, we show that this is due to the inappropriate methods of analysis. In addition essentially all the catalogues show well defined fractal correlations up to their limits and the fractal dimension is $D \simeq 2$. The few exceptions to this result will be discussed and interpreted in detail. The main data of our correlation analysis are collected in Fig.1 (left part) in which we report the *conditional density as a function of scale* for the various catalogues. The relative position of the various lines is not arbitrary but it is fixed by the luminosity function, a part for the cases of IRAS and SSRS1 for which this is not possible. In sec.6 we discuss more in detail the procedure

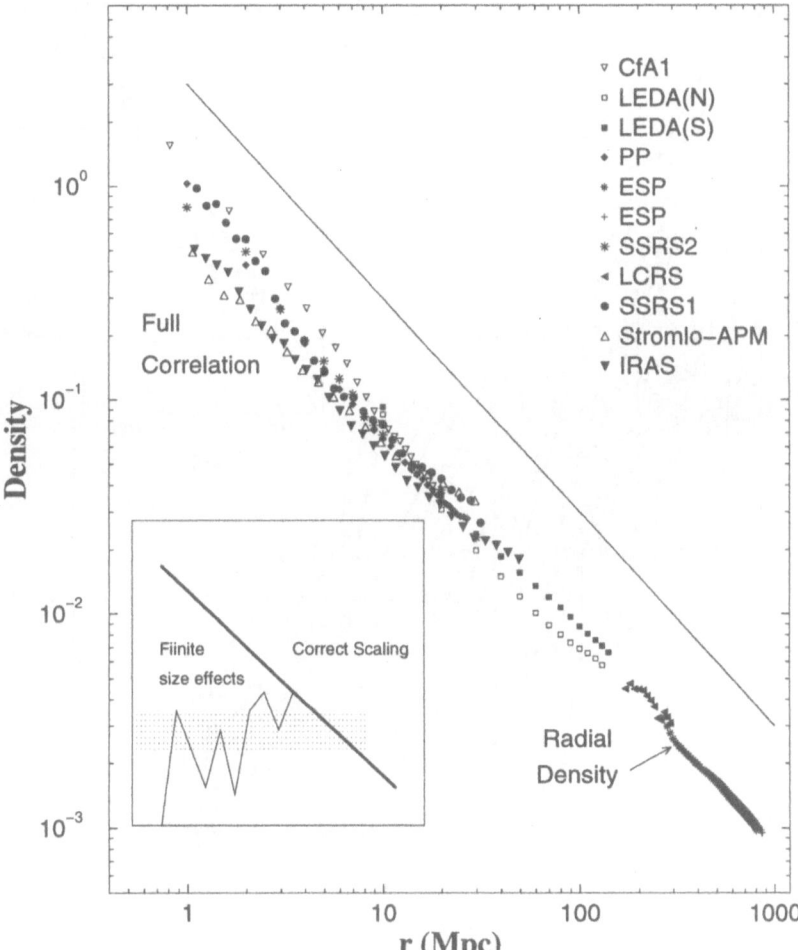

Figure 1. Full correlation for the various available redshift catalogues in the range of distances $0.1 \div 1000h^{-1} Mpc$. A reference line with a slope -1 is also shown (i.e. fractal dimension $D = 2$). Up to $\sim 150h^{-1} Mpc$ the density is computed by the full correlation analysis, while above $\sim 150h^{-1} Mpc$ it is computed through the radial density. For the full correlation the data of the various catalogues are normalized with the luminosity function and they match very well with each other. This is an important test of the statistical validity and consistency of the various data. In the *insert panel* it is shown the schematic behavior of the radial density versus distance computed from the vertex (see text). The behaviour of the radial density allows us to extend the power law correlation up to $\sim 1000h^{-1} Mpc$. However a rescaling is necessary to match the radial density to the conditional density.

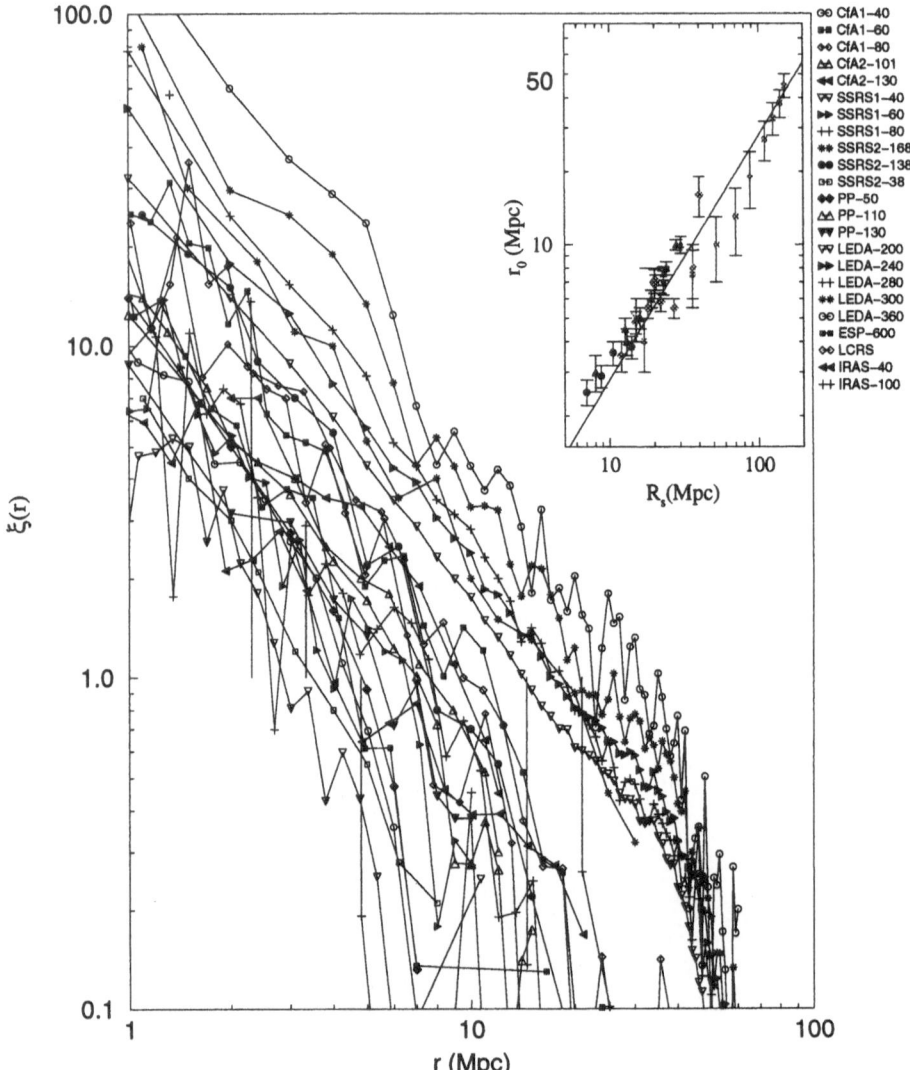

Legend (insert panel):
∞ CfA1–40
∞ CfA1–60
∞ CfA1–80
△△ CfA2–101
◄◄ CfA2–130
▽▽ SSRS1–40
►► SSRS1–60
++ SSRS1–80
∞ SSRS2–168
∞ SSRS2–138
∞ SSRS2–38
◆◆ PP–50
△△ PP–110
▼▼ PP–130
▽▽ LEDA–200
►► LEDA–240
++ LEDA–280
∞ LEDA–300
∞ LEDA–360
■■ ESP–600
∞ LCRS
◄◄ IRAS–40
++ IRAS–100

Figure 2. Usual analysis based on the function $\xi(r)$ of the same galaxy catalogues of Fig.1. This analysis is based on the a priori and untested assumption of the analyticity and homogeneity. These properties are not present in the real galaxy distributions and the results appear therefore rather confusing. This lead to the impression that galaxy catalogues are not good enough and to a variety of theoretical problems like the galaxy cluster mismatch, luminosity segregation, the linear and non linear evolution, etc.. The situation changes completely and it becomes rather clear if one adopts the more general framework that is at the basis of Fig.1. In the *insert panel* we show the dependence of r_0 on R_s for all the catalogs. The linear behaviour is a consequence of the fractal nature of galaxy distribution in these samples.

TABLE 1. The volume limited catalogues are characterized by the following parameters: - $R_d(h^{-1}Mpc)$ is the depth of the catalogue - Ω is the solid angle - $R_s(h^{-1}Mpc)$ is the radius of the largest sphere that can be contained in the catalogue volume. This gives the limit of statistical validity of the sample. - $r_0(h^{-1}Mpc)$ is the length at which $\xi(r) \equiv 1$. - λ_0 is the eventual real crossover to a homogeneous distribution that is actually never observed. The value of r_0 is the one obtained in the deepest sample. The CfA2 and SSRS2 data are not yet available. (Distance are expressed in $h^{-1}Mpc$).

Sample	Ω (sr)	R_d	R_s	r_0	D	λ_0
CfA1	1.83	80	20	6	1.7 ± 0.2	> 80
CfA2	1.23	130	30	10	2.0	?
PP	0.9	130	30	10	2.0 ± 0.1	> 130
SSRS1	1.75	120	35	12	2.0 ± 0.1	> 120
SSRS2	1.13	150	50	15	2.0	?
Stromlo-APM	1.3	100	30	10	2.2 ± 0.1	> 150
LEDA	4π	300	150	45	2.1 ± 0.2	> 150
LCRS	0.12	500	18	6	1.8 ± 0.2	> 500
IRAS 1.2Jy	4π	80	40	4.5	2.0 ± 0.1	$\simeq 25$
ESP	0.006	700	10	5	1.9 ± 0.2	> 800

for the normalization of $\Gamma(r)$ from various catalogues. The properties derived from different catalogues are compatible with each other and show a *power law decay* for the conditional density from $1h^{-1}Mpc$ to $150h^{-1}Mpc$ without any tendency towards homogenization (flattening). This implies necessarily that the value of r_0 (derived from the $\xi(r)$ approach) will scale with the sample size R_s as shown also from the specific data about r_0 of the various catalogues (Coleman *et al.*, 1988) (Sylos Labini *et al.*, 1996b). The behaviour observed corresponds to a fractal structure with dimension $D \simeq 2$. An homogeneous distribution would correspond to a flattening of the conditional density which is never observed.

To check that possible errors in the apparent magnitude do not affect seriously the behavior of $\Gamma(r)$ one can perform the various tests. For example one can change the apparent magnitude of galaxies in the whole catalog by a random factor δm with $\delta m = \pm 0.2, 0.4, 0.6, 0.8$ and 1. We find that the number of galaxies in the VL samples change from 5% up to 15% and that the amplitude and the slope of $\Gamma(r)$ are substantially stable and there are not any significant changes in their behavior. This is because $\Gamma(r)$ measures a global quantity that is very robust with respect to these possible errors. There are several other tests that one can perform, and that are discussed in detail in (Sylos Labini *et al.*, 1997). However we would like to stress that a fractal distribution has a very strong property: it shows power law correlations up to the sample depth. Such correlations *cannot be due* neither by an inhomogeneous sampling of an homogeneous distribution, nor by some selection effects that may occur in the observations. Namely, suppose that a certain kind of sampling reduces the number of galaxies as a function of distance (Davis , 1997). Such an effect in no way can lead to long range correlations, because when one computes $\Gamma(r)$, one makes an average over all the points inside the survey.

It is interesting to compare the analysis of Fig.1 with the usual one, made with the function $\xi(r)$, for the same galaxy catalogues. This is reported in Fig.2 and, from this point of view, the various data appear to be in strong disagreement with the each other. This is due to the fact that the usual analysis looks at the data from the perspective of analyticity and large scale homogeneity (within each sample). These properties are never tested and they are actually not present in the real galaxy distributions, so the result is rather confusing (Fig.2). Once the same data are analyzed within a broader perspective the situation becomes clear (Fig.1) and the data of different catalogues result in agreement with each other. In addition in the insert of Fig.2 we show the dependence of r_0 on R_s for all the catalogs. The linear behaviour is a consequence of the correlation properties of Fig.1 and it provides an additional evidence of fractal behaviour to all scales. In this respect, the proposed luminosity bias effect appears essentially irrelevant while, on the contrary, the linear dependence of r_0 on R_s is very clear.

It is important to remark that analyses like those of Fig.1 have had a profound influence in the field in various ways: first the various catalogues appear in conflict with each other. This has generated the concept of "*not fair sample*" and a strong mutual criticism about the validity of the data between different groups. In other cases the discrepancy observed in Fig.2 have been considered as real physical problems for which various theoretical approaches have been proposed. These problems are, for example, the galaxy-cluster mismatch, luminosity segregation, the richness clustering relation and the linear and non linear evolution of the perturbations corresponding to the "*small*" or "*large*" amplitudes of fluctuations. We can now see that all this problematic is not real and it arises only from a statistical analysis based on inappropriate and to restrictive assumptions that do not find a correspondence in the physical reality. It is also important to note that, even if the galaxy distribution would eventually become homogeneous at some large scale, the use of the above statistical concepts is anyhow inappropriate for the range of scales in which the system shows fractal correlations as those shown in Fig.1.

We would like to stress that a fractal distribution has a very strong property: *it shows global power-law correlations up to the sample depth.* Such correlations *cannot be due* neither to an inhomogeneous sampling of an homogeneous distribution, nor to some selection effects that may occur in the observations. Namely, suppose that a certain kind of sampling reduces the number of galaxies as a function of distance. Such an effect, in no way can lead to long range correlations, because when one computes $\Gamma(r)$, one makes an average over all the points inside the survey. In any case this possible bias could be detected by a difference in the values of $\Gamma(r)$ at different depths, which is not observed (Sylos Labini *et al.*, 1997). We observe instead that all the catalogues, independently on their completeness, show precisely the same correlation properties. The correlations discussed up to now are well defined statistically but limited to the radius of the largest sphere that can be contained in the sample R_s. For example for Las Campanas the depth is very large $R_d \sim 500h^{-1}Mpc$ but R_s is only $20h^{-1}Mpc$ because the sample is very thin. So, it is not surprising that the value of r_0 is also small $(6 \div 7h^{-1}Mpc)$. Given this situation it would be very interesting to find some method that is limited by R_d instead of R_s.

In addition we have studied several cluster catalogs (Abell and ACO) showing that galaxies and clusters are different representations of a single self-similar structure: the correlations of clusters appear to be the continuation of galaxy correlations at larger scales, and clusters have the same fractal dimension $D \approx 2$ (Coleman and Pietronero, 1992) (Sylos Labini *et al.*, 1997) of galaxies (Fig.3).

All these results imply that the previous "correlation lengths" of $5h^{-1}Mpc$ and $25h^{-1}Mpc$, introduced for galaxies and clusters, are spurious and no real correlation length can be defined from the data. Therefore the much discussed mismatch between galaxy and cluster correlations, that is also at the basis of various theories for structure formation, does not actually exist. Cluster correlations correspond just to the continuation of galaxy correlations at larger scales. In the language of Statistical Physics, cluster catalogues are the coarse grained version of galaxy catalogues.

A possible explanation of the shift of r_0 is based on the luminosity segregation effect (Davis *et al.*, 1988) (Park *et al.*, 1994) (Benoist *et al.*, 1996). We briefly illustrate this approach. The fact that the giant galaxies are more clustered than the dwarf ones, i.e. that they are located in the peaks of the density field, has given rise to the proposition that larger objects may correlate up to larger length scales and that the amplitude of the $\xi(r)$ is larger for giants than for dwarfs one. The deeper VL subsamples contain galaxies that are in average brighter than those in the VL subsamples with smaller depths. As the brighter galaxies should have a larger correlation

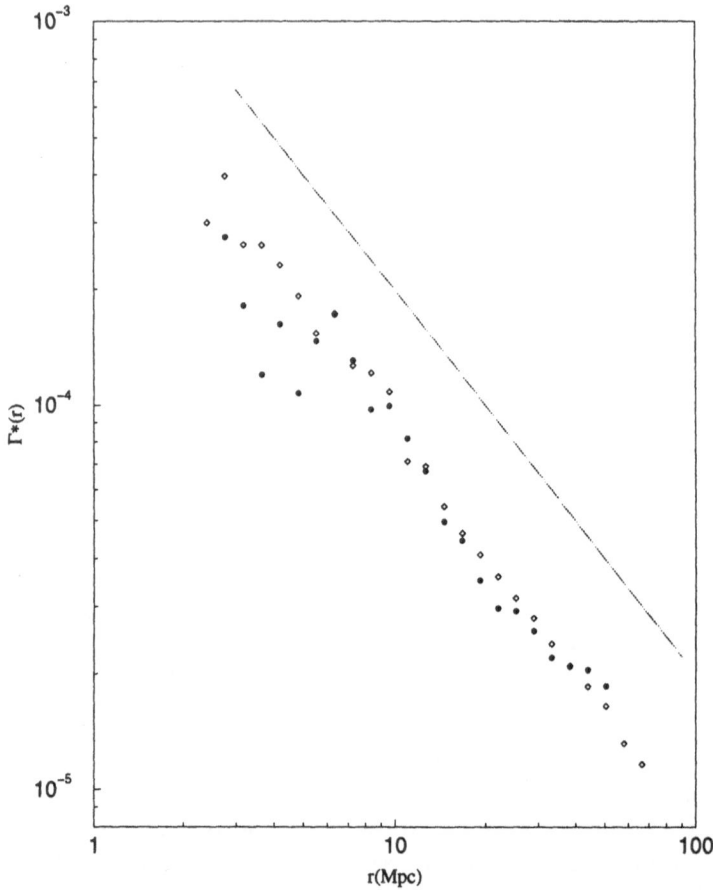

Figure 3. The conditional average density (eq.2) for galaxy clusters: Abell and ACO. The reference line has a slope -1 that corresponds to a fractal dimension $D = 2$.

length the shift of r_0 with sample size could be related, at least partially, with the phenomenon of luminosity segregation.

We would like to stress that as long as $\Gamma(r)$ has a power law decay, r_0 *must be a linear fraction of the sample size*, and that as far as a clear crossover towards homogeneity has been not identified, the "correlation length" r_0 has no physical meaning, just being related to the size of the sample. *Moreover the authors (e.g. (*Davis *et al.*, 1988*) (*Park *et al.*, 1994*) (*Benoist *et al.*, 1996*)) that have introduced the concept do not present any quantitative argument that explains the shift of* r_0 *with sample size.* In addition as we have discussed previously, as far as a clear cut-off towards homogeneity has not been identified, the analysis performed by the $\xi(r)$ function is misleading, i.e. it does not give a physically meaningful result.

We have discussed in detail in (Sylos Labini and Pietronero, 1996) that the observation that the giant galaxies are more clustered than the dwarf ones, i.e. that the massive elliptical galaxies lie in the peaks of the density field, is a consequence of the self-similar behavior of the whole matter distribution. The increasing of the correlation length of the $\xi(r)$ has nothing to do with this effect (Coleman and Pietronero, 1992) (Baryshev *et al.*, 1994).

Finally we would like to stress the conceptual problems of the interpretation of the scaling of r_0 by the luminosity segregation phenomenon. Suppose we have two kind of galaxies of different masses, one of type A and the other of type B. Suppose for simplicity that the mass of the galaxies of type A is twice that of the B. The proposition "galaxies of different luminosities (masses) correlate in different ways" implies that the gravitational interaction is able to distinguish between a situation in which there is, in a certain place, a galaxy of type A or two galaxies of type B placed very near. This seem to be not possible as the gravitational interaction is due the sum of all the masses.

We can go farther by showing the inconsistency of the proposition "galaxies of different luminosities have a different correlation length". Suppose that the galaxies of type A have a smaller correlation length than that of the galaxies of type B This means that the galaxies of type B are still correlated (in terms of the conditional density) when the galaxies of type A are homogeneously distributed. This means that the galaxies of type A should fill the voids of galaxies of type B. This is not the case, as the voids are empty of all types of galaxies, and it seems that the large scale structures distribution is independent on the galaxy morphological types.

6. Normalization of the Density

Here we briefly discuss the procedure of normalization of density ($\Gamma^*(r)$) (Sylos Labini *et al.*, 1997).

A catalog is usually obtained by measuring the redshifts of the all galaxies with apparent magnitude brighter than a certain apparent magnitude limit m_{lim}, in a certain region of the sky defined by a solid angle Ω. An important selection effect exists, in that at every distance in the apparent magnitude limited survey, there is a definite limit which is the absolute magnitude of the fainter galaxy which can be seen at that distance. Hence at large distances, intrinsically faint objects are not observed whereas at smaller distances they are observed. In order to analyze the statistical properties of galaxy distribution, a catalog which does not suffer for this selection effect must be used. In general, it exists a very well known procedure to obtain a sample that is not biased by this luminosity selection effect: this is the so-called *"volume limited"* (VL) sample. A VL sample contains every galaxy in the volume which is more luminous than a certain limit, so that in such a sample there is no incompleteness for an observational luminosity selection effect (Davis and Peebles, 1983), (Coleman and Pietronero, 1992). Such a sample is defined by a certain maximum distance R_{VL} and the absolute magnitude limit M_{VL} given by:

$$M_{VL} = m_{lim} - 5 \log_{10} R_{VL} - 25 - A(z) \tag{23}$$

where $A(z)$ takes into account various corrections (K-corrections, absorption, relativistic effects, etc.), and m_{lim} is the survey apparent magnitude limit.

Of course, given a certain catalog, it is possible to built several VL samples, choosing different limiting distances. To each VL sample we can associate the luminosity factor:

$$\Phi(M_{VL}) = \int_{-\infty}^{M_{VL}} \phi(M) dM \tag{24}$$

that gives the fraction of galaxies per unit volume, present in the sample, with respect the whole density. The normalization of $\Gamma(r)$ in different VL samples can be simply done by dividing their amplitude for the corresponding luminosity factor. Of course such a normalization is parametric, because it depends on the two parameters of the luminosity function δ and M^*. For a reasonable choice of these two parameters we find that the amplitude of the conditional and radial density matches quite well in different VL samples. In Fig.4, we report the $\Gamma^*(r)$ for various VL samples in different surveys after the normalization. The apparent good matching is a proof of validity of analysis and reality of fractal distribution.

7. Samples Validity and Dilution Effects: What is a Fair Sample?

How many galaxies one needs in order to characterize correctly (statistically) the large scale distribution of visible matter ? This fundamental question is addressed in this section, and it will allows us to understand some basic properties of the statistical analysis of galaxy surveys. In such a way, we will be able to clarify the concept of "fair sample", i.e. a sample that contains a statistically meaningful information (Sylos Labini et al., 1996a) (Sylos Labini et al., 1997)

We have discussed in the pervious sections the properties of fractal structures and in particular we have stressed the intrinsic highly fluctuating nature of such distributions. In this perspective it is important to clarify the concept of "fair sample". Often this concept is used as synonymous of a homogeneous sample (see for example (Da Costa et al., 1994)). So the analysis of catalogues along the traditional lines often leads to the

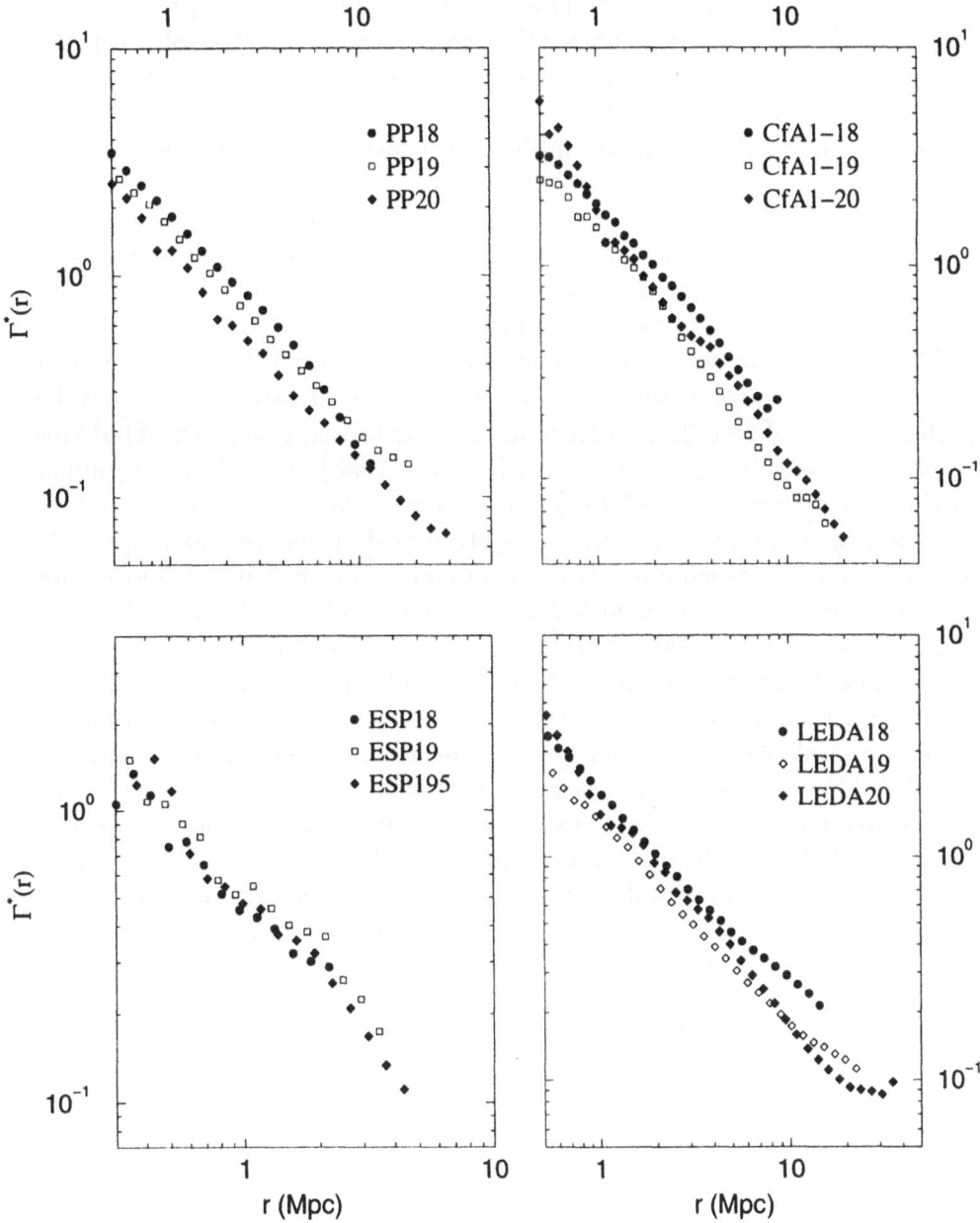

Figure 4. The spatial density $\Gamma^*(r)$ computed in some VL samples of Perseus-Pisces, CfA1, ESP and LEDA and normalized to the corresponding luminosity factor.

conclusion that we still do not have a fair sample and deeper surveys are needed to derive the correct correlation properties. A corollary of this point of view is that since we do not have a fair sample its statistical analysis cannot be taken too seriously.

This point of view is highly misleading because we have seen that self-similar structures never become homogeneous, so any sample containing a self-similar (fractal) structure would automatically be declared "not fair" and therefore impossible to analyze. The situation is actually much more interesting otherwise the statistical mechanics of complex systems would not exist. Homogeneity is a property and not a condition of statistical validity of the sample. A non homogeneous system can have well defined statistical properties in terms of scale invariant correlations, that may be perfectly well defined. The whole studies of fractal structures are about this (Pietronero and Tosatti , 1986) (Erzan *et al.*, 1995). Therefore one should distinguish between a "statistical fair sample", which is a sample in which there are enough points to derive some statistical properties unambiguously and a homogeneous sample, that is a property that can be present or not but that has nothing to do with the statistical validity of the sample itself. We have seen in Sec.3 that even the small sample like CfA1 is statistically fair up to a distance that can be defined unambiguously (i.e $\sim 20h^{-1}Mpc$).

In (Sylos Labini *et al.*, 1996a) we have studied the following question. Given a sample with a well defined volume, which is the minimum number of points that it should contain in order to have a *statistically fair* sample, even if one computes averages over all the points, such as the conditional density and the conditional average density ?

Suppose that the sample volume is a portion of a sphere with a solid angle Ω and radius R, the mass $(N(< R))$ length (R) relation can be written as (Coleman and Pietronero, 1992)

$$N(< R) = B \left(\frac{\Omega}{4\pi} \right) R^D \qquad (25)$$

where D is the fractal dimension or, for the homogeneous case, $D = 3$. The prefactor B is a constant and it is related to the lower cut-off of the fractal structure (Coleman and Pietronero, 1992) (Sylos Labini *et al.*, 1996a). In this letter we consider galaxies of different luminosities having the same clustering properties (i.e. equal fractal dimension): this is a crude approximation and the more complex situation can be described in terms of multifractal (Sylos Labini and Pietronero, 1996).

In principle Eq.25 should refer to *all* the galaxies existing in a given volume. If instead we have a VL sample, we will see only a fraction $N_{VL}(R) = p \cdot N(< R)$ (where $p < 1$) of the total number $N(< R)$. In order to estimate the fraction p it is necessary to know the luminosity function $\phi(L)$ that gives the fraction of galaxies whose absolute luminosity (L) is between L and $L + dL$ (Schecther, 1976) This function has been extensively measured (Da Costa *et al.*, 1994) and it consists of a power law extending from a minimal value L_{min} to a maximum value L^* defined by an exponential cut-off. Therefore we can express the fraction p as

$$0 < p = \frac{\int_{L_{VL}}^{\infty} \phi(L)dL}{\int_{L_{min}}^{\infty} \phi(L)dL} < 1 \qquad (26)$$

where L_{VL} is the minimal absolute luminosity that characterizes the VL sample. The quantity L_{min} is the fainter absolute flux (magnitude M_{lim}) surveyed in the catalog (usually $M_{min} \sim -11 \div -12$).

We have performed several tests in real galaxy samples as well as in artificial distributions with a priori assigned properties. In particular we have eliminated randomly points from the original distribution: such a procedure, for the law of codimension additivity (Mandelbrot, 1982) does not change the fractal dimension, but only the prefactor in Eq.25. In such a way we can control, quantitatively, the behaviour of the conditional density as the value of p decreases, and we are able to conclude that, if the system has self-similar properties, a reliable correlation analysis is only possible if p is substantially larger than $1 \div 2\%$ (Sylos Labini *et al.*, 1996a). Below this value the statistical significance becomes questionable just because the sample is too sparse and large scale correlations are destroyed by this effect.

In relation to the statistical validity it is interesting to consider the IRAS catalogues because they seem to differ from all the other ones and to show some tendency towards homogenization at a relatively small scale Actually the point of apparent homogeneity is only present in some samples, it varies from sample to sample between $\sim 15 \div 25 h^{-1} Mpc$ and it is strongly dependent on the dilution of the sample. Considering that structures and voids are much larger than this scale and that the IRAS galaxies appear to be just where luminous galaxies are it is clear that this tendency appears suspicious. One of the characteristic of the IRAS catalogues with respect to all the other ones is an extreme degree of dilution: this catalogue contains only a very small fraction of all the galaxies. It is important therefore to study what happens to the properties of a given sample if one dilutes randomly the galaxy distribution up to the IRAS limits. A good test can be done by considering the Perseus Pisces catalogue and eliminating galaxies from it. The original distribution shows a well defined fractal behavior. By diluting it to the level of IRAS one observes an artificial flattening of the correlations (Sylos Labini *et al.*, 1996a) (see Fig.5). This effect does not correspond to a real homogenization but it is due to the of dilution. In fact it can be shown that when the dilution is such that the average distance between galaxies becomes comparable with the largest voids (lacunarity) of

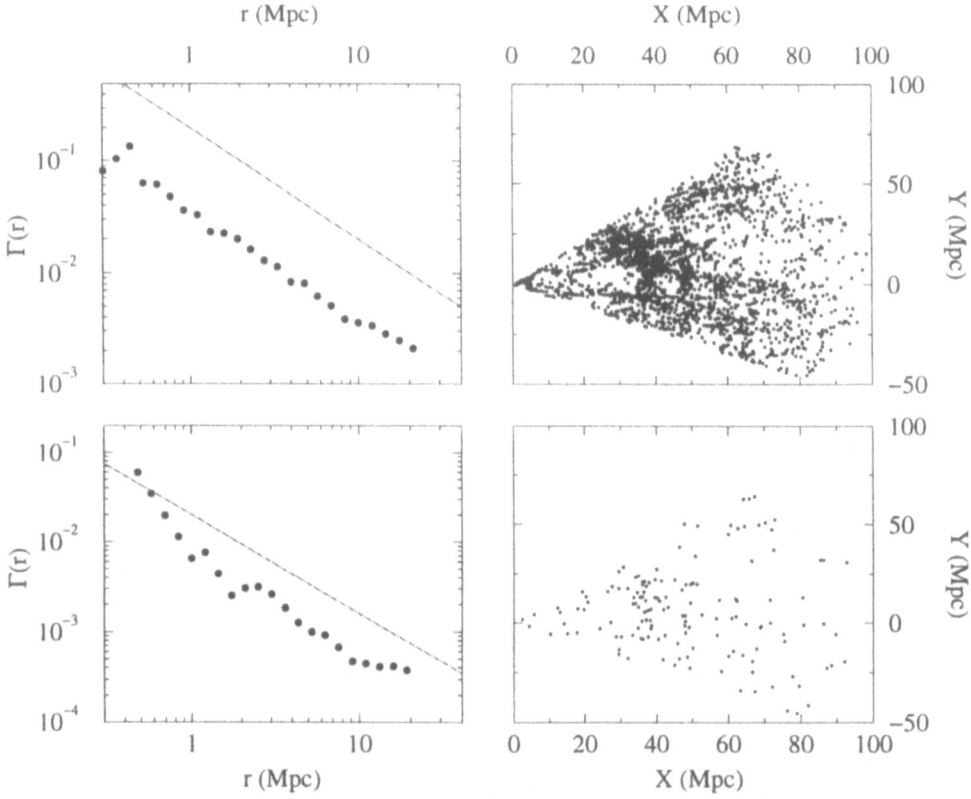

Figure 5. *Top panels:* The conditional density (left) for a volume limited sample of the full Perseus-Pisces redshift survey (right). The percentage of galaxies present in the sample is $\sim 6\%$. The slope is $-\gamma = -1$. *Bottom panels:* In this case the percentage of galaxies is $\sim 1\%$, and the number of galaxies is the same of the IRAS $1.2Jy$ sample in the same region of the sky. We can see that at small scale we have a $1/r^3$ decay just due to the sparseness of the sample, while at large scale the shot noise of the sparse sampling overcomes the real correlations and produces an apparent trend to homogenization. In our opinion, this effect, due to sparseness of this sample, is the origin of the apparent trend towards homogenization observed in some of the IRAS samples.

the original structure there is a loss of correlation and the shot noise of the sparse sampling overcomes the real correlations and produces an apparent trend to homogenization. This allows us to reconcile this peculiarity of the IRAS data with the properties of all the other catalogues. Analogous considerations for other sparse samples like QDOT and the Stromlo-APM samples (Sylos Labini *et al.*, 1997).

8. Conclusions

In summary our main points are:

- The highly irregular galaxy distributions with large structures and voids strongly point to a new statistical approach in which the existence of a well defined average density is not assumed a priori and the possibility of non analytical properties should be addressed specifically.
- The new approach for the study of galaxy correlations in all the available catalogues shows that their properties are actually compatible with each other and they are statistically valid samples. The severe discrepancies between different catalogues that have led various authors to consider these catalogues as *not fair*, were due to the inappropriate methods of analysis.
- The correct two point correlation analysis shows well defined fractal correlations up to the present observational limits, from 1 to $1000h^{-1}Mp$ with fractal dimension $D \simeq 2$. Of course the statistical quality and solidity of the results is stronger up to $100 \div 200h^{-1}Mpc$ and weaker for larger scales due to the limited data. It is remarkable, however, that at these larger scales one observes exactly the continuation of the correlation properties of the small and intermediate scales.
- These new methods have been extended also to the analysis of the number counts and the angular catalogues which are shown to be fully compatible with the direct space correlation analysis. The new analysis of the number counts suggests that fractal correlations may extend also to scales larger that $1000h^{-1}Mpc$.
- The inclusion of the galaxy luminosity (mass) leads to a distribution which is shown to have well defined multifractal properties. This leads to a new, important relation between the luminosity function and that galaxy correlations in space.

Acknowledgments

We thank for useful discussions, suggestions and collaborations L.Amendola, A. Amici, Yu.V. Baryshev, H. Di Nella, R. Durrer, A. Gabrielli.

References

Baryshev, Y., Sylos Labini, F., Montuori, M., Pietronero, L. (1994) *Vistas in Astron.* **38**, 419-500
Benoist C. *et al.*(1996) *Astrophys. J.*, in print
Coleman, P.H. Pietronero, L.,& Sanders,R.H., (1988) *Astron. & Astrophys.* **200** L32-L34
Coleman, P.H. and Pietronero, L., (1992) *Phys.Rep.* **231**, pp.311-391
Davis, M., Peebles, P. J. E. (1983) *Astrophys. J.* **267**, 465-482
Davis M. *et al.*, (1988) *Astrophys. J.* **333**, L9-L12

126

Da Costa L.N., *et al.*(1994) *Astrophys. J.* **424** L1-L4

Davis M., (1997)(astro-ph/9610149) *in the Proc. of the Conference "Critical Dialogues in Cosmology"* N. Turok ed.

Di Nella H., Montuori M., Paturel G., Pietronero L., and Sylos Labini F., (1996) *Astron.Astrophys.Lett.* **308**, L33-L36

Erzan A, Pietronero L., Vespignani A. (1995) *Rev. Mod. Phys.* **67**, 554

Mandelbrot, B. (1982) *The Fractal Geometry of Nature,* Freeman, New York

Park, C., Vogeley, M.S., Geller, M., Huchra, J. (1994) *Astrophys. J.* **431** 569

Peebles, P.E.J., (1993) *Principles of physical Cosmology* Princeton Univ.Press.

Pietronero L., and Tosatti E. Eds (1986) *Fractal in Physics* North-Holland, Amsterdam

Pietronero, L. *Physica A* **144**, 257-284

Schecther P., (1976) *Astrophys. J.*, **203** 297

Sylos Labini, F., (1994), *Astrophys. J.*, **433**, 464-467

Sylos Labini, F., Gabrielli, A., Montuori, M., Pietronero,L. (1996) **266** 195-242

Sylos Labini, F., Montuori, M., Pietronero,L. (1996) *Physica A* **230**, 336-358

Sylos Labini, F., Pietronero,L. (1996) *Astrophys. J.* **469**, 28-42

Sylos Labini, F. Amendola, L. (1996) *Astrophys. J.* **468** L1-L4

Sylos Labini, F., Montuori, M.and Pietronero, L. (1996), preprint

DARK MATTER: An Introduction

D.N. SCHRAMM

The University of Chicago
5640 South Ellis Avenue, Chicago, IL 60637, USA

This lecture series is to provide an overview of the basic dark matter arguments and the role of the relevant cosmological parameters. In particular, the current situation on the age of the universe and the Hubble constant are reviewed since they provide the framework for the basic density arguments. It is shown that the Big Bang Nucleosynthesis constraints on the cosmological baryon density, when compared with dynamical arguments, demonstrate that the bulk of the baryons are dark and also that the bulk of the matter in the universe is non-baryonic. The recent extragalactic deuterium observations as well as the other light element abundances are examined in detail. Comparison of nucleosynthesis baryonic density arguments with other baryon density arguments is made. Discussion of the interface of density and age arguments is also presented.

1. Introduction

One of the most exciting topics in all the physical sciences is the apparent fact that the bulk of the matter of our universe is not only not seen but seems to be made out of a different substance than baryons. This paper reviews the key arguments regarding the existence of this "dark matter." The conclusion that we need the bulk of the universe to be in the form of non-baryonic dark matter will be emphasized since this seems to require new physics beyond the standard particle model. The review will open with a discussion of the age of the universe since that provides a framework for our later density discussion and also because there is much action on the age front. The review will then turn to Big Bang Nucleosynthesis and its comparison to other density determinations.

2. Age of the Universe

The problem of estimating the age of the universe is longstanding. For example, in 1650, Bishop James Ussher[1] determined, by a technique of summing the Biblical begats and making other corrections and connections based on the then available historical and astronomical records, that the universe began in 4004 BC, at the moment that would correspond to sunset in Jerusalem on the evening before October 23. This would correspond to 4 PM U.T. on October 22.

D. N. Schramm and P. Galeotti (eds.), Generation of Cosmological Large-Scale Structure, 127–152.
© 1997 *Kluwer Academic Publishers.*

This early determination illustrates a key point which we will also apply to more modern techniques. Namely, while Bishop Ussher was able to obtain a result with reported accuracy of about 8 significant figures, his systematic errors are considerably larger. (Even his intrinsic error is larger than the accuracy of his result indicates, since the Jewish calendar, using essentially the same technique, obtains an age that is over 200 years off from Ussher's.)

Today the age of the universe can be estimated by three independent means:

1. Dynamics (Hubble age and deceleration)
2. Oldest Stars (globular clusters)
3. Radioactive Dating (nucleocosmochronology)

We will see that despite much activity on the dynamical technique, the best age bounds are still those derived from nuclear arguments - namely #2 and #3. Each of these gives a lower bound of $t \gtrsim 10$ Gyr as plotted in Figure 1. Furthermore, the age of the disk also bounds the age of the universe ($t_{disk} \sim 10 Gyr$), as does radioactive dating of the Earth-Meteorite system at $t_{ss} = 4.6 \pm 0.1 Gyr$.

2.1. THE AGE FROM DYNAMICS

The use of the Hubble constant to determine an age is the most quoted and least accurate of all the age determination methods. The point is that it is not really determining an age but only a dynamic timescale. For perspective let us note that in the past decade astronomers have published values ranging from $H_0 \sim 100$ km/sec/Mpc down to values near $H_0 \sim 40$ km/sec/Mpc. The higher values tended to come from people using empirical techniques like Tulley-Fisher, whereas the smaller values came primarily from people using supernovae. In principle, supernovae are better understood physically, but some astronomical calibrations inevitably creep in. However, few hidden-variables should creep in since the physics is in reasonable shape, unlike the empirical technology. A critical question tends to be the accuracy of intermediate distance calibrators and the correction for infall into the Virgo cluster. Most of us can't see anything wrong at face value with the Tulley-Fisher techniques other than a possible susceptibility to the so-called Malmquist bias. However, many physicists have a certain fondness for the use of Type-I supernovae as standard candles. Type I's seem to be due to the detonation of a C-O white dwarf star converting its C-O to Fe. Such a model has a physical relationship between its luminosity and basic nuclear quantities that can be measured in the lab. Current best-fit models (c.f. Nomoto) tend to convert about $0.7 M_\odot$ of C-O, which yields $H_0 \sim 60$ km/sec/Mpc. However, even in the extreme where the entire $1.4 M_\odot$ Chandrasekhar mass is burned, H_0 is never below

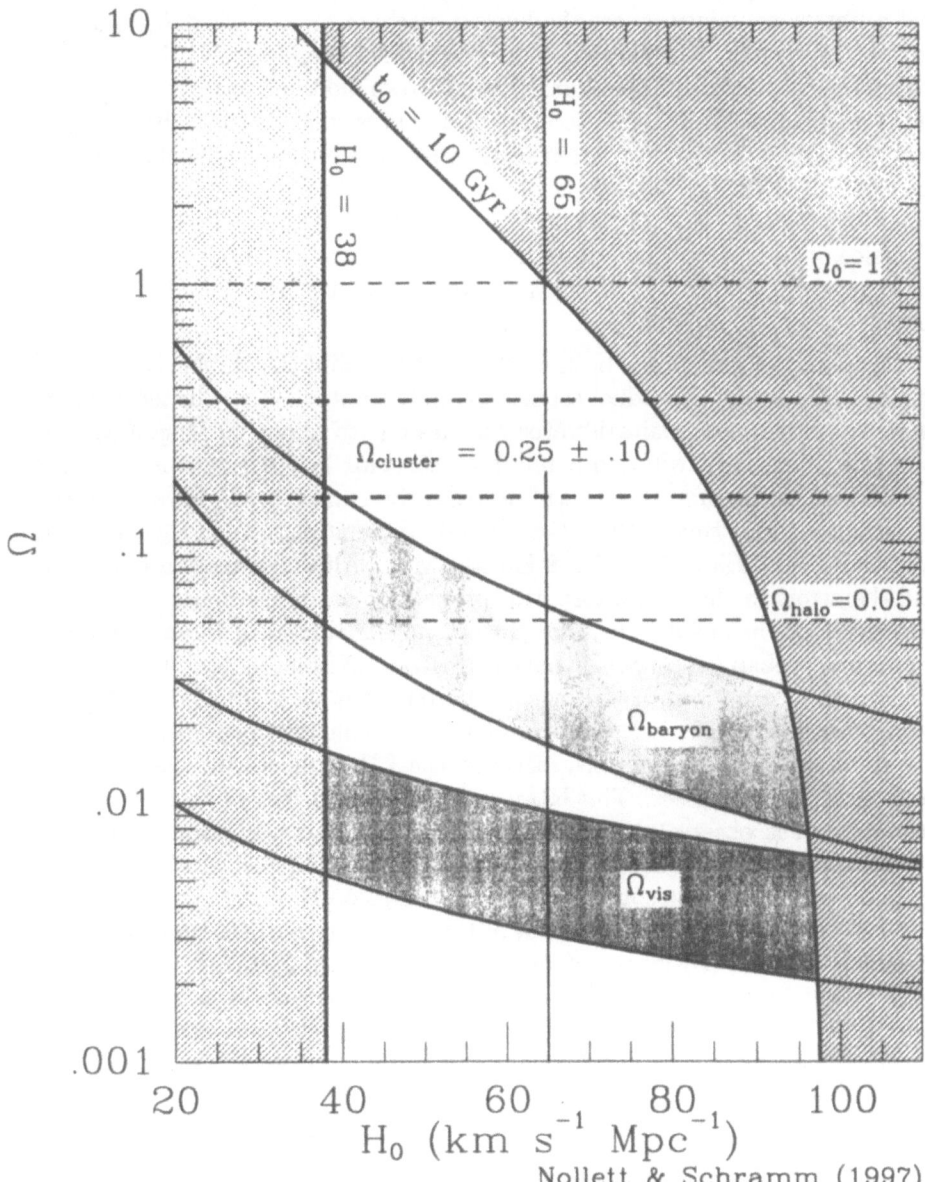

Figure 1: An updated version of $H_0 - \Omega$ diagram of Gott, Gunn, Schramm and Tinsley[59] showing that Ω_b does not intersect $\Omega_{VISIBLE}$ for any value of H_0 and that $\Omega_{TOTAL} > 0.1$, so non-baryonic dark matter is also needed[84].

$\sim$ 40 km/sec/Mpc (see also Nugent et al.[2]). Sandage and Tammann's[3] empirical calibrations, which ignore the nuclear mechanism, now yield $H_0 \sim 58 \pm 7$ km/sec/Mpc after using HST-measured cepheids to calibrate M101, which fall within the theoretically allowed range and correspond to almost complete burning of a Chandrasekhar core. Recently, Riess, Press and Kirshner[4] have argued that there may be some variation in type IA light curves which shifts H_0 up to $\sim 66 \pm 3$. Kirshner[5] also argues that the expanding photosphere of type II supernova implies $H_0 \sim 73$. While selecting between 40 and 75 is still a matter of choice, it does seem that values less than 40 can be reliably excluded. Why these numbers tend to be systematically lower than the Tulley-Fisher numbers remains to be fully understood.

Most recently there has been much publicity about the Hubble Space Telescope (HST) seeing individual cepheid variable stars in Virgo Cluster[6] galaxies as well as other potential calibrator galaxies out to about 20 Mpc. Over the next few years, HST will find many more cepheids in other galaxies in Virgo so that part of the uncertainty will decrease. Freedman et al.[7] have been very conscientious in listing both statistical and systematic errors in their recently quoted value of $H_0 = 73 \pm 7 \pm 8$ km/sec/Mpc. Although most astronomers add the errors in the quadrature, it is probably more realistic not to add systematic errors in quadrature since the second derivatives of these systematic errors are probably not well behaved. In fact, in some cases, the distributions may even be bi-modal. Hence, a better estimate is 73 ± 15. Even this large error does not include the possibility that cepheids themselves may have a systematic shift in luminosity between the LMC (where the calibration is done) and other galaxies. This latter point has now been supported with the new Hipparcus observations of Feist and Catchpole[8] which seem to increase the cepheid scale by $\sim$10% but is still preliminary. Historically, Hubble got $H_0 \sim 500$ due to using cepheids calibrated from Pop II objects and applied to Pop I in other galaxies. While much of the metalicity effect is now taken into account, the observed trends of cepheids in M31 may hint that there is still some residual effect. Thus, while some systematic errors will be reduced with more HST detections of cepheids in other galaxies, some systematic errors will remain (including potential differential reddening between the southern LMC direction and the northern Virgo direction).

With all of these systematics, it is clear at the present time that the SN technique and the Tulley-Fisher techniques are not really in conflict. Recently, the first principles technique of using gravitational lensing, and the time delay between two images have entered the fray. In particular, Kundic et al.[9], using the Q50 0957 $\pm$ 561 A, B system found a time delay of 417 ± 3 days which yields $H_0 = 63 \pm 12$ at 95% confidence Furthermore, Schechter et al.,[10] using

PG1115 + 080 found a 24 ± 3 days time delay, which Keeton and Kochonek[11] said yields $H_0 = 60 \pm 17$. Perhaps we are finally seeing a convergance at

$$H_0 = 65 \pm 10 \text{ km/sec/Mpc}.$$

This range would include all recent determinators. Converting to years, this yields

$$\frac{1}{H_0} = 15 \pm^3_2 \text{ Gyr}.$$

While such a convergance may be a bit premature, it is, nonetheless, not a driving factor in the age discussion.

Age, t_u, is related to H_0 by:

$$t_u = \frac{f(\Omega)}{H_0}$$

where, for standard matter-dominated models, with the cosmological constant $\Lambda = 0$,

$$f(\Omega) = \begin{cases} 1 & \Omega = 0 \\ 2/3 & \Omega = 1 \\ \sim 0.5 & \Omega = 4 \end{cases}.$$

From dynamics alone we can put an upper limit on Ω by limiting the deceleration parameter q_o. From limits on the deviations of the redshift-magnitude diagrams at high redshift, we know that $q_o \lesssim 2$ (for zero cosmological constant $\Omega = 2q_o$). Thus, we can argue that $\Omega \lesssim 4$ or that $f(\Omega) \gtrsim 0.5$. Therefore, from dynamics alone, with no further input, we can conclude only that

$$6 \lesssim t_u \ (Gyr) \lesssim 18$$

Since the lower bound here could almost be obtained from the age of the earth, it is clear that the dynamical technique is not overly restrictive, even with our convergence assumption.

Even high values of H_0 can be consistent with high ages by invoking the cosmological constant, $\lambda = \Lambda/3H_0^2$. Even $\lambda \sim 0.4$ allows high H_0 to be consistent with a flat universe. It is interesting to speculate that $\Lambda \neq 0$ can be produced by a late-time vaccum phase transition of the type proposed by Hill, Schramm and Fry[12]. However, such models do require a fair degree of tuning.

2.2. THE AGE FROM THE OLDEST STARS

Globular cluster dating is an ancient and honorable profession. The basic age comes from determining how long it takes for low mass stars to burn their core hydrogen and thereby move off the main sequence. The central temperature of such stars is determined by their composition and the degree of mixing. While there has certainly been some static as to what is the dispersion between the age of the youngest versus the oldest globular cluster in a given calculation, there is a surprising convergence on the age of the oldest clusters. Since the age of the very oldest cluster is the critical cosmological question, it is really somewhat of a red herring as to how much less the youngest cluster may be. The convergence on the age of the oldest does require a consistency of assumptions about primordial helium and metalicity (including O/Fe). Difference between different groups can be explained away once agreement is made on these assumptions. For example, Sandage's[13] oldest ages of $\sim$ 18 Gyr and Iben and Renzini's[14] of $\sim$ 16 Gyr are consistent if the same helium is used. (Lower helium yields higher ages. Iben assumed the Pagel[15] value of y = 0.23.) Another decrease of a billion years occurs if O/Fe is assumed high as current observations show for extreme Pop II.

Another effect is the fact that these old stable stars will have some gravitational settling of their helium which will also shorten the ages about 1 Gyr relative to calculations where core helium enrichment is purely due to nuclear burning. All of these assumptions give a standard model[16,17,18,19] for the oldest globulars of $\sim$ 14 $\pm$ 1 Gyr where the $\pm$1 is only the difference between different groups using the same standard assumptions. Additionally, Feist and Catchpole[8] have argued from Hipparcus data that the RR Lyrae stars in the LMC should be recalibrated, which shifts the age downward to a central value of around 12.5 Gyr. In addition to the calculational errors, there are also uncertainties in composition/opacity, uncertainties in distance/turnoff luminosity, and uncertainties in reddening/surface temperature at turnoff which increase the statistical error from $\pm$1 to $\pm$2 Gyr. Then, there are systematic uncertainties due to model assumptions: the helium abundance, settling, O/Fe, etc. For example, helium abundances might even be enhanced from the Big Bang (BBN) value due to helium production accompanying the extreme Pop II metal production and perhaps preferential helium in cluster formation[20]. Also note that the current best fit BBN helium is actually closer to 0.25 than 0.23. Shi et al.[20] showed that assumptions about He could lower the best fit age by as much as 2 Gyr without violating any other constraint (e.g. Y must be $\leq$ 0.28 to fit RR Lyrae blue edge). Furthermore, there are recent suggestions from the first Keck spectroscopic temperature determinations of globular cluster stars that the true temperatures are as much as 200 K hotter than the photometric determinations. This could also shift the age downward by as much as 2 Gyr. Furthermore, Shi et al.[20] (see also Shi[21]) have shown that mass loss due to the variable strip crossing the main sequence near the cluster turnoff could also shift the age down by 1 to 2 Gyr. However, these combined effects do not add linearly. No matter what, low mass stars can burn their hydrogen only so fast. We estimate that systematics add an additional ±2 Gyr which should not be added in quadrature with the ±2 Gyr statistical uncertainty, since most of the systematic effects are binary assumptions rather than selections from smooth, well behaved distributions. Thus, we conclude that $t_{GC} = 13 \pm 2 \pm 2$ Gyr.

One can use the standard solar model to get a quick estimate of an extreme globular age. The main line pp-chain is the main energy generation mechanism for the Sun and the globular clusters. The basic pp part of the solar model is now well confirmed by the calibrated GALLEX and SAGE solar neutrino experiments. Since the Sun has a much higher metalicity than the oldest globular clusters, and presumably has higher helium content and is at least as massive, if not more massive, it is paramount that the calculated main sequence lifetime of 10 Gyr for our Sun will always be a lower bound on the oldest globular cluster lifetimes. This 10 Gyr is also consistent with Shi et al.shiet95 and with an independant study by Chaboyer[22]. Thus, it is reasonable to conclude that shifting the "best" fit age for the oldest globulars down to 11 Gyr cannot be excluded. But an extreme lower bound at 10 Gyr is not able to be broken.

Note that the time delay for cluster formation does not change this limit, since it is certainly possible to hypothesize an isocurvature model where globular clusters are the first objects[23] to form after recombination (their Jeans mass at that time is the globular cluster mass). Their Kelvin-Helmholtz time is only $\sim 10^7$ yr, so in principle, they could be present as early as 10^8 yr after the Big Bang. (Of course, standard CDM models extend this to several Gyr.)

2.3. NUCLEOCOSMOCHRONOLOGY

Nucleocosmochronology is the use of abundance and production ratios of radioactive nuclides coupled with information on the chemical evolution of the Galaxy to obtain information about time scales over which the solar system elements were formed. Typical estimates for the Galaxy's (and Universe's) minimum age as determined from cosmochronology are of the order of 9.6 Gyr (e.g. Meyer and Schramm[24]). In recent years questions about the role of β-delayed fission in estimating actinide production ratios as well as uncertainties in ^{187}Re decay due to thermal enhancement and the discussion of Th/Nd abundances in stars have obfuscated some of the limits one can obtain. In par-

Table 1: AGE OF OLD THINGS IN THE UNIVERSE
(Age of Universe is Greater Than Age of Oldest Things)

Globular Clusters
$t_{GC} = 13 \pm 2 \pm 2Gyr$
≥ 10 Gyr
Long Lived Radioactive Isotopes (Nucleocosmochronology)
$t_{NC} \gtrsim 10$ Gyr
Solar System
$t_{SS} = 4.6 \pm 0.1$ Gyr

ticular, we note that the formalism of Schramm and Wasserburg[25] as modified by Meyer and Schramm[24] continues to provide firm bounds on the mean age of the heavy elements (see also recent reprint of Wasserburg and Busso[26]). In fact, Th/U provides a firm lower limit to the age and Re/Os, a firm upper limit. These limits are based solely on nuclear physics inputs and abundance determinations. To extend these mean age limits to a total age limit requires some galactic evolution input. However, as Reeves and Johns[27] first showed, and as Meyer and Schramm[24] developed further, one can use chronometers to constrain Galactic evolution models and thereby further restrict the age from the simple mean age limits of Schramm and Wasserburg[25]. To try to push further on such ranges and give ages to ± 1 Gyr accuracy, as some authors have done, always necessitates making some very explicit assumptions about Galactic evolution beyond the pure chronometric arguments. At the present time such model-dependent ages are not fully justified and should probably not be used as arguments to question (or support) cosmological models, but pure, nuclear derived lower bounds are very useful. In particular, the Meyer and Schramm[24] lower bound of $t_{NC} > 9.6$ Gyr which involves the mean age and the nuclear constrains on maximal evolutionary effects is a very firm bound.

2.4 AGE SUMMARY

The age situation at the present time can be summarized by Figure 1 and by Table 1. We see there that an $\Omega = 1$ universe is consistent with $t > 10$ Gyr as long as $H_0 \lesssim 65$ km/sec/Mpc. If uncertainties on H_0 (including bounds on systematics) ever exclude 65, then one would require $\Lambda_o \neq 0$ to achieve the flat universe favored by inflation models.

Naively, we expect gravitational microphysics on the Planck scale, M_p to determine the scale of Λ_o. An effective $\lambda_o \sim 1$ requires $\rho_\Lambda \sim 10^{-121} M_p^4$. This seems like remarkable tuning. Of course, some late-time transition on the fraction of an eV scale could substitute for M_p if the early $\rho_\Lambda \sim M_p^4$ effects could be surpressed to more than 121 orders of magnitude. Because these problems seem awkward to avoid, most physicists think $\Lambda = 0$.

As an anthropic aside, if it were ever shown that $\Lambda_o \neq 0$, then we may have to appeal to the following anthropic argument (ugh!). While particle physics prefers a large value for $\Lambda_o \sim M_p^4$, the only values consistent with an old universe have to have $\Lambda_o < 10^{-121} M_p^4$. Thus, our existence plus particle theory would make the maximum value consistent with our existence the most likely value. (Hopefully, a better motivated physics explanation for Λ_o will eventually be found.)

To repeat the main conclusion: at present there is no age problem, even for $\Omega = 1, \Lambda = 0$ models, since the real uncertainties including systematics allow completely consistent age values.

3. Big Bang Nucleosynthesis

The study of the light element abundances has undergone a recent burst of activity on many fronts. New results on each of the cosmologically significant abundances have sparked renewed interest and new studies. The bottom line remains: primordial nucleosynthesis has joined the Hubble expansion and the microwave background radiation as one of the three pillars of Big Bang cosmology. Of the three, Big Bang Nucleosynthesis probes the universe to far earlier times (~ 1 sec) than the other two and led to the interplay of cosmology with nuclear and particle physics. Furthermore, since the Hubble expansion is also part of alternative cosmologies such as the steady state, it is BBN and the microwave background that really drive us to the conclusion that the early universe was hot and dense.

Recent heroic observations of ^{6}Li, Be and B, as well as ^{2}D, ^{3}He and new ^{4}He determinations, have all gone in the direction of strengthening the basic picture of cosmological nucleosynthesis. Theoretical calculations of cosmic ray production of ^{6}Li, Be and B have fit the observations remarkably well, thus preventing these measurements from disturbing the standard scenario[28]. The recent reports of D/H in quasar absorption systems at redshift $Z \sim 3$ are particularly interesting and will be discussed[29,30,31] since BBN requires that fragile deuterium be found in primitive material. However, the possible variation of D/H in different lines of sight at $Z \sim 3$ argues that perhaps hydrogen cloud interlopers may be immitating deuterium, at least in lines of sight having higher apparent D/H. New work by Tytler, Burles, and Kirkman[31] on a previous high D/H system seems to support this point of view. Furthermore, recent theoretical calculations have confirmed that quark-hadron inspired inhomogenous

Big Bang Nucleosyntheis does not significantly alter the basic conclusions of standard BBN. We will also briefly discuss the possible impact on BBN of the recent ROSAT and ASCA x-ray satellite results on clusters of galaxies and the recent halo microlensing results. This summary will attempt to put it all together within an historical framework. The bottom line that emerges is how dramatically robust BBN is and how it gives a dramatically tight constraint on the density of baryons in the universe.

Let us now briefly review the history. This agreement works only if the baryon density is well below the cosmological critical value. This summary draws on the reviews of Walker et al.[32]; Schramm[33]; and Copi, Schramm and Turner[34].

It should be noted that there is a symbiotic connection between BBN and the 3K background dating back to Gamow and his associates, Alpher and Herman. The initial BBN calculations of Gamow's group[35] assumed pure neutrons as an initial condition and thus were not particularly accurate, but their inaccuracies had little effect on the group's predictions for a background radiation.

Once Hayashi[36] recognized the role of neutron-proton equilibration, the framework for BBN calculations themselves has not varied significantly. The work of Alpher, Follin and Herman[37] and Taylor and Hoyle[38], preceeding the discovery of the 3K background, and of Peebles[39] and Wagoner, Fowler and Hoyle[40], immediately following the discovery, and the more recent work of our group of collaborators[32,34,28,41,42,43,44,45] all do essentially the same basic calculation, the results of which are shown in Figure 1. As far as the calculation itself goes, solving the reaction network is relatively simple by the standards of explosive nucleosynthesis calculations in supernovae, with the changes over the last 25 years being mainly in terms of more recent nuclear reaction rates as input, not as any great calculational insight, although the current Kawano code[45] is somewhat streamlined relative to the earlier Wagoner code[40]. In fact, the earlier Wagoner code is, in some sense, a special adaptation of the larger nuclear network calculation developed by Truran[46] for work on explosive nucleosyntheis in supernovae. With the exception of Li yields and non-yields of Be and B, to which we will return, the reaction rate changes over the past 25 years have not had any major affect (see Yang et al.[44] and Krauss and his collaborators[47,48] or Copi, Schramm and Turner[34] for discussion of uncertainties). The one key improved input is a better neutron lifetime determination[49]. There has been much improvement in the $t(\alpha, \gamma)$ ^{7}Li reaction rate[50] but as the width of the curves in Figure 2 shows, the ^{7}Li yields are still the poorest determined, both because of this reaction and even more because of the poorly measured ^{3}He (α, γ) ^{7}Be.

With the exception of the effects of elementary particle assumptions, to which we will also return, the real excitement for BBN over the last 25 years has not really been in redoing the basic calculation. Instead, the true action is focused on understanding the evolution of the light element abundances and using that information to make powerful conclusions. In the 1960's, the main focus was on ^{4}He which is very insensitive to the baryon density. The agreement between BBN predictions and observations helped support the basic Big Bang model but gave no significant information, at that time, with regard to density. In fact, in the mid-1960's, the other light isotopes (which are, in principle, capable of giving density information) were generally assumed to have been made during the T-Tauri phase of stellar evolution[51], and so, were not then taken to have cosmological significance. It was during the 1970's that BBN fully developed as a tool for probing the universe. This possibility was in part stimulated by Ryter et al.[52] who showed that the T-Tauri mechanism for light element synthesis failed. Furthermore, ^{2}D abundance determinations improved significantly with solar wind measurements[53,54] and the interstellar work from the Copernicus satellite[55]. (Recent HST observations reported by Linsky et al.[56] have compressed the ^{2}D error bars considerably.) Reeves, Audouze, Fowler and Schramm[57] argued for cosmological ^{2}D and were able to place a constraint on the baryon density excluding a universe closed with baryons. Subsequently, the ^{2}D arguments were cemented when Epstein, Lattimer and Schramm[58] proved that no realistic astrophysical process other than the Big Bang could produce significant ^{2}D. This baryon density was compared with dynamical determinations of density by Gott, Gunn, Schramm and Tinsley[59]. See Figure 1 for an updated $H_0 - \Omega$ diagram.

In the late 1970's, it appeared that a complimentary argument to ^{2}D could be developed using ^{3}He. In particular, it was argued[60] that, unlike ^{2}D, ^{3}He was made in stars; thus, its abundance would increase with time. Unfortunately, recent data on ^{3}He in the interstallar medium[61] has shown that ^{3}He has been constant for the last 5 Gyr. Thus, low mass stars are not making a significant addition, contrary to these previous theroetical ideas. Furthermore, Rood, Bania and Wilson[62] have shown that interstellar ^{3}He is quite variable in the Galaxy, contrary to expectations for a low mass star-dominated nucelus. However, the work on planetary nebulae shows that at least some low mass stars do produce ^{3}He. Nonetheless, the current observational situation clearly shows that arguments based on theoretical ideas about ^{3}He evolution should be avoided (c.f. Hata et al.). Since ^{3}He now seems not to have a well behaved history, simple ^{3}He or ^{3}He + D inventory arguments are misleading at best. One is not free to go to arbitrary low baryon densities and high primordial D and ^{3}He, since processing of D and ^{3}He in massive stars also produces metals which are constrained[63,64] by the metals in the hot intra-cluster gas, if not the Galaxy.

It was interesting that other light elements led to the requirement that ^{7}Li be near its minimum of ^{7}Li/H $\sim 10^{-10}$, which was verified by the Pop II Li measurements of Spite and Spite[65,66,67], hence yielding the situation emphasized by Yang et al[44]. that the light element abundances are consistent over nine orders of magnitude with BBN, but only if the cosmological baryon density, Ω_b, is constrained to be around 6% of the critical value (for $H_0 \simeq 50$ km/sec/Mpc). The Li plateau argument was further strengthened with the observation of ^{6}Li in a Pop II star by Smith, Lambert and Nissen[68]. Since ^{6}Li is much more fragile than ^{7}Li, and yet it survived, no significant nuclear depletion of ^{7}Li is possible[69,70,71]. This observation of ^{6}Li has now been verified by Hobbs and Thorburn[72], and a detection in a second Pop II star has been reported. Lithium depletion mechanisms are also severly constrained by the recent work of Spite et al.[73] showing that the lithium plateau also is found in Pop II tidally locked binaries. Thus, meridonal mixing is not causing lithium depletion. Recently Nollett et al.[74] have discussed how ^{6}Li itself might eventually become another direct probe of BBN depending on the eventual low energy measurement of the ^{2}D $(\alpha\gamma)$ ^{6}Li cross section on spectroscipy improvements for extreme metal-poor dwarfs.

Another development back in the 70's for BBN was the explicit calculation of Steigman, Schramm and Gunn[75] showing that the number of neutrino generations, N_ν, had to be small to avoid overproduction of ^{4}He. (Earlier work[38,76,77] had commented about a dependence on the energy density of exotic particles but had not done an explicit calculation probing N_ν.) This will subsequently be referred to as the SSG limit. To put this in perspective, one should remember that the mid-1970's also saw the discovery of charm, bottom and tau, so that it almost seemed as if each new detector produced new particle discoveries, and yet, cosmology was arguing against this "conventional" wisdom. Over the years, the SSG limit on N_ν improved with ^{4}He abundance measurements, neutron lifetime measurements, and with limits on the lower bound to the baryon density, hovering at $N_\nu \lesssim 4$ for most of the 1980's and dropping to slightly lower than 4 just before LEP and SLC turned on.[28,32,78,79] This was verified by the LEP results[80] where now the overall average is $N_\nu = 2.99 \pm 0.02$. A recent examination of the cosmological neutrino limit by Copi et al.[81] in the light of the recent ^{3}He and D/H work shows that the BBN limit remains between 3 and 4 for all reasonable assumption options.

The recent apparent convergence of the extra-galactic D/H measurements towards the lower values[82,83] D/H $\sim 3 \times 10^{-5}$ may eventually collapse the Ω_B band in figure 2 to a relatively narrow strip on the high Ω_B side. However,

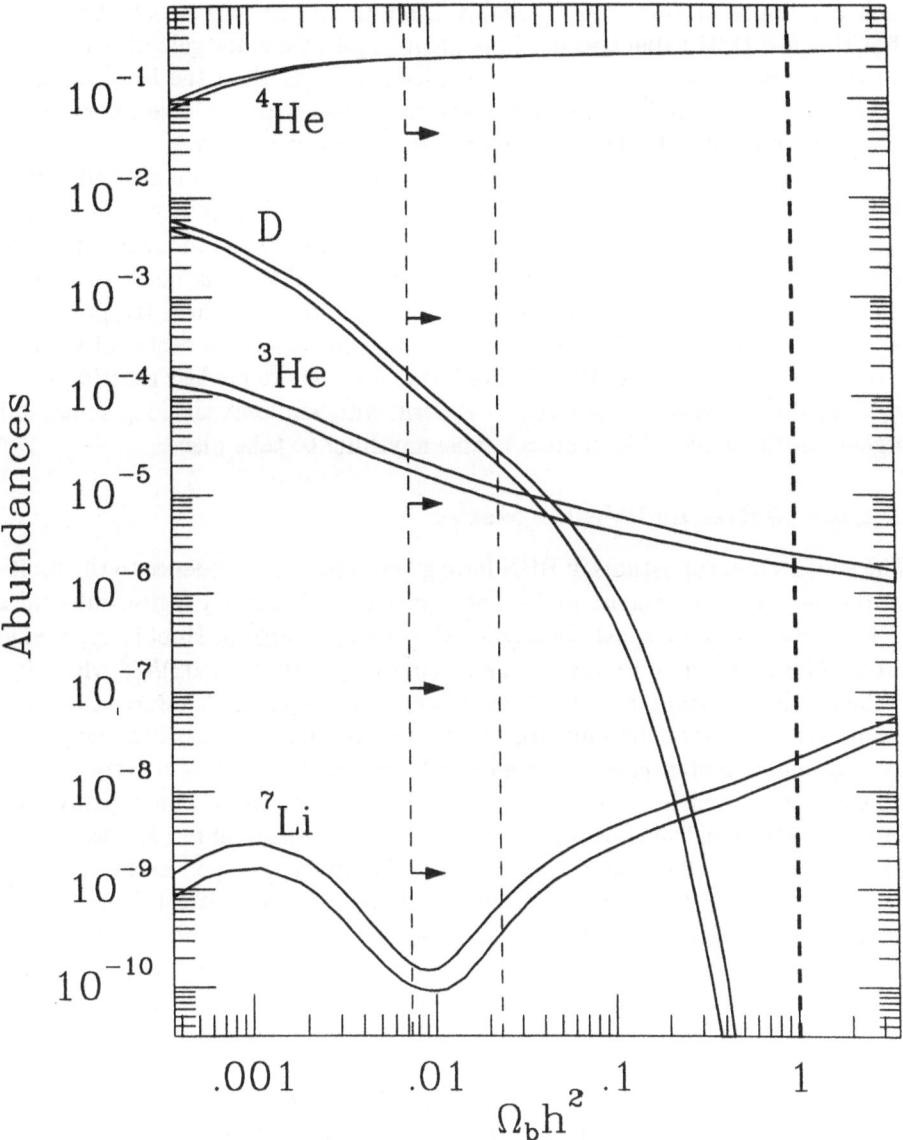

Figure 2: Big Bang Nucleosynthesis abundance yields versus baryon density (Ω_b) and $\eta \equiv \frac{n_b}{n_\gamma}$ for a homogeneous universe. ($h \equiv H_0/100$ km/sec/Mpc; thus, the concordant region of $\Omega_b h^2 \sim 0.015$ corresponds to $\Omega_b \sim 0.06$ for $H_0 = 50$ km/sec/Mpc.) Figure is from Copi, Schramm and Turner[34]. Note concordance region is slightly larger than Walker et al.[32] due primarily to inclusion of possible systematic errors on Li/H. The width of the curves represents the uncertainty due to input of nuclear physics in the calculation.

such a collapse at present is probably a bit premature. It should be noted that this low D/H value also has important implications for galactic evolution since the present ISM value is not significantly lower than the high Z value. This would seem to favor infall models or models with variable initial mass functions to explain heavy element production in our Galaxy.

The power of homogeneous BBN comes from the fact that essentially all of the physics input is well determined in the terrestrial laboratory. The appropriate temperature regimes, 0.1 to 1 MeV, are well explored in nuclear physics laboratories. Thus, what nuclei do under such conditions is not a matter of guesswork, but is precisely known. In fact, it is known for these temperatures far better than it is for the centers of stars like our sun. The center of the sun is only a little over 1 keV, thus, below the energy where nuclear reaction rates yield significant results in laboratory experiments, and only the long times and higher densities available in stars enable anything to take place.

4. Dark Matter and Visible Matter

The success and robustness of BBN have given renewed confidence to the limits on the baryon density constraints. Let us convert this density regime into units of the critical cosmological density for the allowed range of Hubble expansion rates. This is shown in Figure 1. In particular, $\Omega_b = (0.08\pm0.02)h_{50}^{-2}$ where h_{50} is the Hubble constant in units of 50 km/sec/Mpc. Figure 1 also shows the lower bound on the age of the universe of 10 Gyr from both nucleochronology and from globular cluster dating[84] and a lower bound on H_0 of 40 from extreme type IA supernova models with pure 1.4 $M_\odot$ carbon white dwarfs being converted to ^{56}Fe. The constraint on Ω_b means that the universe *cannot be closed with baryonic matter*. (This point was made over twenty years ago[57] and has proven to be remarkably strong.) If the universe is truly at its critical density, then nonbaryonic matter is required. This argument has led to one of the major areas of research at the particle-cosmology interface, namely, the search for non-baryonic dark matter. In fact, from the lower bound on Ω_{TOTAL} from cluster dynamics of $\Omega_{TOTAL} > 0.1$, it is clear that non-baryonic dark matter is required unless $H_0 < 50$. The need for non-baryonic matter is strengthened on even larger scales[85]. Figure 1 also shows the range of $\Omega_{VISIBLE}$ and shows that there is no overlap between Ω_b and $\Omega_{VISIBLE}$. Hence, the bulk of the baryons are dark.

The estimate of the Ω_{VIS} in Figure 1 can be obtained from noting that stellar material tends to have mass-to-light ratios in solar units as shown below

$$\frac{M}{L}\Big|_* \sim 2-8 \; \frac{M_\odot}{L_\odot}$$

for blue band light. (The number is greater than one since mass is dominated by low mass stars and light by high mass ones, and initial mass functions favor low mass ones.) The uncertainty comes from the location of the low mass cutoff. Recent observations from HST[86] argue that the cutoff is above the red dwarf limit of several tenths $M_\odot$ since these objects were not found to be ubiquitous. To obtain a density from an M/L one needs to multiply by the average luminosity density, $\mathcal{L}_B$, in that wave band. Kirshner et al.[87] have determined :

$$\mathcal{L}_B \simeq 1.8 \pm 0.2 \times 10^8 h \frac{L_\odot}{Mpc^3}$$

where h is the Hubble constant in units of 100 km/sec/Mpc. One can then obtain Ω_{VIS} by dividing by

$$\rho_{crit} \equiv \frac{3H_0^2}{8\pi G} = 2.8 \times 10^{11} h^2 \frac{M_\odot}{Mpc^3} = 2 \times 10^{-29} h^2 g/cm^3$$

Hence,

$$\Omega_{VIS} = \frac{\frac{M}{L}|_B \cdot \mathcal{L}_B}{\rho_{crit}} = \frac{0.003 \pm 0.002}{h}.$$

Note that Ω_{VIS} obtained in this way is inversely proportional to H_0. (One can also estimate Ω_{VIS} from the dynamics of the shining regions of galaxies and obtain a similar value but independent of H_0.)

Another interesting conclusion[59] regarding the allowed range in baryon density is that it is in agreement with the density implied from the dynamics of single galaxies, *including their dark halos*. The MACHO and EROS[88,89] reports of halo microlensing may well indicate that at least some of the dark baryons are in the form of brown dwarfs in the halo. However, Gates, Gyuk and Turner[90,91], and Alcock et al.[88] show that the observed distribution of MACHOs favors less than 50% of the halo being in the form of MACHOs, but a 100% MACHO halo cannot be completely excluded yet.

5. The Need for Non-Baryonic Dark Matter

The arguments requiring some sort of dark matter fall into separate and quite distinct areas. These arguments are summarized in Figures 1 and 3. First are the arguments using Newtonian mechanics applied to various astronomical systems that show that there is more matter present than the amount that is shining. It should be noted that these arguments reliably demonstrate that galactic halos seem to have a mass ~ 10 times the visible mass.

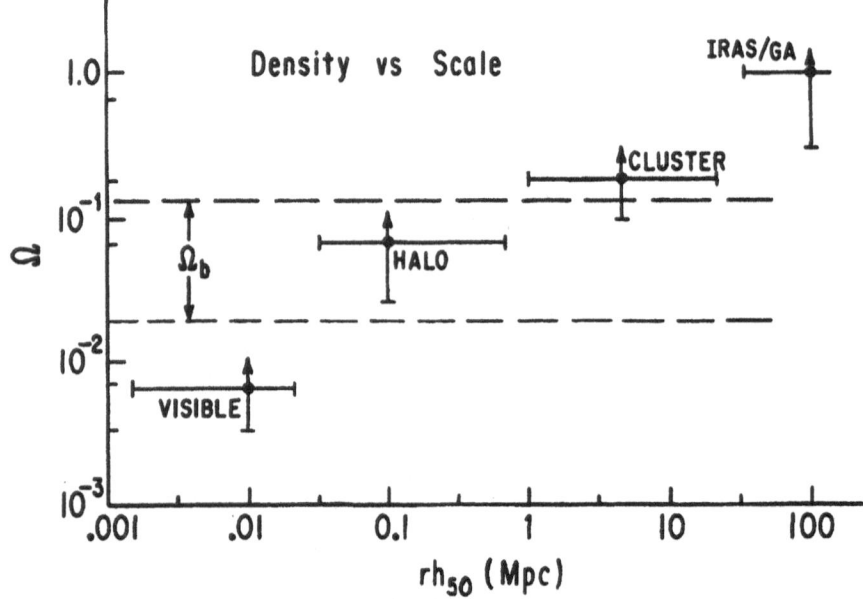

Figure 3: Implied densitites versus the scale of the measurements.

For dynamical estimates of Ω one estimates the mass from

$$M \sim \frac{v^2 r}{G}$$

where v is the relative velocity of the objects being studied, r is their separation distance, and G is Newton's constant. The proportionality constant out front depends on orientation, relative mass, etc. For large systems such as clusters, one uses averaged quantities. For single galaxies v would represent the rotational velocity and r the radius of the star or gas cloud. Note that since v is measured by relative redshift, it has no H_0 dependence, whereas distances will scale with $1/H_0$. Also note that the luminosity of a system will scale as the surface area and hence is proportional to $1/H_0^2$. Hence, for dynamically measured systems,

$$\frac{M}{L}\Big|_{DYN} \propto H_0.$$

Again,

$$\Omega_{DYN} = \frac{\frac{M}{L}\big|_B^{DYN} \cdot \mathcal{L}_B}{\rho_{crit}}$$

which now is independent of H_0. It is this technique which yields the halo and cluster points shown on Figure 3 and the cluster bound on Ω shown on Figure 3. It should be noted that the value of $\Omega_{CLUSTER} \sim 0.2$ is also obtained in those few cases where alignment produces giant gravitational-lens arc. Recent work using weak gravitational lensing by Kayser[92] also supports large Ω. The IRAS/Great Attractor point on Figure 3 comes from the large scale flow arguments. As Davis and Nusser[85] show, if the large scale velocity flows measured from the IRAS survey are due to gravity, then $\Omega_{IRAS} \gtrsim 0.2$. Similar arguments have been obtained using the Great Attractor study or the Potent[93] technique. All imply $\Omega > \Omega_{BARYON}$, hence the need for non-baryonic dark matter. However, there is still considerable uncertainty of the exact value of Ω determined in this way as discussed by Szalay[94] at this school. But all groups agree it is greater than 0.3. However, as Figure 1 illustrates, except for $H_0 < 50$, $\Omega_{CLUSTER}$ already required $\Omega_{TOTAL} > \Omega_{BARYON}$ and hence the need for non-baryonic dark matter.

An Ω of unity is, of course, preferred on theoretical grounds since that is the only long-lived natural value for Ω, and inflation[95,96] or something like it provided the early universe with the mechanism to achieve that value and thereby solve the flatness and smoothness problems. Note that our need for exotica is not dependent on the existence of dark galatic halos and that high values of H_0 (see Figure 1) increase the need for non-baryonic dark matter.

Non-baryonic matter can be divided into two major categories[97] for cosmological purposes: hot dark matter (HDM) and cold dark matter (CDM). Hot dark matter is matter that is relativistic until just before the epoch of galaxy formation, the best example being low mass neutrinos with $m_\nu \sim 20$eV. Cold dark matter is matter that is moving slowly at the epoch of galaxy formation. Because it is moving slowly, it can clump on very small scales, whereas HDM tends to have more difficulty in being confined on small scales and be CDM as well. Examples of CDM could be the lightest super-symmetric particle which is presumed to be stable and might also have masses of several GeV. Following Michael Turner, all such weakly interacting massive particles are called "WIMPS." Axions, while very light, would also be moving very slowly[98] and thus would clump on small scales and be CDM as well. Or, one could also go to non-elementary particle candidates, such as planetary mass blackholes or quark nuggets of strange quark matter, possibly produced at the quark-hadron transition[99,100]. Another possibility would be any sort of massive toplogical remnant left over from some early phase transition. Note that CDM would clump in halos, thus requiring the dark baryonic matter to be out between galaxies, whereas HDM would allow baryonic halos. The MACHO and EROS events may eventually require at least some CDM to fill out the halo. Obviously, mixed models with some HDM and some CDM have even more flexibility and have thus become quite popular as data constraints increase.

Some baryonic dark matter must exist since we know that the lower bound from Big Bang Nucleosynthesis is greater than the upper limits on the amount of visible matter in the universe. If the baryonic dark matter is not in the halo, it could be in hot intergalactic gas, hot enough not to show absorption lines in the Gunn-Peterson test, but not so hot as to be seen in the x-rays. The exciting report by Jakobsen et al.[101] (see also Davidsen[102]) of a Gunn-Peterson effect observed with HST for He-II at high Z showed that at least some hot IGM exists (and verifies that He seems primordial). Recent analyses by Davidson[7] show that the universe at high Z probably had the bulk of its baryons in the Lyman-α gas clouds.

Another possible hiding place for the dark baryons would be failed galaxies, large clumps of baryons that condensed gravitationally but did not produce stars. Such clumps are predicted in galaxy formation scenarios that include large amounts of biasing where only some fraction of the clumps shine. Evidence for some hot gas is found in clusters of galaxies from the ROSAT and ASCA satellites. In particular, Mushotzky[103] and White et al.[104] have discussed how certain observed rich clusters have $M_{HOTGAS}/M_{TOTAL} \sim 1/5$, which, with $\Omega_{CLUSTER} \sim 0.2$ and $\Omega_b \sim 0.05$, would imply no conflict with BBN, but that clusters are not fair samples of the baryon to non-baryon ratio in the universe. Apparent variation[101] in M_{HOT}/M_{TOTAL} for small groups relative to clusters seems to support the point of view that megaparsec scales are not always fair samples. If true, this is difficult to reconcile with cold dark matter models but can be fit with topological defects and HDM or with mixed models[105].

While the exact nature of the seeds to stimulate the clumping has yet to be fully determined, COBE's discovery of large scale fluctuations in the microwave background probably shows that some sort of seeds did indeed exist and that the basic scenario is probably on the right track.

6. Seeds and Large Scale Structure

To form structure in the universe requires some sort of "seed" in addition to the baryonic matter and the non-baryonic dark matter. The types of seeds

Table 2: TWO GENERIC SEED FAMILIES

Density Fluctuations	Topological Seeds
Quantum Fluctuations from Inflation	Topological Defects such as strings, textures or domain walls
Gaussian	Non-Gaussian
Adiabatic	Iso-Curvature or almost equivalently isothermal

tend to fall into two broad categories:

1. density fluctuations;
2. topological seeds.

Density fluctuations are presumably produced via quantum fluctuations near the end of the inflation epoch and topological seeds such as cosmic strings or textures are topological defects again left over from some vacuum phase transition. Thus, in either case a cosmological phase transition seems to be required. Table 1 shows other names and properties these two seed classifications sometimes go by.

Since COBE only was able to observe anisotropies on angular scales $\gtrsim 7°$, it was not able to directly probe the scales that eventually led to the observed large scale structure.

The horizon size at the decoupling of the microwave radiation corresponds only to an angular size of less than 2°, so COBE could not probe the causal processes that led to structure. Many new microwave experiments are now probing the angular scales of $\sim 30'$ to $\sim 2°$ to try to differentiate between the models of Table 2 and various CDM/HDM assumptions. These experiments are being done in balloons and at the South Pole and at various other remote terrestrial locations such as Saskatoon, Canada. At present, there is much ambiguity among the different experiments, so no single picture emerges. Non-gaussian models may even claim to predict that different directions should see different results, but the error bars are too large to claim this has been proven yet. Many experiments do find a few times higher signal at $\sim 1°$ than COBE finds for larger scale (COBE $\frac{\Delta T}{T} \sim 10^{-5}$) which is what the fluctuation model predicts. Detailed analysis by Jungman, Kosowsky, Kamionkowski and Spergel[106] and by Scott and White[107] show that if the fluctuations are gaussian then the details of the accoustic peaks near 1° and less will determine H_0, Ω_T, Ω_B, Λ_0, etc. Thus, the next generation satellites, MAP and PLANCK (formerly called COBRAS/SAMBA), may be able to give us precise answers to most of the traditional cosmological questions. The intercomparison of statistically significant galaxy distribution data and microwave anisotropy data, plus direct dark matter searches and accelerator searches will hopefully resolve the origin of structure in the universe by the end of this decade.

7. Alternative Measures of Ω_B

In addition to the BBN arguments presented above in section 3, it is interesting that we are now obtaining new alternative probes of Ω_B. In particular, the height of the first doppler peak in the MWB anisotropy, mentioned in section 6 above, is directly related to $\Omega_B h^2$. Since H_0 is also related to the ratio of the first and third peak heights, one might eventually be able to deconvolve the two quantities using just the MWB anisotropy. The correct preliminary analyses based on Sasketoon, South Pole and balloon data overlap with the BBN $\Omega_B h^2$ values on the high side; thus, they are more consistent with the low primordial D/H values. That this overlap exists is remarkable independent support for BBN arguments.

Eventually, with MAP and PLANCK, we should have a very definitive test here.

Another way of probing Ω_B is via the Lyman-α clouds. The recent work of Davidson[7], showing that the presence of ionized intergalactic helium and clouds at high redshift, is well interpreted by baryonic gas with Ω_B again overlapping with the high side of the BBN range.

The x-ray gas in clusters might also provide some clues here, but at present it seems to have so many uncertainties that all that can be said is that it is worth watching. However, even at present with the appropriate caveats, it too leans towards the high side of the Ω_B range from BBN.

8. Conclusion

It seems clear that we need the bulk of the matter in the universe to be in some non-baryonic form. A possible mixed dark matter solution may be motivated by the need for neutrino masses from the solar neutrino problem. The best fit solar ν mass hints at $m_{\nu_\mu} \sim 10^{-3}$ eV which in a simple see-saw model may imply $m_{\nu_\tau} \sim 10$ eV, that is, great hot dark matter.

Particularly exciting are the experimental searches for non-baryonic dark matter. These searches include accelerator searches for supersymmetry (or other Weakly Interactiong Massive Particles, WIMPS) and for neutrino oscillations. They also include direct underground searches for WIMPS and axions and even satellite searches for WIMP annihilation products.

Hopefully, with all this activity, we will find the dark matter within a decade.

9. Acknowledgements

I would like to thank my recent collaborators, David Dearborn, Brian Fields, Dave Thomas, Gary Steigman, Brad Meyer, Keith Olive, Angela Olinto, Bob Rosner, Michael Turner, George Fuller, Karsten Jedamzik, Rocky Kolb, Grant Mathews, Bob Rood, Jim Truran and Terry Walker for many useful discussions. I would further like to thank Poul Nissen, Jeff Linsky, Julie Thorburn, Doug Duncan, Lew Hobbs, Evan Skillman, Bernard Pagel and Don York for valuable discussion regarding the astronomical observations.

This work is supported by the NASA and the DoE(nuclear) at the University of Chicago, and by the DoE and NASA grant NAG5-2788 at Fermilab.

10. References

1. Ussher, J. (1658) *The Annals of the World* (E. Tyler for J. Crook, London).
2. Nugent, P., Branch, D., Boran, E., Fisher, A., Vaughn, T, and Hauschildt, D. (1995) *Phys. Rev. Lett.* **75**, 394-397.
3. Sandage, A. and Tamman, G. 1995 Carnegie Institute preprint.
4. Riess, A.G., Press, W.H., and Kirshner, R.P. (1995) *Astrophys. J. Lett.*, **438**, L17-L20.
5. Schmidt, B., Kirshner, R., and Eastman, R. (1992) *Astrophys. J.*, **395**, 366-386.
6. Freedman, W.L. et al. (1994) *Nature* **371**, 757-762.
7. Freedman, W.L. *et al.* (1997) in Proc. of the Conference on Critical Dialogues in Cosmology, June 1996, Princeton University, ed. N.Turok, in press.
8. Feast, M.W., Carter, B.S., Roberts, G., Marang, F. and Catchpole, R.M. (1997) *MNRAS*, **285**, 317-338.
9. Kundic, T. *et al.* (1997) astro-ph/9610162 and *Astrophys. J.*, **482**, 631-635.
10. Schechter, P.L. (1996) *IAU Symposia* **173**, 263-264.
11. Keeton, C.R. and Kochonek, C.S. (1997) astro-ph/9611216, *Astrophys. J.*, **482**, 604-620.

148

12. Hill, C., Schramm, D.N., and Fry, J. (1989) *Comm. Nuc. Part. Phys.*, **19**, 25-39.
13. Sandage, A. and Tamman, G. (1995) in *Current Topics in Astrofundamental Physics: The Early Universe*, eds. N. Sanchez and A. Zichichi (Kluwer, Dordrecht) pp. 403-443.
14. Iben, I. and Renzini, A. (1984) *Physics Reports*, **105**, 330-406.
15. Pagel, B. (1989) NORDITA preprint.
16. Pagel, B. and Jimenez, R. (1997) *Physics Reports*, in preparation.
17. Chaboyer, B., Demarque, P., Kernan, P.J., and Krauss, L. (1996) *Science*, **271**, 957-961.
18. Sandage, A. (1993) *Astrophys. J.*, **106**, 719-723.
19. Mazzitelli, I and d'Antonna, F. (1997) *Astron. and Astrophys.*, **302**, 382-400.
20. Shi, X., Schramm, D.N., and Dearborn, D. (1995) *Phys. Rev. D*, **50**, 2414-2420.
21. Shi, X. (1995) *Astrophys. J.*, **446**, 637-645.
22. Chaboyer, B. (1994) CITA preprint CITA-94-52, and (1995) *Astrophys. J.*, **444**, L9-L12.
23. Lee, S., Schramm, D.N., and Mathews, G. (1995) *Astrophys. J.* **449**, 616-622.
24. Meyer, B.S. and Schramm, D.N. (1986) *Astrophys. J.* **311**, 406-417.
25. Schramm, D.N., and Wasserburg, G.J. (1970) *Astrophys. J.* **162**, 57-69.
26. Wasserburg, G. and Busso, M. (1996) Caltech abstract, LPSXXVIIz.
27. Reeves, H. and Johns, O. (1976) *Astrophys. J* **206**, 958-962.
28. Olive, K., Schramm, D.N., Steigman, G. and Walker, T. (1990) *Phys. Lett. B* **236**, 454-460.
29. Songaila, A., Cowie, L.L., Hogan, C.J., and Rugers, M. (1990) *Nature* **368**, 599-604.
30. Carswell, R.F. et al. (1994) *MNRAS* **268**, L1-L12.
31. D. Tytler, Burles, S., and Kirkman, D. (1996) astro-ph/9612121.
32. Walker, T., Steigman, G., Schramm, D.N., Olive, K., and Kang, H.S. (1991) *Astrophys. J.* **376**, 51-69.
33. Schramm, D.N. (1994) in *Evolution of the Universe and Its Observational Quest*, Proc. of Yamada Conf. XXXVII, Tokyo, June 1993, ed. Sato, K. (Universal Academic Press, Tokyo), pp. 61-74. See also Schramm, D.N. (1995) in *The Light Element Abundances*, Proc. of ESO/EIPC Workshop, Elba, May 1994, ed. P. Crane (Springer-Verlag, Heidelberg), 51-72.
34. Copi, C., Schramm, D.N. and Turner, M.S. (1994) *Science* **267**, 192-199.
35. Alpher, R.A., Bethe, H., and Gamow, G. (1948) *Phys. Rev.* **73**, 803-804.

36. Hayashi, C. (1950) *Prog. Theor. Phys.* **55**, 224-235.

37. Alpher, R.A., Follin, J.W., and Herman, R.C. (1953) *Phys. Rev.* **92**, 1347-1361.

38. Taylor, R. and Hoyle, F. (1964) *Nature* **203**, 1108-1110.

39. Peebles, P.J.E. (1966) *Phys. Rev. Lett.* **16**, 410-413.

40. Wagoner, R., Fowler, W.A., and Hoyle, F. (1967) *Astrophys. J.* **148**, 3-49.

41. Schramm, D.N. and Wagoner, R.V. (1977) *Ann. Rev. of Nuc. Sci.* **27**, 37-74.

42. Olive, K., Schramm, D.N., Steigman, G., Turner, M.S., and Yang, J. (1981) *Astrophys. J.* **246**, 557-568.

43. Boesgaard, A.M. and Steigman, G. (1985) *Ann. Rev. of Astron. and Astrophys.* **23**, 319-378.

44. Yang, J., Turner, M.S., Steigman, G., Schramm, D.N., and Olive, K. (1984) *Astrophys. J.* **281**, 493-511.

45. Kawano, L., Schramm, D.N., and Steigman, G. (1988) *Astrophys. J.* **327**, 750-754.

46. Truran, J., Doctoral Thesis, Yale University (1965); Truran, J.W., Cameron, A.G.W., and Gilbert, A. (1966) *Can. Jour. of Phys.* **44**, 563-592.

47. Krauss, L.M. and Romanelli, P. (1990) *Astrophys. J.* **358**, 47-59.

48. Kernan, P. and Krauss, L. (1994) *Phys. Rev. Lett.* **72**, 3309-3312.

49. Mampe, W., Ageron, P., Bates, C., Pendlebury, J.M., and Steyerl, A. (1996) *Phys. Rev. Lett.* **63A**, 593-596; Mampe, W. *et al.* (1993) *JETP Lett.***57**, 82-87.

50. Schramm, D.N. and Mathews, G. (1995) in *Particle and Nuclear Astrophysics in the Next Millenium: Proc. Snowmass Summer Study*, eds. Kolb, E.W. and Peccei, R.D. (World Scientific, Singapore) 479-497.

51. Fowler, W.A., Greenstein, J. and Hoyle, F. (1962) *Geophys. R.A.S.* **6**, 148-220.

52. Ryter, C., Reeves, H., Gradstajn, E. and Audouze, J. (1970) *Astron. and Astrophys.* **8**, 389-397.

53. Geiss, J. and Reeves, H. (1971) *Astron. and Astrophys.* **18**, 126-132.

54. Black, D. (1971) *Nature* **234**, 148-149.

55. Rogerson, J. and York, D. (1973) *Astrophys. J.* **186**, L95-L98.

56. Linsky, J., Brown, A., Gayley, K., Diplas, A., Savage, B., Ayres, T., Landsman, W., Shore, S. and Heap, S. (1993) *Astrophys. J.* **402**, 694-709.

57. Reeves, H., Audouze, J., Fowler, W.A., and Schramm, D.N. (1973) *Astrophys. J.* **179**, 909-930.

150

58. Epstein, R., Lattimer, J. and Schramm, D.N. (1976) *Nature* **263**, 198-202.

59. Gott, J.R., Gunn, J., Schramm, D.N., and Tinsley, B.M. (1974) *Astrophys. J.* **194**, 543-553.

60. Rood, R.T., Steigman, G., and Tinsley, B.M. (1976) *Astrophys. J.* **207**, L57-L60.

61. Gloeckler, G. and Geiss, J. (1996) *Nature* **381**, 210-212.

62. Rood, R.T., Bania, T. and Wilson, J. (1992) *Nature* **355**, 618-620.

63. Copi, C.J., Schramm, D.N., and Turner, M.S. (1995) *Astrophys. J.* **455**, L95-L98.

64. Scully, S.T., Cassé, M., Olive, K.A., Schramm, D.N., Truran, J. and Vangioni-Flam, E. (1996) *Astrophys. J.* **462**, 960-968.

65. Spite, J. and Spite, M. (1982) *Astron. and Astrophys.* **115**, 357-366.

66. Rebolo, R., Molaro, P. and Beckman, J. (1988) *Astron. and Astrophys.* **192**, 192-205.

67. Hobbs, L. and Pilachowski (1988) *Astrophys. J.* **326**, L23-L26.

68. Smith, V.V., Lambert, D.L. and Nissen, P. (1982) *Astrophys. J.* **408**, 262-276.

69. Olive, K. and Schramm, D.N. (1992) *Nature* **360**, 4349-442.

70. Steigman, G., Fields, B., Olive, K., Schramm, D.N., and Walker, T. (1993) *Astrophys. J.* **415**, L35-L38.

71. Lemoine, M., Schramm, D.N., Truran, J.W. and Copi, C.J. (1997) *Astrophys. J.* **478**, 554-562.

72. Hobbs, L. and Thorburn, J. (1994) *Astrophys. J. Lett.* **428**, L25-L28.

73. Spite, M., Nissen, P.E. and Spite, F. (1996) *Astron. and Astrophys.* **307**, 172-183.

74. Nollett, K., Lemoine, M. and Schramm, D.N. (1997) *Phys. Rev. C.* **56**, 1144-1151.

75. Steigman, G., Schramm, D.N., and Gunn, T. (1977) *Phys. Lett.* B **66**, 202-204.

76. Schvartzman, V.F. (1969) *JETP Letters* **9**, 184-186.

77. Peebles, P.J.E. (1971) *Physical Cosmology* (Princeton University Press).

78. Schramm, D.N. and Kawano, L. (1989) *Nuc. Inst. and Methods A* **284**, 84-88.

79. Pagel, B. (1990) in Proc. of 1989 Rencontres de Moriond.

80. ALEPH, L3, OPAL, DELPHI results (1993) 1993 Lepton-Photon meeting at Ithaca, NY.

81. Copi, C., Schramm, D.N., and Turner, M.S. (1997) *Phys. Rev.* D **55**, 3389-3393.

82. Tytler, D. (1997) in *Proc. of the 18th Texas Symposium on Relativisitic*

Astrophysics, in press (World Scientific, Singapore).

83. Hogan, C. (1997) in *Proc. of the 18th Texas Symposium on Relativisitic Astrophysics*, in press (World Scientific, Singapore).

84. Shi, X., Schramm, D.N., Dearborn, D., and Truran, J.W. (1995) *Comments on Astrophysics*, **17**, 343-360.

85. Davis, M. and Nusser, A. (1995) in Proc. Maryland Symposium on Dark Matter, in press; see also Davis, M. (1997) in *Proc. 18th Texas Symposium on Relativistic Astrophysics*, in press (World Scientific, Singapore).

86. Flynn, C., Gould, A. and Bahcall, J. (1996) *Astrophys. J. Lett.* **466**, L55-L58.

87. Kirshner, R., Oemler, A. and Schechter, P.L. (1979) *Astronom. J.* **84**, 951-959.

88. Alcock, C. et al. (1993) *Nature* **365**, 621-623.

89. Aubourg, E. et al. (1993) *Nature* **365**, 623-625.

90. Gates, E., Gyuk, G. and Turner, M.S. (1995) *Phys. Rev. Lett.* **74**, 3724-3727.

91. Alcock, C. et al. (1997) *Astrophys. J.* **486**, 697-726.

92. Kayser, N. (1995) in Proc. of the Texas Symposium on Relativistic Astrophysics, Munich, December 1994, in press.

93. Bertschinger, E., Dekel, A., Faber, S.M., Dressler, A. and Burstein, D. (1990) *Astrophys. J.* **364**, 370-395; Dekel, A., Bertschinger, E., Yahil, A., Strauss, M., Davis, M., and Huchra, J. (1993) *Astrophys. J.* **412**, 1-21.

94. Szalay, A. (1995) in Proc. of the Australian National University Summer School on Cosmology, Canberra, Australia, January 1994 (World Scientific, Singapore).

95. Guth, A. (1981) *Phys. Rev.* D **23**, 347-356.

96. Linde, A. (1990) *Particle Physics and Inflationary Cosmology* (Harwood Academic Publishers, N.Y).

97. Bond, R. and Szalay, A. (1992) in Proc. Texas Relativistic Astrophysical Symposium, Austin, Texas.

98. Turner, M.S., Wilczek, F. and Zee, A. (1983) *Phys. Lett.* B**125**, 35-38; **125**, 519-522.

99. Crawford, M. and Schramm, D.N. (1982) *Nature* **298**, 538-540.

100. Alcock, C. and Olinto, A. (1988) *Ann. Rev. Nuc. Part. Phys.* **38**, 161-184.

101. Jakobsen, P., Boksenberg, A., Deharveng, J.M., Greenfield, P., Jedrzewski, R. and Paresce, F. (1994) *Nature* **370**, 35-39.

102. Davidsen, A. (1997) in in *Proc. 18th Texas Symposium on Relativistic Astrophysics*, in press (World Scientific, Singapore).

103. Mushotsky, R. (1993) in Relativistic Astrophysics and Particle Cosmology: Texas PASCOS 92, eds. Akerlof, C. W. and Srednicki, M. A. *Annals of the N.Y. Academy of Sciences* **688**, 184-190.
104. White, S.D.M., Navarro, J.F., Evrard, A.E. and Frenck, C.S. (1993) *Nature* **366**, 429-433.
105. Strickland, R. and Schramm, D.N. (1997) *Astrophys. J.*, **481**, 571-577.
106. Jungman, G., Kosowsky, A., Kamionkowski, M. and Spergel, D.N. (1996) *Phys. Rev. Lett.* **76**, 1007-1010; *Phys. Rev. D* **54**, 1332-1344.
107. Scott, D. and White, M. (1995) *Gen. Rel. and Grav.* **27**, 1023-1030.

Ten Things Everyone Should Know About Inflation

Michael S. TURNER

Enrico Fermi Institute, The University of Chicago
5640 S. Ellis Ave., Chicago, IL 60637, USA

Theoretical Astrophysics, Fermi National Accelerator Laboratory
Batavia, IL 60510-0500, USA

These lecture notes are organized into ten lessons that summarize the status of inflationary cosmology.

1 Inflation is a Bold and Expansive Paradigm

The hot big-bang cosmology is very successful. It provides a physical description of the Universe from about 10^{-2} sec onward.[1] However, it raises fundamental questions about initial conditions: The origin of the smoothness and flatness of our Hubble volume, the small (one part in 10^5) density inhomogeneities needed to seed all the structure seen in the Universe today, and the tiny baryon asymmetry that results in the existence of matter today.

Inflation explains how a region of size much, much greater than our Hubble volume could have become smooth and flat[2] as well as the origin of the density inhomogeneities needed to seed structure.[3] With regard to the smoothness and flatness, inflation is a temporary fix: It does not guarantee that the observable Universe in the exponentially distant future will be isotropic and homogeneous.[4]

Models of inflation are based upon well defined, albeit speculative physics – usually the semi-classical evolution of a weakly coupled scalar field. The physics is speculative because a) there is no evidence for the existence of even a single fundamental scalar field and b) the energy scale associated with inflation is typically much greater than $1\,\mathrm{TeV}$ and in most models around $10^{14}\,\mathrm{GeV}$.

I believe that it is fair to say that inflation has revolutionized the way cosmologists view the Universe. It leads to the current working hypothesis for an extension of the standard cosmology: The Inflation/Cold Dark Matter Paradigm. This paradigm has the potential to extend the standard cosmology back to times as early as 10^{-32} sec and address almost all the pressing questions in cosmology. The key elements of this paradigm are: flat Universe, nonbaryonic dark matter in the form of slowly moving elementary particles (cold dark matter), and nearly scale-invariant, adiabatic density perturbations. As I will emphasize, the inflation/cold dark matter paradigm is highly testable and a

D. N. Schramm and P. Galeotti (eds.), Generation of Cosmological Large-Scale Structure, 153–192.
© *1997 Kluwer Academic Publishers.*

flood of observations are doing so. At the outside, within the next decade this paradigm will have been falsified or more firmly established.

There are even grander implications of inflation, albeit very difficult to test.[5] Cosmologists have long used the Copernican principle to argue that the entire Universe must be smooth because of the smoothness of our Hubble volume. In the post-inflation view, our Hubble volume is smooth because it is a small part of a region that underwent inflation, and thus it need not reflect the large-scale features of the Universe as a whole. On the largest scales the structure of the Universe is likely to be very rich: Different regions may have undergone different amounts of inflation, beginning at different times; some regions may not have undergo inflation and may have collapsed to black holes; other regions may be governed by different realizations of the laws of physics because they evolved into different vacuum states of equivalent energy. It is likely that most of the volume of the Universe is still undergoing inflation and that inflationary patches are being constantly produced (eternal inflation). In this case, "the age of the Universe" is a meaningless concept: Our expansion age merely measures the time back to the end of our inflationary event.

If inflation is correct, it will be a major advance in our understanding of the origin and evolution of our Hubble volume and it will open a new window on physics beyond the standard model of particle physics. It is possible that inflation is just plain wrong, and over the years other explanations have put forth to address the dilemma of the initial data. For example, Penrose has suggested the smoothness and flatness of the Universe has to do with the nature of initial singularities.[6] It has, however, been shown that any microphysical solution to the horizon and flatness problems must involve the two key elements of inflation – superluminal expansion and entropy production – suggesting to me that inflation or something closely related is likely to be the correct explanation.[7]

2 There is No Standard Model of Inflation

It would be nice if there were a standard model of inflation, but there isn't. Because inflation involves physics beyond the standard model of particle physics and is probably to tied to fundamental physics at energies of $\mathcal{O}(10^{14}\,\mathrm{GeV})$ this is not surprising. What is important, is that inflationary models make three robust predictions (see next Section) which allow the paradigm to be decisively tested. Moreover, cosmological measurements should also be able to discriminate between different models (see final Section).

The two required elements of any inflationary model are: superluminal expansion (i.e., accelerated expansion, $\ddot{R} > 0$) and massive entropy production.[7]

They are usually achieved by means of the classical evolution of a scalar field rolling down its potential-energy curve. During the first part of its evolution, the field rolls so slowly that its potential energy density is nearly constant; this drives a nearly exponential expansion (superluminal expansion). During the late part of its evolution, the scalar field rapidly oscillates about the minimum of its potential and the decay of these oscillations eventually leads to the production of particles and the reheating of the Universe (entropy production). The entropy produced is the heat that today is the Cosmic Background Radiation (CBR). Because of the massive entropy produced, any initial baryon asymmetry is diluted to a level much, much less than the observed $\mathcal{O}(10^{-10})$ baryon number per photon and baryogenesis after inflationary is mandatory. The basic mechanics of inflation are well understood and summarized elsewhere.[8]

All models of inflation have one feature in common: the scalar field responsible for inflation has a very flat potential-energy curve and is very weakly coupled. This typically implies a dimensionless coupling of the order of 10^{-14}. Such a small number, like other small numbers in physics (e.g., the ratio of the weak to Planck scales $\approx 10^{-17}$ or the ratio of the mass of the electron to the W/Z boson masses $\approx 10^{-5}$), runs counter to one's belief that a truly fundamental theory should have no tiny parameters, and cries out for an explanation.[a]

Models of inflation range from the very simple (e.g., chaotic inflation[9]) to those that attempt to be part of a grander scheme (e.g., models that make contact with speculations about physics at very high energies – grand unification[10], supersymmetry[11,12,13], preonic physics[14], or supergravity[15].) Some have attempted to link inflation with superstring theory;[16] others have focussed on the naturalness issue, trying to explain the small dimensionless number associated with inflation.[17]

While the scale of the vacuum energy that drives inflation is typically of the order of $(10^{14}\,\text{GeV})^4$, a model of inflation at the electroweak scale, vacuum energy $\approx (1\,\text{TeV})^4$, has been proposed.[18] Multiple epochs of inflation are also possible.[19] Inflation has been considered in the context of alternative theories of gravity. The most successful is first-order inflation,[20,21] where gravity is described by Jordan – Brans – Dicke theory (or a similar theory of gravity)

---

[a]It is sometimes stated that inflation is unnatural because of the small coupling of the scalar field responsible for inflation; while the small coupling certainly begs explanation, inflationary models are not unnatural in the technical sense as the small number is always stabilized against the effect of quantum corrections. In some models, the small number in the inflationary potential is related to other small numbers in particle physics: for example, the ratio of the electron mass to the weak scale or the ratio of the unification scale to the Planck scale. Explaining the origin of the small number associated with inflation is both a challenge and an opportunity.

and inflation is triggered by a strongly first-order phase transition (e.g., GUT symmetry breaking) of the kind originally envisioned by Guth.[2]

There are certainly details of inflation that are both model-dependent and not completely understood. For example, the basics of reheating were laid out early on;[22] however, important details are still under study today.[23] While the physics issues such as reheating and model building are important and interesting, they do not affect the basic predictions of inflation that are crucial to its testing. In the end, observations may give the best guidance about models and even physics issues.

3   Inflation Makes Three Robust Predictions

Inflation theorists are very inventive and there are probably no set of predictions that are common to all models of inflation. However, a theory without definite predictions is not testable – and is hardly a theory at all (Mach's principle provides an interesting case in point). The philosopher of science Karl Popper argued that the status of a scientific theory is tied to its vulnerability – strong theories constantly subject themselves to falsification. I believe that inflation is a strong theory in the sense of Popper and that it makes three predictions which allow it to be falsified. They are:

1. **Flat universe.** This is perhaps the most fundamental prediction of inflation. Through the Friedmann equation it implies that the total energy density is always equal to the critical energy density; it does not however predict the form (or forms) that the critical density takes on today or at any earlier or later epoch.

2. **Nearly scale-invariant spectrum of gaussian density perturbations.** These density perturbations (scalar metric perturbations) arise from quantum-mechanical fluctuations in the field that drives inflation;[3] they begin on very tiny scales (of the order of 10^{-23} cm) and are stretched to astrophysical size by the tremendous growth of the scale factor during inflation (factor of e^{60} or greater). Scale invariant refers to the fact that the fluctuations in the gravitational potential are independent of length scale; or equivalently that the horizon-crossing amplitudes of the density perturbations are independent of length scale. While the shape of the spectrum of density perturbations is common to all models, the overall amplitude is model dependent. Achieving density perturbations that are consistent with the observed anisotropy of the CBR and large enough to produce the structure seen in the Universe today requires a horizon crossing amplitude of around $(\delta\rho/\rho)_H \approx 2 \times 10^{-5}$. This is the most severe

constraint on inflationary models and leads to the small dimensionless number associated with inflation.

3. **Nearly scale-invariant spectrum of gravitational waves.** These gravitational waves (tensor metric perturbations) arise during inflation from quantum-mechanical fluctuations in the metric itself and today have wavelengths from $\mathcal{O}(1\,\text{km})$ to the size of the present Hubble radius and beyond.[24] Scale invariant here refers to the fact that gravitational waves of all wavelength cross the horizon with the same dimensionless strain amplitude. Once again, the overall amplitude is model dependent (proportional to the inflationary vacuum energy). The uniformity of the CBR provides a cosmological upper bound to the overall amplitude, but unlike density perturbations, there is no cosmological lower bound to the amplitude of gravity-wave perturbations.

There are other interesting consequences of inflation that are not generic. For example, in models of first-order inflation, where reheating occurs through the nucleation and collision of vacuum bubbles, there is an additional, much larger amplitude, but narrow-frequency-band spectrum of gravitational waves ($\Omega_{\text{GW}} h^2 \sim 10^{-7}$).[25] Large-scale primeval magnetic fields of interesting size can be seeded during inflation.[26] It is also possible to produce topological defects during or near the end of inflation[27] or isocurvature perturbations in a matter component (axions,[28] baryons,[29] or something else[30]). Such auxiliary predictions are interesting, but are not part of the core predictions that can be used to falsify inflation. On the other hand, they could prove very helpful in establishing inflation.

4   Can Inflation Lead to an Open Universe?

Yes, BUT!!

Whether or not flatness is a generic prediction of inflation has been the topic of much debate recently. I believe that flatness should be taken as a firm prediction of inflation and I will explain why. If there is one episode of inflation, solving the "horizon" problem and solving the "flatness" problem (maintaining Ω very close to unity until the present epoch) are linked geometrically by the simple expression[31]

$$|\Omega_0 - 1| \lesssim \left(\frac{H_0^{-1}}{d_{\text{Patch}}}\right)^2 \tag{1}$$

where d_{Patch} is the present size of the inflationary patch that our Hubble volume resides within, which is assumed to have size H^{-1} at the beginning of inflation. If we make no assumption about the smoothness of the Universe on superHubble scales before inflation or about our location within our inflationary region, solving the horizon problem requires $H_0^{-1} \ll d_{\text{Patch}}$ and this implies $|\Omega_0 - | \ll 1$.

Open inflation requires that this linkage be evaded and that the amount of inflation be tuned to a specific value. The number of e-folds of inflation N is determined by the shape of the scalar-field potential,

$$N = \frac{8\pi}{m_{\text{Pl}}^2} \int_{\phi_i}^{\phi_f} \frac{V(\phi)d\phi}{V'} . \tag{2}$$

The value of N required to achieve a given value of Ω today depends upon the reheating temperature after inflation, the value of Ω before inflation, and the temperature today,

$$N = \frac{1}{2} \ln\left[\frac{|\Omega_i - 1|}{|\Omega_0 - 1|}\right] + \ln\left[\frac{T_{\text{RH}}T_0}{m_{\text{Pl}}H_0}\right] . \tag{3}$$

The amount of inflation needed is linked to both the initial state and the epoch of our existence and invites one to invoke the anthropic principle. I see this as a major step backward and counter to the spirit of the inflationary program.

In any case, the simplest way to evade Eq. (1) is to assume that the smooth patch that inflates has an initial size that is ten or even hundred times larger than the Hubble radius. The more elegant way is to assume two epochs of inflation, the first ending with the nucleation of a bubble and second tied to the slow roll of a scalar field.[32] The open Universe resides within the bubble nucleated by the first episode of inflation (which looks like an open universe[33]) and is reheated by the second, slow-roll epoch of inflation.

Open, double inflation in the context of eternal inflation can trade tuning for a distribution of values of Ω_0. My hunch is that the distribution is likely to be very strongly peaked, either around $\Omega_0 = 0$ or $\Omega_0 = 1$, rather than uniform. The recent work of Vilenkin and Winitzki suggests that this is the case.[34]

The scientific question of the flatness of the Universe will be answered, probably within the next five years by using the fine-scale anisotropy of the CBR. If the Universe is found to be flat, I will score it as an important victory for inflation. If the flatness prediction is falsified I will consider it a major defeat. If the Universe is found not to be flat, but other tests of inflation prove successful (e.g., CBR anisotropy and/or gravitational waves), I will be willing to take another look at open inflation.

5   Inflation Implies Particle Dark Matter and Maybe More

While inflation predicts a flat, critical-density Universe, it sheds no light on the form that the critical density should take. Cosmological observations have narrowed the possibilities. Denote the fraction of critical density contributed of all forms of energy density by Ω_0; inflation predicts $\Omega_0 = 1$. Big-bang nucleosynthesis constrains the baryon density tobe well below the critical density: $0.007h^{-2} \leq \Omega_B \leq 0.024h^{-2} < 0.10$ (for $h > 0.5$),[35] which implies that most of the critical density must be in a form other than baryons. When the primeval deuterium abundance is pinned down by a definitive determinations of D/H in high-redshift hydrogen clouds, the baryon density will be pegged to a precision of around 10% or so. Tytler and his collaborators have made a very strong case for a primeval deuterium abundance of $D/H \simeq 2.5 \times 10^{-5}$,[36] which implies that $\Omega_B h^2 \simeq 0.024$ or $\Omega_B \simeq 0.05(0.7/h)^2$.

It is also known that: most of the matter is dark (luminous matter contributes less than 1% of the critical density, $\Omega_{\text{LUM}} \simeq 0.003h^{-1}$) and the fraction of critical density in matter that clusters exceeds 30%, $\Omega_M > 0.3$.[37] Thus, it follows that at least 25% of the critical density should be in the form of nonbaryonic matter in the form of particles, $\Omega_{\text{nbparticles}} = \Omega_M - \Omega_B \gtrsim 0.25$. Particle physics has provided three very good candidates whose relic abundance (if they exist) should be of the order of the critical density: an axion of mass around 10^{-5} eV;[38] a neutralino of mass between $10\,\text{GeV}$ and $500\,\text{GeV}$;[39] and a neutrino of mass of the order of $10\,\text{eV}$.[40]

All are well motivated: the axion is a prediction of the most promising solution to the strong-CP problem; the neutralino is predicted by supersymmetric extensions of the standard model; and essentially all attempts to unify the forces and particles of Nature lead to the prediction that neutrinos have small, but nonzero, masses. In fact, these three particle dark matter candidates are so well motivated that we should probably take seriously the possibility that more than one might contribute significantly to the matter density today.

Finally, it is possible that there is another, even more exotic component, which is smoothly distributed and contributes up to 70% of the critical density, $\Omega_X = \Omega_0 - \Omega_M \lesssim 0.7$. The fact that evidence for $\Omega_M \sim 1$ is still lacking and that a case is mounting for $\Omega_M \sim 0.3$,[37] suggests that inflationists should consider the possibility of a smooth component seriously. Candidates for such include vacuum energy, tangled strings, and rolling scalar fields.[41]

While Occam's Razor argues against a smorgasbord, Nature might enjoy a more interesting meal, and inflation gives no guidance.

6 Large-scale Structure from Quantum Fluctuations

This is perhaps the most striking prediction of inflation, and I believe, the motivation for Stephen Hawking's description of the COBE DMR discovery of CBR anisotropy[42] as "the most important discovery of all time." I believe Hawking said this thinking the COBE discovery might prove to be the first evidence that the density perturbations that seeded all structure in the Universe arose from quantum fluctuations during inflation.

Recall, scale invariant refers to density perturbations that cross the horizon with the same amplitude, independent of length scale. Different scales cross the horizon at different times, so the spectrum of density perturbations today is not independent of scale.

Gaussian means that the density contrast, $\delta\rho(\mathbf{x}, t)/\bar{\rho}$, is a gaussian random field, described fully by its two-point correlation function, or equivalently by the power spectrum, which is the Fourier transform of the correlation function and the square of the Fourier transform of the density contrast.

Both scale invariant and gaussian are generic predictions as they are linked to central features of inflation. Because the scalar field that drives inflation is very weakly coupled, it behaves like a free field and its fluctuations are gaussian.[3] The density perturbations are proportional to the scalar-field fluctuations and hence they too should be gaussian. The deviation of the fluctuations from scale invariance is related to the steepness of the scalar potential; since the scalar field responsible for inflation must take the 60 or so Hubble times to evolve to the minimum of its potential in order to solve the horizon/flatness problems its potential cannot be too steep.

The relationship between the inflationary potential and the power spectrum of density perturbations today $(P(k) \equiv \langle|\delta_k|^2\rangle)$ is given by

$$P(k) = \frac{1024\pi^3}{75} \frac{k}{H_0^4} \frac{V_*^3}{m_{\text{Pl}}^6 V_*'^2} \left(\frac{k}{k_*}\right)^{n-1} T^2(k)$$

$$n - 1 = -\frac{1}{8\pi}\left(\frac{m_{\text{Pl}} V_*'}{V_*}\right)^2 + \frac{m_{\text{Pl}}}{4\pi}\left(\frac{m_{\text{Pl}} V_*'}{V_*}\right)'$$

$$\frac{dn}{d\ln k} = -\frac{1}{32\pi^2}\left(\frac{m_{\text{Pl}}^3 V_*'''}{V_*}\right)\left(\frac{m_{\text{Pl}} V_*'}{V_*}\right)$$

$$+\frac{1}{8\pi^2}\left(\frac{m_{\text{Pl}}^2 V_*''}{V_*}\right)\left(\frac{m_{\text{Pl}} V_*'}{V_*}\right)^2 - \frac{3}{32\pi^2}\left(m_{\text{Pl}}\frac{V_*'}{V_*}\right)^4$$

$$T(q) = \frac{\ln(1 + 2.34q)/2.34q}{[1 + 3.89q + (16.1q)^2 + (5.46q)^3 + (6.71q)^4]^{1/4}}, \tag{4}$$

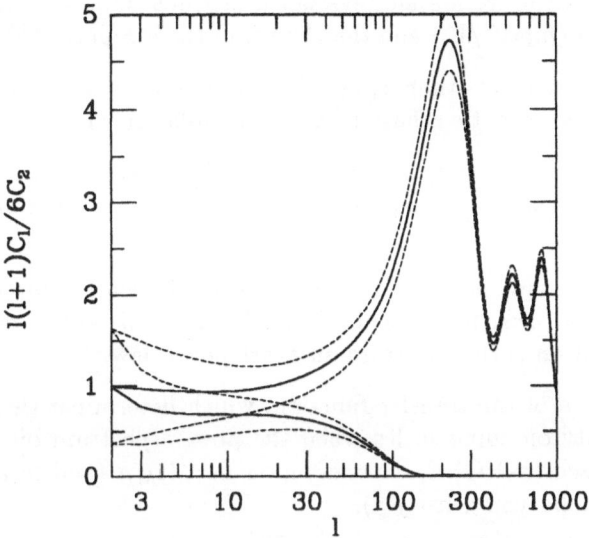

Figure 1: Angular power spectra ($C_l \equiv \langle|a_{lm}|^2\rangle$) of CBR anisotropy for gravity waves (lower curves) and density perturbations (upper curves), normalized to the quadrupole anisotropy; broken lines indicate sampling variance. Temperature fluctuations measured on angular scale θ are approximately, $(\delta T/T)_\theta \sim \sqrt{l(l+1)C_l/2\pi}$ with $l \sim 200°/\theta$.

where $V(\phi)$ is the inflationary potential, prime denotes $d/d\phi$, V_* is the value of the scalar potential when the scale k_* crossed outside the horizon during inflation, $T(k)$ is the transfer function which accounts for the evolution of the mode k from horizon crossing until the present, $q = k/h\Gamma$, and $\Gamma \simeq \Omega_M h$ is the "shape" parameter.[43] It is very convenient to chose $k_* = H_0$ so that V_* is the value of the inflationary potential when the scale that fixes the CBR quadrupole crossed outside the horizon. These expressions are given to lowest-order in the deviation from scale invariance, and assume a matter-dominated Universe today; the next-order corrections have been calculated[44] and the analogous expressions which include the possibility of a cosmological constant have been written down.[45]

There are several important things to take note of:

1. The overall amplitude of the power spectrum depends upon the combination $V_*^3/V_*'^2$.

2. The quantity $n - 1$, which measures the deviation from scale invariance, is generally not equal to zero.[46]

3. The deviation from scale invariance depends upon the steepness of the potential $(m_{Pl}V'_*/V_*)$ and the change in the steepness.[47,48]

4. Typically n is less than 1, and for many models is in the range 0.94 to 0.96[47] (e.g., chaotic inflation[9] and new inflation[49]).

5. There are inflationary models where n is as larger than 1 (e.g., hybrid inflation[50]) and as small as 0.7 (e.g., power-law inflation and natural inflation).

6. Generally, the spectrum of perturbations is nearly, but not exactly, a power law: $dn/d\ln k$ is typically of the order of -10^{-3}; only for power-law inflation is the spectrum an exact power law.[51]

7. The shape of the transfer function, which determines the level of inhomogeneity on small scales when the power spectrum on large scales is normalized by COBE, depends upon $\Gamma = \Omega_M h$ (and to a lesser extent upon the baryon density[52]).

Density perturbations give rise to CBR anisotropy which can be computed very precisely.[53] CBR anisotropy probes the power spectrum at early times ($z \simeq 1100$), when the perturbations were still in the linear regime and astrophysical effects were minimal. Thus, it is one of the most important tests and powerful probes of inflation. Expanding the CBR temperature on the sky in spherical harmonics,

$$\frac{\delta T(\theta, \phi)}{T} = \sum_{lm} a_{lm} Y_{lm}(\theta, \phi), \tag{5}$$

the anisotropy is fully characterized by its angular power spectrum $C_l \equiv \langle|a_{lm}|^2\rangle$, shown in Fig. 1. The ensemble average for the variance of the multipoles, $\langle|a_{lm}|^2\rangle$, is related to the power spectrum (as described elsewhere[53]); note, isotropy in the mean implies $\langle a_{lm}\rangle = 0$. The variance of multipole l is dominated by modes of wavenumbers $k \simeq lH_0/2$. CBR anisotropy on large-angular scales ($l \ll 100$) arises almost solely from the Sachs-Wolfe effect and to good approximation can be computed analytically[8,54]

$$C_l = \frac{H_0^4}{2\pi} \int_0^\infty (u/u_{EQ})|j_l(u)|^2 P(u/u_{EQ}) du/u, \tag{6}$$

where $u = k\tau_0$ and $u_{EQ} = k_{EQ}\tau_0$. The variance of the quadrupole anisotropy provides a handy means of normalizing the power spectrum

$$S \equiv \frac{5\langle|a_{2m}|^2\rangle}{4\pi} \simeq 2.2 \frac{V_*/m_{Pl}^4}{(m_{Pl}V'_*/V_*)^2}. \tag{7}$$

The detection of CBR anisotropy by the COBE DMR was a major advance as it allowed the spectrum of density perturbations to be normalized on very large scales ($k \sim H_0$). For precisely scale-invariant density perturbations and no gravitational-wave contribution to CBR anisotropy the normalization procedure is easy to describe: S is equated to the COBE determination of the variance of the quadrupole anisotropy, $Q_{COBE} = (17\mu\,\text{K}/2.2728\,\text{K})^2 \simeq 3.8 \times 10^{-11}$,[55] which then implies

$$\frac{V_*/m_{Pl}^4}{(m_{Pl}V_*'/V_*)^2} = 1.8 \times 10^{-11} \tag{8}$$

Bunn et al.[56] have done a careful analysis of the COBE four-year data which takes account of the fact that the COBE normalization for S depends upon n as well as the possible contribution of gravitational waves to CBR anisotropy. (The "pivot point" for the COBE data is $l \sim 15$; that is, the COBE determinations of C_{15} and n are almost uncorrelated.) This leads to the more accurate normalization

$$\frac{V_*/m_{Pl}^4}{(m_{Pl}V_*'/V_*)^2} = 1.7 \times 10^{-11}\frac{\exp[-2.02(n-1)]}{\sqrt{1+\frac{2}{3}\frac{T}{S}}} \tag{9}$$

where T is the tensor contribution to the variance of the CBR quadrupole. The 1σ error is 15%. (Bunn et al. have also generalized this result to allow for the possibility of a cosmological constant.[56])

The Bunn et al. normalization can also be expressed in terms of the horizon-crossing amplitude for the comoving scale $k = H_0$:

$$\delta_H(k = H_0) \equiv \left[\frac{k^{3/2}|\delta_k|}{\sqrt{2\pi^2}}\right]_{k=H_0} = 1.9 \times 10^{-5}\frac{\exp[-1.01(n-1)]}{\sqrt{1+\frac{2}{3}\frac{T}{S}}}. \tag{10}$$

7   Gravitational Waves: The Smokin' Gun

The inflation-produced gravitational waves are the smokin' gun signature of inflation and crucial to learning about the inflationary potential. Both a flat Universe[57] and scale-invariant density perturbations (so called Harrison–Zel'dovich spectrum[58]) were discussed as features of any "sensible cosmology" long before inflation. The nearly scale-invariant spectrum of gravitational waves which arises from quantum mechanical fluctuations excited in the space-time metric during inflation is a very important prediction of inflation that sets it apart from just any other sensible cosmology. Detecting these gravitational waves will be very challenging.

Unlike the scalar perturbations, which must have an amplitude of around 10^{-5} to seed structure formation, there is no astrophysical clue as to the amplitude of the tensor perturbations. They can be characterized by their power spectrum today[59]

$$P_T(k) \equiv \langle |h_k|^2 \rangle = \frac{8}{3\pi} \frac{V_*}{m_{\text{Pl}}{}^4} \left(\frac{k}{k_*} \right)^{n_T - 3} T_T^2(k)$$

$$n_T = -\frac{1}{8\pi} \left(\frac{m_{\text{Pl}} V_*'}{V_*} \right)^2$$

$$\frac{dn_T}{d\ln k} = \frac{1}{32\pi^2} \left(\frac{m_{\text{Pl}}{}^2 V''}{V} \right) \left(\frac{m_{\text{Pl}} V'}{V} \right)^2 - \frac{1}{32\pi^2} \left(\frac{m_{\text{Pl}} V'}{V} \right)^4$$

$$= -n_T[(n-1) - n_T]$$

$$T_T(k) \simeq \left[1 + \frac{4}{3} \frac{k}{k_{\text{EQ}}} + \frac{5}{2} \left(\frac{k}{k_{\text{EQ}}} \right)^2 \right]^{1/2}, \quad (11)$$

where $T_T(k)$ is the transfer function for gravity waves and describes the evolution of mode k from horizon crossing until the present, $k_{\text{EQ}} = 6.22 \times 10^{-2}\,\text{Mpc}^{-1}\,(\Omega_M h^2/\sqrt{g_*/3.36})$ is the scale that crossed the horizon at matter-radiation equality, Ω_M is the fraction of critical density in matter, and g_* counts the effective number of relativistic degrees of freedom (3.36 for photons and three light neutrino species). The quantity $k^{3/2}|h_k|/\sqrt{2\pi^2}$ corresponds to the dimensionless strain (metric perturbation) on length scale $\lambda = 2\pi/k$.

Like density perturbations, gravity waves lead to CBR anisotropy which can be fully described by an angular power spectrum. The gravity-wave angular power spectrum is related to the power spectrum and can be computed very accurately;[60] it is shown in Fig. 1. The following analytical expression is accurate to about 10%,

$$C_l = 36\pi^2 \frac{\Gamma(l+3)}{\Gamma(l-1)} \int_0^\infty F_l(u)(u/u_{\text{EQ}})^3 P_T(u/u_{\text{EQ}})\, du/u$$

$$F_l(u) = -\int_{(\tau_{\text{LS}}/\tau_0)u}^u dy \left(\frac{j_2(y)}{y} \right) \left(\frac{j_l(u-y)}{(u-y)^2} \right) \quad (12)$$

$$(13)$$

where $u = k\tau_0$, $\tau_0 = 2H_0^{-1}$ is the conformal age of the Universe today, and $\tau_{\text{LS}} = \tau_0/\sqrt{(1 + z_{\text{LS}})}$ is the conformal age at last scattering. The tensor contribution to the variance of the CBR quadrupole is a convenient normalization

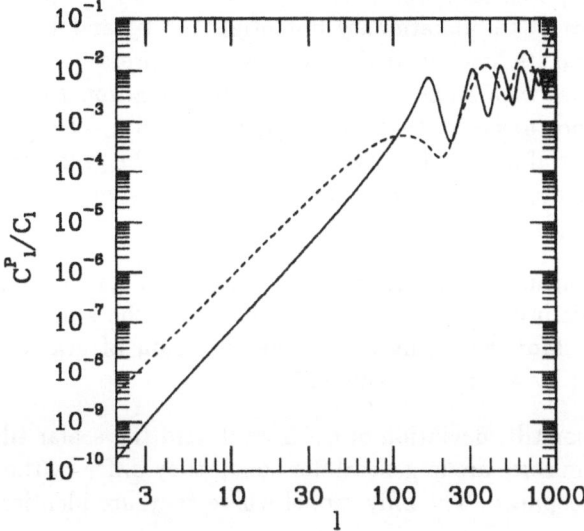

Figure 2: Polarization angular power spectra for gravity waves (broken) and density perturbations (solid). The polarization of the CBR anisotropy is roughly $\sqrt{C_l^P/C_l}$.

for the spectrum:

$$T \equiv \frac{5\langle|a_{2m}|^2\rangle}{4\pi} = 0.61(V_*/m_{\mathrm{Pl}}^4).\tag{14}$$

The predicted variance of the CBR quadrupole anisotropy is $T + S$.

There are several important things to take note of:

1. The contribution of gravity waves to the variance of the CBR quadrupole
   is proportional to the value of the vacuum energy that drives inflation,
   and if T can be determined, the energy scale of inflation can be determined.

2. Using the COBE four-year results as an upper bound, $T < Q_{\mathrm{COBE}}$,
   it follows that, $V_* < 6 \times 10^{-11} m_{\mathrm{Pl}}{}^4$, or equivalently, $V_*^{1/4} < 3.4 \times 10^{16}\,\mathrm{GeV}$. This indicates that inflation, if it occurred, involved energies
   much smaller than the Planck scale. (To be more precise, inflation could
   have begun at a much higher energy scale, but the portion of inflation
   relevant for us, i.e., the last 60 or so e-folds, occurred at an energy scale
   much smaller than the Planck energy. In chaotic inflation[9], inflation is
   supposed to begin at the Planck energy density.)

3. The four potential observables, S, T, $n-1$ and n_T, depend upon three properties of the inflationary potential, V_*, V'_* and V''_*. Thus, the potential and its first two derivatives can be expressed in terms of the four observables with an additional consistency relation, $T/S = -7n_T$, which is an important test of inflation.[61] If S, T, and $n-1$ can be determined, the potential and its first two derivatives can be "reconstructed"; in addition, if n_T can also be measured, the consistency of inflation can be tested.[62]

4. The amplitude of the gravity-wave spectrum and its "tilt" (deviation from scale invariance) are related: the larger the amplitude, the greater the tilt. Moreover, this means the spectrum of gravity waves can be described by a single parameter.

5. The tensor tilt, deviation of n_T from 0, and the scalar tilt, deviation of $n-1$ from zero, are in general not equal; they differ by the rate of change of the steepness. The only model where they are identical is power-law inflation.

6. The variation of the tensor index with scale, $dn_T/d\ln k$, is typically $\mathcal{O}(10^{-3})$.

There are two basic approaches to detecting the tensor perturbations: CBR anisotropy and/or polarization and direct detection of gravity waves. The first approach probes the spectrum at very long wavelengths, $\lambda \sim H_0^{-1}/100 - H_0^{-1} \sim 10^{26}\,\mathrm{cm} - 10^{28}\,\mathrm{cm}$, while the second probes much shorter wavelengths, $\lambda \sim 10^8\,\mathrm{cm} - 10^{14}\,\mathrm{cm}$. If some information (detections/upper limits) could be obtained at both wavelengths, both T and n_T could be measured or at least constrained.

While the scalar and tensor angular power spectra are very different (see Fig. 1), sampling variance sets a fundamental limit to how well the two can be separated from measurements of the one CBR sky we have access to. For multipole l, $2l+1$ multipole amplitudes can be used to determine the variance; "the variance of the variance" is

$$\frac{\langle (C_l^{\text{estimate}} - C_l)^2 \rangle^{1/2}}{C_l} = \sqrt{\frac{2}{2l+1}}, \tag{15}$$

where C_l^{estimate} is the estimate based upon CBR measurements; sampling variance is shown in Fig. 1. Using anisotropy alone, sampling variance implies that the tensor contribution can be reliably separated only if $T/S \geq 0.1$.[63] The tensor and scalar perturbations led to different levels of polarization of

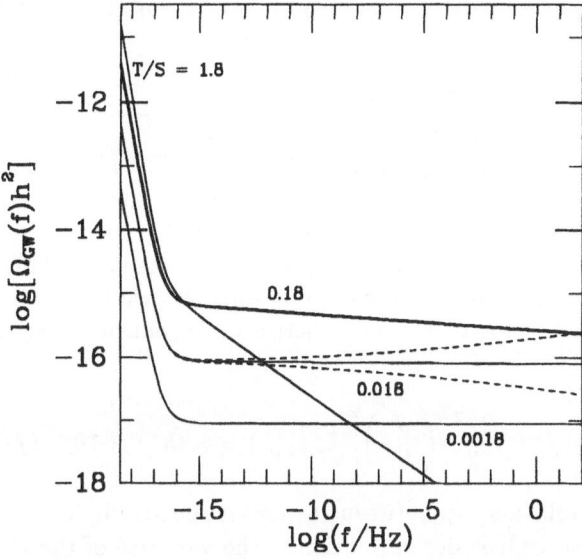

Figure 3: Spectral energy density in gravity waves produced by inflation; for $T/S = 0.018$, $dn_T/d\ln k = -10^{-3}$, 0, 10^{-3}. $T/S = 0.18$ (heavy curve) maximizes the energy density at $f = 10^{-4}$ Hz. Curves are from Eq. (16) using $H_0 = 60\mathrm{km\,s}^{-1}\,\mathrm{Mpc}^{-1}$, $\Omega_M = 1$, and $g_* = 3.36$.

the anisotropy; see Fig. 2. The sampling-variance limit based upon CBR polarization is about a factor of five smaller,[63] but requires that polarization on large-angular scales be measured at less than a fraction of a percent. Recently, it has been pointed out that scalar and tensor perturbations excite different patterns of polarization,[64] which could allow sampling variance to be evaded. In any case, the potential of polarization remains to be seen: the signal is small (maximum polarization is a few percent of the anisotropy); CBR polarization has yet to be detected; and the severity of polarization foregrounds are yet to be determined.

Earth-based laser interferometers which operate in the 10 Hz to kHz range are being built in the US (LIGO) and in Europe (VIRGO). A space-based interferometer is being planned by ESA (LISA) and ideas for a smaller mission are being discussed in the US (OMEGA). Space-based interferometers could operate at frequencies as low as 10^{-4} Hz.

It is straightforward to go from the power spectrum to the fraction of critical density contributed by gravity waves per log frequency interval

$$\Omega_{\rm GW}(f) \equiv \frac{1}{\rho_{\rm crit}} \frac{d\rho_{\rm GW}}{d\ln k} = \frac{\Omega_M^2 \, V_*/m_{\rm Pl}^4}{(k/H_0)^{2-n_T}} \left[1 + \frac{4}{3} \frac{k}{k_{\rm EQ}} + \frac{5}{2} \left(\frac{k}{k_{\rm EQ}} \right)^2 \right], \qquad (16)$$

where $f = k/2\pi$ and $\Omega_{\rm GW}(f)$ is shown in Fig. 3. The long plateau, frequencies greater than $f_{\rm EQ} = k_{\rm EQ}/2\pi \sim 10^{-15}\,{\rm Hz}$, reflects the scale invariance of the gravitational-wave spectrum. The rise for smaller frequencies, as f^2, traces to the fact that the longest wavelength modes crossed the horizon during the matter-dominated epoch. The energy density in gravitational waves can also be expressed in terms of the *rms* metric perturbation or strain, $h_{rms}^2(k) \equiv k^3 |h_{\bf k}|^2/2\pi^2$,

$$\Omega_{\rm GW}(f) = \frac{2\pi^2}{3} \left(\frac{f}{H_0} \right)^2 h_{rms}^2(k) = 6.3 h^{-2} \times 10^{-7} \, (f/{\rm Hz})^2 \qquad (17)$$

Using the relationship between the tensor spectral index n_T and the amplitude T, and the COBE determination of the variance of the CBR quadrupole anisotropy, Eq. (16) can be rewritten in terms of n_T (or T/S) alone.[65] Doing so, it then follows that on the "long plateau" ($f \gg 10^{-15}\,{\rm Hz}$)

$$\Omega_{\rm GW}(f) h^2 = 5.1 \times 10^{-15} \, (g_*/3.36) \frac{n_T}{n_T - 1/7}$$

$$\times \exp[n_T N + \frac{1}{2} N^2 (dn_T/d\ln k)], \qquad (18)$$

where $N \equiv \ln(k/H_0) \simeq 33 + \ln(f/10^{-4}{\rm Hz}) + \ln(0.6/h)$ and I have also allowed for the possible variation of the tensor index n_T which is typically -10^{-3}.

The importance of the amplitude – tilt relationship can be seen in Fig. 3. Sadly, tilt goes in the direction of pushing $\Omega_{\rm GW}$ down as the amplitude T/S is increased. There is a bright side: $\Omega_{\rm GW}(f \sim {\rm Hz})$ is maximized for $T/S \simeq 0.18$, which is close to the value predicted by chaotic inflation and exceeds the sampling-variance limit to the detection of tensor perturbations using CBR anisotropy alone. While there are essentially no inflationary models where $|n - 1| \ll 0.1$, there are many models where $-n_T \ll 0.1$ (e.g., natural inflation and new inflation). Because of the amplitude – tilt relationship, the gravity-wave background in these models is very small.

The range of T/S accessible to a gravity-wave detector operating at $f = 10^{-4}\,{\rm Hz}$ (appropriate for LISA) and $f = 100\,{\rm Hz}$ (appropriate for LIGO and VIRGO) is shown as a function of the detector energy sensitivity in Fig. 4.[65] A sensitivity of $\Omega_{\rm GW}(f)h^2 \sim 10^{-15}$ is needed for a serious search for inflation-produced gravity waves. With its initial strain detectors, the earth-based LIGO

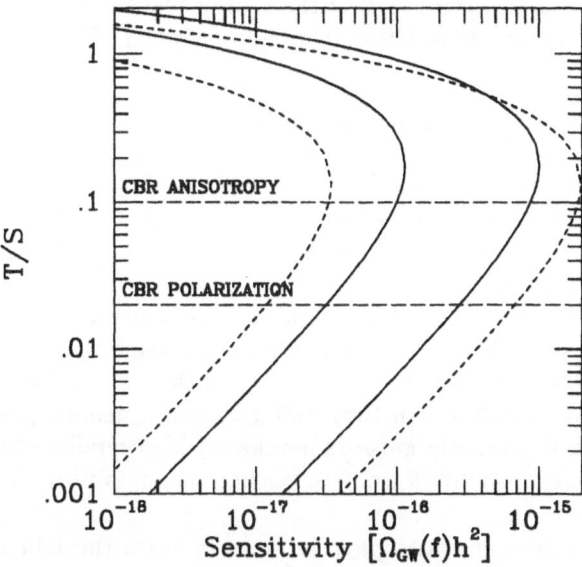

Figure 4: The range of T/S probed (interval interior to parabola) as a function of energy sensitivity for $f = 10^{-4}\,\mathrm{Hz}$ (solid curves) and $f = 100\,\mathrm{Hz}$ (broken curves). The "pessimistic" (left) parabola assumes $dn_T/d\ln k = -10^{-3}$ and the "optimistic" (right) parabola assumes $dn_T/d\ln k = 10^{-3}$. Also shown are the sampling-variance limiting sensitivities for CBR anisotropy and polarization.

should be able to detect a stochastic background of gravity waves with 90% confidence provided $\Omega_{\mathrm{GW}}(f \sim 100\,\mathrm{Hz})h^2 \geq 2.8 \times 10^{-6}$; with advanced strain detectors this should improve to 2.8×10^{-11}.[66,67] Unfortunately, this misses the mark by four orders of magnitude. (If one were to ignore the relation between tilt and amplitude and assume $n_T \equiv 0$, LIGO would only miss by three orders of magnitude.[67])

Because the energy density in gravity waves is proportional to strain squared times frequency squared, a detector operating at lower frequency has better energy-density sensitivity for fixed strain sensitivity. Earth-based detectors cannot cannot operate at lower frequencies because of seismic noise, but space-based detectors can. The design study for the space-based LISA indicates an energy sensitivity of around $\Omega_{\mathrm{GW}}(f)h^2 \sim 10^{-13}$ at $f = 10^{-4}\,\mathrm{Hz}$,[68] which is more promising, but still misses by two orders of magnitude. (There is also a worrisome background of coalescing white-dwarf binaries, which could dominate inflation at frequencies greater than around $10^{-4}\,\mathrm{Hz}$.[69])

Detection of the inflation-produced gravity waves presents a very great challenge. But the possible payoffs are commensurately large: A smokin' gun test of inflation; a determination of the energy scale of inflation (through T); and a consistency test of inflation (through n_T and T/S).

8   CDM: A Testable, Ten Parameter Theory

CBR anisotropy has been detected on angular scales ranging from 100° to a fraction of degree at the level of about $\delta T/T \sim 10^{-5}$ (see Fig. 5). This provides strong support for the notion that structure formed by the gravitational amplification of small primeval density inhomogeneities. One of the pressing problems in cosmology is the formulation of a detailed and coherent picture of structure formation. The two key elements in any such theory are the quantity and composition of matter in the Universe and the nature of the perturbations that seed the formation of structure. Inflation makes definite predictions about both, and with the rapidly growing number of high-quality observations that bear on the issue, structure formation has become an important testing ground for inflation.

Recall, inflation and astrophysical data indicate the following about the quantity and composition of matter in the Universe: baryons contribute a small fraction of the critical density, $\Omega_B = (0.007 - 0.024)h^{-2}$; particle dark matter plus baryons contribute at least 30% of the critical density; and there may be a smooth component that brings the total to the critical density. The inflationary prediction concerning the seed perturbations is sharper: almost scale-invariant gaussian density perturbations, whose horizon-crossing amplitude is determined by COBE to be about 2×10^{-5}.

Particle dark matter can be classified by its velocity dispersion at the epoch of matter – radiation equality: cold, $v_{rms} \ll 1$; hot, $v_{rms} \sim 1$; and warm, v_{rms} not too much smaller than 1. Neutrinos and neutralinos were once in thermal equilibrium and their velocity dispersion is set by the temperature at matter – radiation equality ($T_{EQ} \simeq 6h^2 \, \text{eV}$) and is inversely proportional to their mass. Neutrinos are light and therefore hot; neutralinos are heavy and therefore cold. Axions are cold in spite of their small mass because they were never in thermal equilibrium and were produced by a coherent, rather than thermal, process.[38] Around the time of matter – radiation equality, density perturbations on small scales can be damped by the freestreaming of dark matter particles from regions of high density to regions of low density; for neutrinos this is a significant effect and perturbations on scales less than about 30 Mpc are strongly damped. For cold dark matter the freestreaming scale is much less than 1 Mpc and most likely uninteresting. For warm dark matter the freestreaming scale is around 1 Mpc (essentially by definition) and has interesting consequences.

For hot dark matter structure must form "top down" – superclusters form

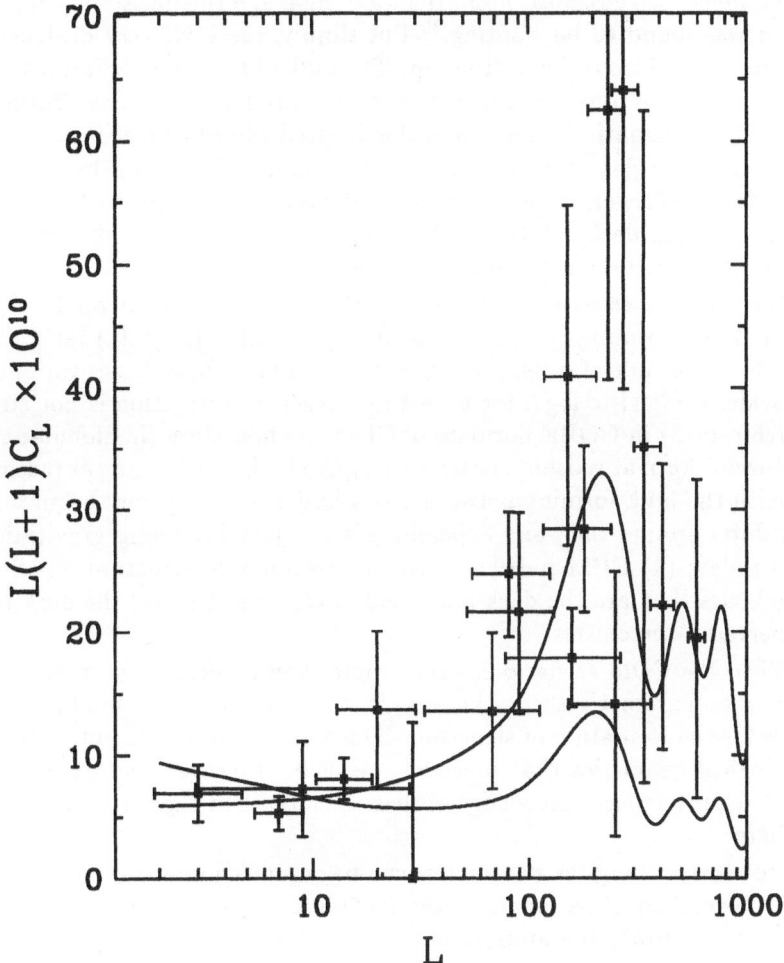

Figure 5: Summary of CBR anisotropy measurements and predictions for two CDM models. Plotted are the squares of the measured multipole amplitudes ($C_l = \langle |a_{lm}|^2 \rangle$) versus multipole number l. The relative temperature difference on angular scale θ is given roughly by $\sqrt{l(l+1)C_l/2\pi}$ with $l \sim 200^\circ/\theta$. The theoretical curves are standard CDM (upper curve) and CDM with $n = 0.7$ and $h = 0.5$ (lower curve).

and fragment into galaxies. More than a decade ago this possibility was studied and was found to be wanting.[70] Put simply, there is every evidence that structure formed from the bottom up: The bulk of the galaxies formed at redshifts from two to four and superclusters are only forming today. Warm dark matter is problematic because subgalactic-sized objects must form from the fragmentation of galaxies; the abundance of high-redshift (up to redshifts of almost five) hydrogen clouds is difficult at best to reconcile with this fact.[71] That leaves cold dark matter as the unique "prediction" of inflation. As we shall see, this prediction has been very successful.

Here are the essential features of CDM:[72,73] (1) it is hierarchical, with smaller things forming first and larger things forming (slightly) later; (2) because the amplitude of density perturbations on very small scales varies slowly with scale, $k^{2/3}|\delta_k| \propto \log k$ for $k \gg k_{EQ}$, structure formation is not strongly hierarchical; (3) in COBE normalized CDM the first stars (in globular-cluster size objects) form at redshifts of ten or so, galaxies begin forming at redshifts of five (with the bulk forming between $z \sim 4$ and $z \sim 2$), clusters begin forming at redshifts around one, and superclusters are just becoming gravitationally bound today; (4) CDM particles form the cosmic infrastructure on all scales – in galaxies they are the dark halos and in clusters they are the dark matter that pervades the cluster.

When the CDM scenario emerged more than a decade ago many referred to it as a no-parameter theory because it was so specific compared to previous models for the formation of structure. This was an overstatement as there are cosmological quantities that must be specified. However, the data available did not require precise knowledge of these quantities to begin testing the CDM paradigm.

Broadly speaking the parameters can be organized into two groups.[74] First are the cosmological parameters: the Hubble constant h; the density of ordinary matter, $\Omega_B h^2$; the amplitude of the scalar perturbations, S, the scalar power-law index, n, and the rate at which it varies, $dn/d\ln k$; the amplitude of the tensor perturbations, T/S, and tensor power-law index, n_T. The inflationary parameters fall into this category because there is no standard model of inflation; on the other hand, once determined they can be used to discriminate among models of inflation.

The second group specifies the composition of invisible matter in the Universe: radiation, dark matter, and vacuum energy. Radiation refers to relativistic particles: the photons in the CBR, three massless neutrino species (assuming none of the neutrino species has a mass), and possibly other undetected relativistic particles (some particle-physics theories predict the existence of additional massless particle species). At present relativistic particles contribute almost nothing to the energy density in the Universe, $\Omega_R \simeq 4.2 \times 10^{-5}h^{-2}$; early on – when the Universe was smaller than about 10^{-5} of its present size – they dominated the energy content; the level of radiation today is important as it determines when the transition from radiation domination to matter domination took place and thereby the shape of the transfer function (through Γ).

Dark matter could include other particle relics besides CDM. For example, each neutrino species has a number density of $113\,\mathrm{cm}^{-3}$, and a neutrino species of mass $5\,\mathrm{eV}$ would account for about 20% of the critical density ($\Omega_\nu = m_\nu/90h^2\,\mathrm{eV}$). Predictions for neutrino masses range from $10^{-12}\,\mathrm{eV}$ to several MeV, and there is some experimental evidence that at least one of the neutrino species has a small mass.[75] Finally, there is the cosmological constant. Introduced and then abandoned by Einstein to prevent the expansion of the Universe, and resurrected by Bondi, Gold and Hoyle in 1948 to address an age crisis, it is still with us. In the modern context it corresponds to an energy density associated with the quantum vacuum. At present, there is no reliable calculation of the value that the cosmological constant should take,[76] and so its existence must be regarded as a logical possibility, with its value to be determined by observations. (As mentioned earlier, there are even more exotic possibilities for the smooth component.[41])

The original no-parameter CDM model, often referred to as standard CDM,[72] is characterized by simple choices for the cosmological and the invisible matter parameters: precisely scale-invariant density perturbations ($n = 1$), $h = 0.5$, $\Omega_B = 0.05$, $\Omega_{\mathrm{CDM}} = 0.95$; no radiation beyond the photons and the three massless neutrinos; no dark matter beyond CDM; and zero cosmological constant. Standard CDM served its purpose well as the DOS 1.0 of cosmology: it focussed attention on a specific CDM model.

While inflation models predict that the shape of the spectrum is approximately scale-invariant, the overall amplitude depends on the particular inflationary model. For standard CDM the overall amplitude was fixed by comparing the predicted level of inhomogeneity with that seen today in the distribution of bright galaxies. Galaxy-number fluctuations in spheres of radius $8h^{-1}$ Mpc are unity:

$$\sigma_r^2 \equiv \int_0^\infty \frac{dk}{k} \frac{k^3 P(k)}{2\pi^2} \left(\frac{3j_1(kr)}{kr} \right)^2 = 1 \qquad (19)$$

for $r = 8\,h^{-1}\mathrm{Mpc}$. This normalization ($\sigma_8 = 1$) corresponds to the assumption that light, in the form of bright galaxies, traces mass. Choosing σ_8 to be less than one means that light is more clustered than mass and is a biased tracer of mass. There is some evidence that bright galaxies are somewhat more clumped than mass with biasing factor $b \equiv 1/\sigma_8 \simeq 1 - 2$; e.g., the number of fluctuations of IRAS galaxies on the $8h^{-1}$ Mpc scale is less than one:, $\sigma_8(\text{IRAS}) = 0.69 \pm 0.04$,[77] implying that infrared selected galaxies are less clustered than optically bright galaxies.

As discussed earlier, COBE changed the normalization procedure; given the values of the CDM parameters and normalizing to COBE σ_8 can be computed. Further, an independent means of determining σ_8, based upon the abundance of rich clusters, has been developed; comparing this value to the that computed from the COBE normalized spectrum now provides a check/constraint. But I am getting ahead of myself.

Is a ten parameter theory testable? With sufficient high-quality data the answer is yes. The standard model of particle physics has at least nineteen parameters; not only has it been tested, but most of the parameters have been determined, many to better than 1% precision. In the next two Sections I hope to make the case that inflation/cold dark matter can be tested with the same decisiveness. Much of my case will rely upon CBR anisotropy; if 2000 or so multipoles can be measured to a precision close to that dictated by sampling variance, I believe this ambitious goal is achievable.

9 Status of Inflation: So Far, So Good

The testing of inflation necessarily focuses on its three robust predictions and their consequences.

9.1 Flatness

The first prediction is a flat Universe with nonbaryonic dark matter. There is strong evidence coming from a number of directions that Ω_M is at least 0.3.[37] This makes nonbaryonic dark matter inescapable since the big-bang nucleosynthesis upper bound is $\Omega_B < 0.024h^{-2} < 0.10$ (for $h > 0.5$) and is half way (on a logarithmic scale) to the simplest realization of a flat Universe, $\Omega_M = 1$; see Fig. 6. In the case that $\Omega_M \approx 0.3$ it is possible that the bulk of the closure density resides in a smooth component.

Testing the flatness prediction has an even brighter future. The position of the first acoustic (Doppler) peak is sensitive to Ω_0, $l_{\text{peak}} \simeq 220/\sqrt{\Omega_0}$. The current data, while certainly not definitive, put a smile on my face: Hancock infers $\Omega_0 = 0.7^{+1.0}_{-0.4}$.[78] It is likely that even before MAP is launched in 2000, ground-based and balloon-based measurements will determine the position of the first acoustic peak well enough to peg Ω_0 to 10%.

Measurements of the deceleration of the Universe using the magnitude – redshift diagram for SNeIa constrain a nearly orthogonal combination, Ω_M –

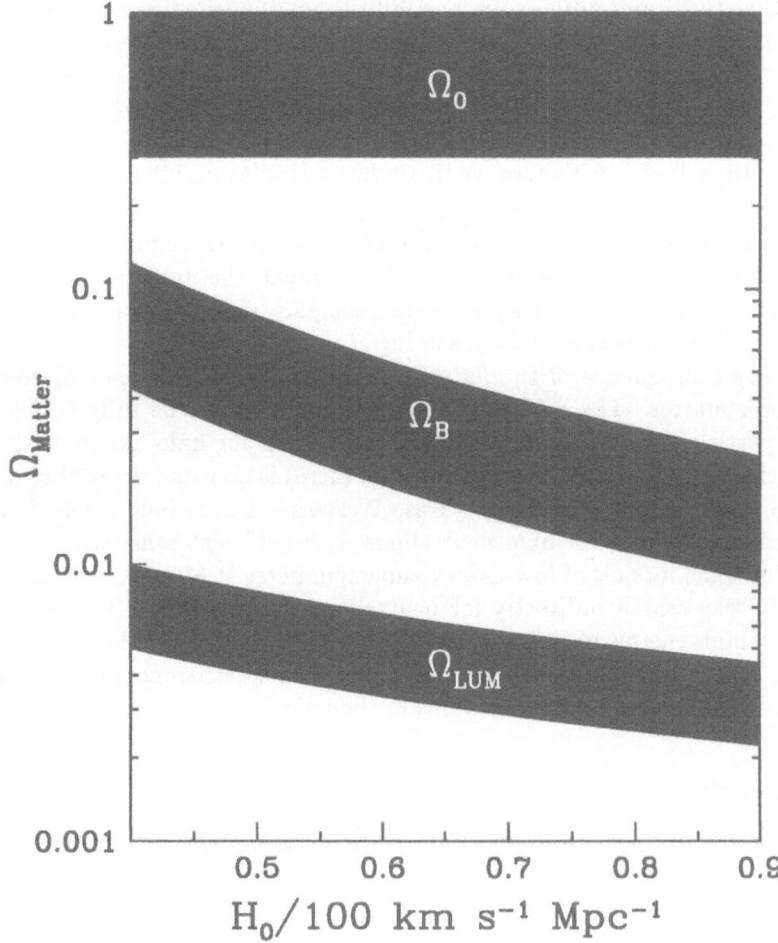

Figure 6: Summary of knowledge of Ω_M. The lowest band is luminous matter, in the form of bright stars and associated material; the middle band is the big-bang nucleosynthesis determination of the density of baryons; the upper region is the estimate based upon the peculiar velocities of galaxies and other dynamical methods.[37] The gaps between the bands illustrate the two dark matter problems: most of the ordinary matter is dark and most of the matter is nonbaryonic.

Ω_Λ; together, they can determine both Ω_M and Ω_Λ. (One should keep in mind that there are more exotic possibilities for the smooth component.[41,80]) Assuming a flat Universe and using their first seven SNeIa, Perlmutter et al[79] derive the bound $\Omega_\Lambda < 0.51$ (95%); or without the assumption of flatness, $-0.4 < \Omega_M - \Omega_\Lambda < 2.7$ (95%). Perlmutter's group, the Supernova Cosmology Project, now has a total of more than fifty SNeIa at redshifts $z \sim 0.4 - 0.8$, and the High-Redshift Supernova Team has a similar number; more definitive results should be coming soon.

The combination of the age and Hubble constant can, in principle, determine or at least constrain Ω_M and Ω_Λ. At the moment, the uncertainties preclude any firm conclusions. Taken at face value, the data seem to favor a cosmological constant (if the Universe is flat); see Fig. 7.

A key consequence of the flatness prediction is the existence of nonbaryonic dark matter. The cold dark matter scenario won't be fully tested until CDM particles are detected. A large-scale search for halo axions with sufficient sensitivity to detect them (if they are there) is now underway,[81] and soon, CDMS, the Stanford – Berkeley – Case Western – UCSB bolometric detector, will began searching for halo neutralinos with sufficient sensitivity to detect them for some models of low-energy supersymmetry.[82] SuperKamiokande and MACRO can search indirectly for neutralinos that annihilate in the sun and produce high-energy neutrinos, and Ting's AMS, which will be flown on the shuttle a year from now, will be able to search for positrons and antiprotons produced by neutralino annihilations in the halo.

9.2 Gravity waves

The search for inflation-produced gravitational waves was summarized in Section 7.

9.3 Density Perturbations/Cold Dark Matter

Figure 5 summarizes the status of testing the second robust prediction of inflation through CBR anisotropy: The measurements are generally consistent with the inflationary prediction. The power-law index is constrained to be 1.1 ± 0.2, which is well within the range predicted by inflation, and when COBE is used to normalize the spectrum, there are CDM models that are consistent with all the other observations.

There is much data that can be used to test the cold dark matter scenario. To focus the discussion, I will consider four "families" of models, distinguished by their invisible matter content: standard invisible matter content (CDM);

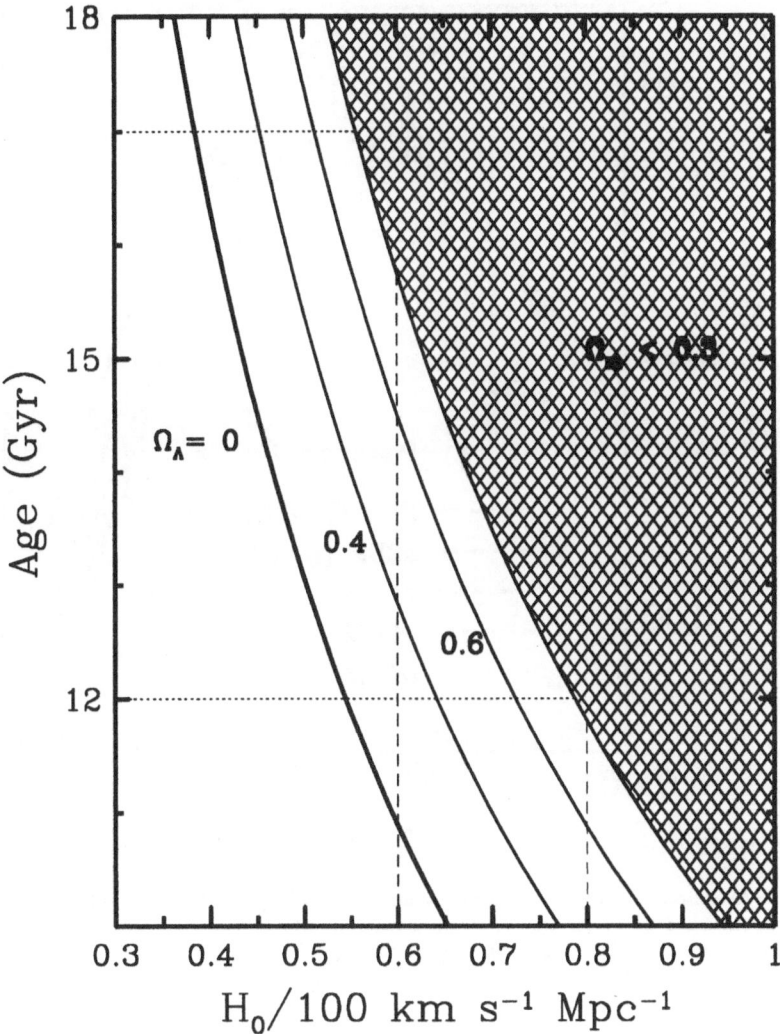

Figure 7: The relationship between age and H_0 for flat-universe models with $\Omega_M = 1 - \Omega_\Lambda$. The cross-hatched region is ruled out because $\Omega_M < 0.3$. The dotted lines indicate the favored range for H_0 and for the age of the Universe (based upon the ages of the oldest stars).

178

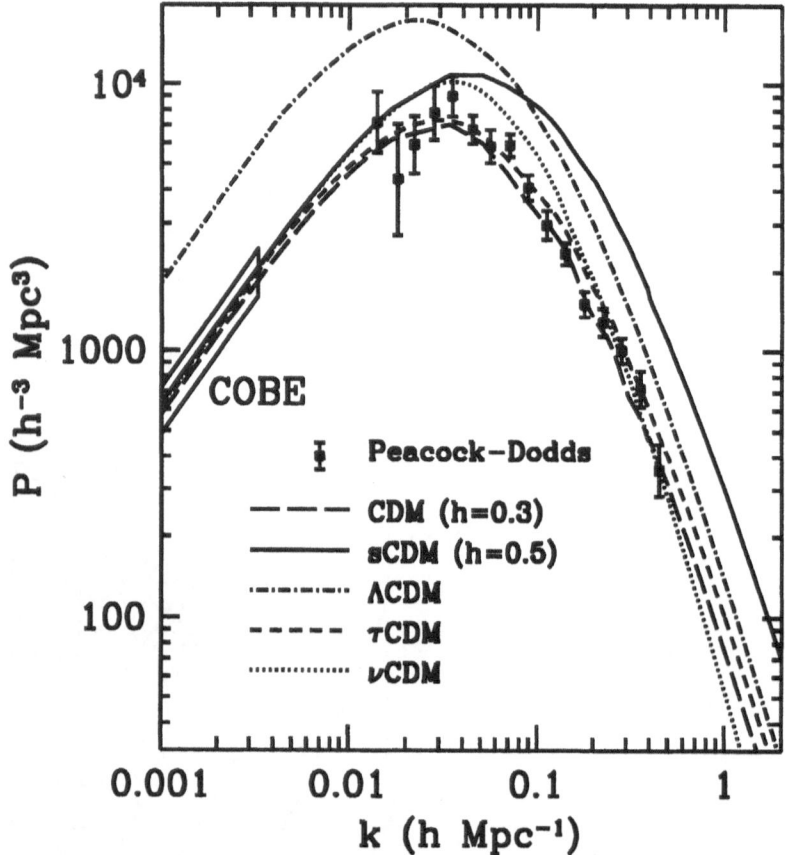

Figure 8: Measurements of the power spectrum, $P(k) = |\delta_k|^2$, and the predictions of different COBE-normalized CDM models. (COBE constrains the power spectrum at wavenumbers k around $2h \times 10^{-3}$ Mpc as indicated by rectangle.) The points are from several redshift surveys as compiled by Peacock and Dodds;[83] the models are: ΛCDM with $\Omega_\Lambda = 0.6$ and $h = 0.65$; standard CDM (sCDM), CDM with $h = 0.35$; τCDM (with the energy equivalent of 12 massless neutrino species) and νCDM with $\Omega_\nu = 0.2$ (unspecified parameters have their standard CDM values). The offset between a model and the points indicates the level of biasing. Note, ΛCDM does not pass through the COBE rectangle because a cosmological constant alters the relation between the power spectrum and CBR anisotropy.

extra radiation (τCDM); small hot dark matter component (νCDM); and cosmological constant (ΛCDM). There are of course other possibilities: extra radiation + cosmological constant, or a more exotic smooth component, which has been analyzed elsewhere.[80] Within each family, the five cosmological parameters (h, $\Omega_B h^2$, n, T/S and n_T) must be specified. Once specified, the power spectrum can be COBE-normalized and the expected level of inhomogeneity in the Universe today computed.

In assessing the viability of CDM models I will summarize work done in collaboration with Dodelson and Gates;[74] others have done similar work.[84] We began with three robust observational constraints on the power spectrum: the shape of the power spectrum; the power on cluster scales; and the early formation of objects. The first constraint, the shape of the power spectrum on scales from a few Mpc to a few 100 Mpc (see Fig. 8), comes from redshift surveys of the distribution of bright galaxies today.[83] In the absence of an understanding of the relationship between the distribution of light, which is what these surveys determine, and of mass, the bias factor is left as a free parameter. Models whose power spectra deviates from the measured power spectrum (as compiled by Peacock and Dodds[83]) by more than two sigma (value of χ^2 whose likelihood is less than 5%) were rejected. (Very roughly, this constrains the shape parameter: $\Gamma = \Omega_M h = 0.25 \pm 0.1$.)

The abundance of x-ray emitting clusters is sensitive to the level of inhomogeneity on scales around $8h^{-1}$ Mpc and thus provides a good means of inferring the value of σ_8. We used $0.5 \leq \sigma_8 \leq 0.8$ for models with $\Omega_M = 1$[85] and let this range scale with $\Omega^{-0.56}$ for models with a cosmological constant ($\Omega_\Lambda = 1 - \Omega_M$).[86]

The formation of objects at high redshift (early structure formation) probes the power spectrum on small scales. At redshifts of two to four, hydrogen clouds, detected by their absorption features in the spectra of high-redshift quasars ($z \sim 4 - 5$), contribute a fraction of the critical density, $\Omega_{\text{clouds}} \simeq (0.001 \pm 0.0002)h^{-1}$.[87] Insisting that the predicted level of inhomogeneity is sufficient to account for this leads to a lower limit to the power on small scales ($\lambda \sim 0.2h^{-1}$ Mpc).

Figure 9 summarizes the viable models. Models with standard invisible-matter content must lie in a region that runs diagonally from smaller Hubble constant and larger n to larger Hubble constant and smaller n. That is, higher values of the Hubble constant require more tilt. As is well appreciated,[88,89] standard CDM is outside of the allowed range – so much for DOS 1.0, onto Windows 95! Current measurements of CBR anisotropy on the degree scale, as well as the COBE four-year anisotropy data, preclude n less than about 0.7, which implies that the largest H_0 consistent with the simplest CDM models is

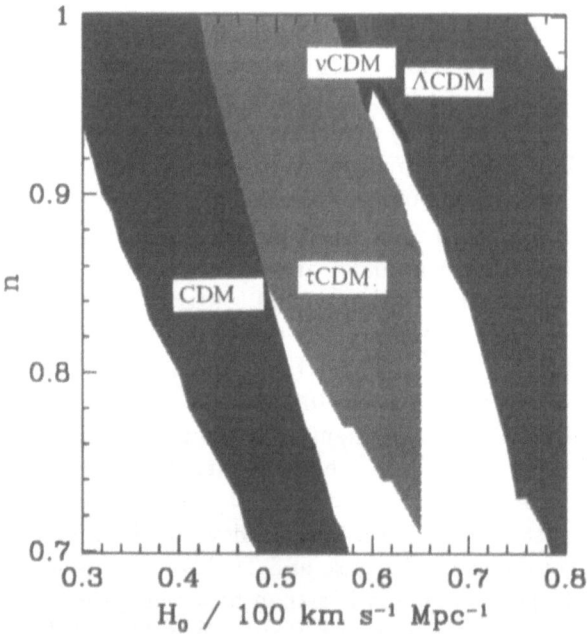

Figure 9: Acceptable values of the cosmological parameters n and h for CDM models with standard invisible-matter content (CDM), with 20% hot dark matter (νCDM), with additional relativistic particles (the energy equivalent of 12 massless neutrino species, denoted τCDM), and with a cosmological constant that accounts for 60% of the critical density (ΛCDM). A model is considered viable if it passes the three tests for *any* value of $\Omega_B h^2$ between 0.01 and 0.02 and any level of gravitational waves. The τCDM models have been truncated at a Hubble constant of $65 \, \mathrm{km \, s^{-1} \, Mpc^{-1}}$ because a larger value would result in a Universe that is younger than 10 Gyr.

slightly less than $60 \, \mathrm{km \, s^{-1} \, Mpc^{-1}}$.

If the invisible-matter content is nonstandard, higher values of H_0 can be accommodated. With tilt and hot dark matter, H_0 as large as $65 \, \mathrm{km \, s^{-1} \, Mpc^{-1}}$ is consistent with the constraints. The introduction of a cosmological constant permits H_0 as large as $80 \, \mathrm{km \, s^{-1} \, Mpc^{-1}}$, and additional radiation allows a Hubble constant as large as the age constraint permits (we assumed $t_0 \geq 10$ Gyr, which requires $H_0 \leq 65 \, \mathrm{km \, s^{-1} \, Mpc^{-1}}$).

In passing I mention that a similar analysis has been carried out for open-inflation models and they do not fare nearly as well.[90] The only viable models have $n > 1.1$ or $\Omega_0 > 0.5$. Cold dark matter seems to be weighing in on the side of a flat Universe.

A host of other observations test CDM. Some are more controversial and/or open to interpretation. They tend to distinguish the cosmological-constant family of models from the other three families. This is because models with standard invisible matter, extra radiation, or a hot dark matter component are all matter dominated today and have the same kinematic properties – age for a given Hubble constant and distance to an object of given redshift. The introduction of a cosmological constant leads to an older Universe and greater distance to an object at fixed redshift.

Together, the Hubble constant and age of the Universe, have great leverage. Determinations of the Hubble constant based upon a variety of techniques (Type Ia and II supernovae, IR Tully-Fisher and fundamental-plane methods) have converged on a value between $60 \, \mathrm{km \, s^{-1} \, Mpc^{-1}}$ and $80 \, \mathrm{km \, s^{-1} \, Mpc^{-1}}$.[91] This corresponds to an expansion age of less than 11 Gyr for a flat, matter-dominated model; for ΛCDM, the expansion age can be significantly higher, as large as 16 Gyr for $\Omega_\Lambda = 0.6$ (see Fig. 7). On the other hand, the ages of the oldest globular clusters indicate that the Universe is between 13 Gyr and 17 Gyr old; further, these age determinations, together with the those for the oldest white dwarfs and the long-lived radioactive elements, provide an ironclad case for a Universe that is at least 10 Gyr old.[92,93,94] At face value, the age/Hubble constant combination favor ΛCDM. But again, I want to stress that, within the uncertainties in both the age and Hubble constant, all of the models are viable.

Clusters are large enough that the baryon fraction should reflect its "universal value," $\Omega_B/\Omega_M = (0.007 - 0.024)h^{-2}/(1 - \Omega_\Lambda)$. Most of the (observed) baryons in clusters are in the hot, intracluster x-ray emitting gas. From x-ray measurements of the flux and temperature of the gas, baryon fractions in the range $(0.04 - 0.10)h^{-3/2}$ have been determined;[95,96,97] further, a recent detailed analysis and comparison to numerical models of clusters in CDM indicates an even smaller scatter, $(0.07 \pm 0.007)h^{-3/2}$.[98] ¿From the cluster baryon fraction and Ω_B, Ω_M can be inferred: $\Omega_M = (0.25 \pm 0.15)h^{-1/2}$, which for the lowest Hubble constant consistent with current determinations ($h = 0.6$) implies $\Omega_M = 0.32 \pm 0.2$. Unless one of the assumptions underlying this analysis is wrong, ΛCDM is strongly favored.

E. Turner emphasized the frequency of gravitational lensing of distant QSOs as an important cosmological test.[99] The underlying principle is simple: the comoving distance to fixed redshift depends upon the cosmology – it is largest for a cosmological constant, and in a matter-dominated Universe decreases with increasing Ω_M – and the probability of lensing increases with comoving distance (more lenses along the line of sight). For a flat Universe, Kochanek has obtained the upper limit $\Omega_\Lambda < 0.65$ (95%), and for a matter-dominated Universe $0.25 < \Omega_M < 2$ (95%).[100] Neither result is decisive.

ΛCDM is consistent with all the observations discussed here as well as others; see Fig. 10. For this reason, it is the strawman CDM model.[101] The parameters for this best fit model are: $\Omega_\Lambda \sim 0.5 - 0.65$ and $h \sim 0.6 - 0.7$. One should keep in mind that the introduction of a cosmological constant is a big step, one which twice before proved to be a misstep, and raises a fundamental question – the origin of the implied vacuum energy, about $10^{-8}\,\text{eV}^4$.[76]

ΛCDM's hold on the title of best-fit CDM model is by no means unshakeable: should the Hubble constant turn out to be less than $60\,\text{km}\,\text{s}^{-1}\,\text{Mpc}^{-1}$ and should a flaw be found in the cluster-baryon-fraction argument, the other models become very viable and are theoretically more attractive. Bartlett et al have pointed out that if the Hubble constant is around $30\,\text{km}\,\text{s}^{-1}\,\text{Mpc}^{-1}$, then CDM with $n \approx 1$ is consistent with all the measurements discussed above.[102] Lineweaver et al have analyzed the existing CBR anisotropy data and conclude that it favors a Hubble constant of around this value.[103] The rub is squaring this "determination" of H_0 with the multitude of other determinations that indicate a value almost twice as large. Appealing to a difference between the local value of the expansion rate and the global value is of little help – in the context of CDM, the difference, which arises due to cosmic and sampling variance, is expected to be only 10% or so.[104]

In finishing this status report I would like to emphasize three things. First, inflation is currently consistent with all the observational data, which is no mean feat. Second, cold dark matter is consistent with a large body of high-quality cosmological data, ranging from measurements of CBR anisotropy to our growing understanding of the evolution of galaxies and clusters of galaxies. This too is no mean feat; at the moment, the only other potentially viable paradigm for structure formation is topological defects + nonbaryonic dark matter. These models, when COBE normalized, appear to be in great jeopardy as they predict very little power on small scales (high bias). Finally, the quantity of high-quality data that bear on inflation and cold dark matter is growing rapidly, leading me to believe that inflation/cold dark matter will be decisively tested soon.

10   The Future: Precision Testing and More

Inflation is a bold attempt to build upon the success of the standard cosmology and extend our understanding of the Universe to times as early as $10^{-32}\,\text{sec}$ after the bang. Its three robust predictions – flat Universe, nearly scale-invariant spectrum of density perturbations, and nearly scale-invariant spectrum of gravitationally waves – are the keys to its testing. In addition, much can be learned about the specific, underlying model of inflation if other

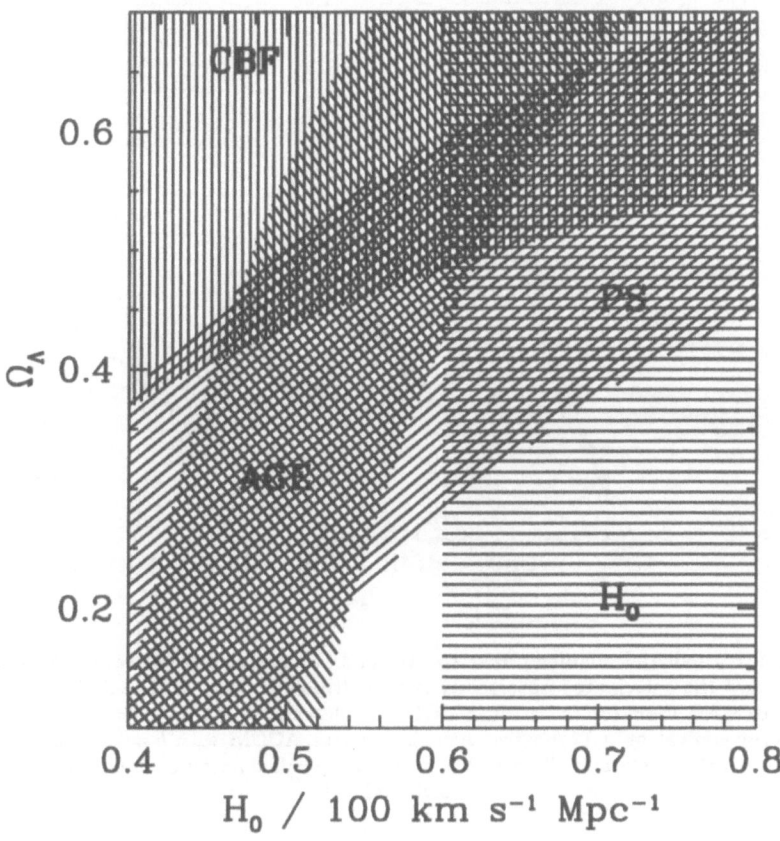

Figure 10: Summary of constraints projected onto the $H_0 - \Omega_M$ plane: (CBF) comes from combining the BBN limit to the baryon density with x-ray observations of clusters; (PS) arises from the power spectrum; (AGE) is based on age determinations of the Universe; (H_0) indicates the range currently favored for the Hubble constant. The constraint $\Omega_\Lambda < 0.66$ has been implicitly taken into account since the Ω_Λ axis extends only to 0.7. The darkest region indicates the parameters allowed by all constraints.

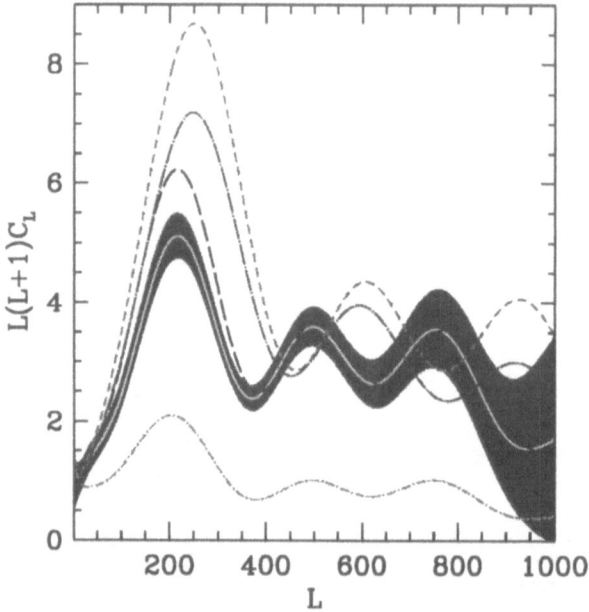

Figure 11: Predicted angular power spectra of CBR anisotropy for several viable CDM models and the anticipated uncertainty (per multipole) from a CBR satellite experiment similar to MAP. From top to bottom the CDM models are: CDM with $h = 0.35$, τCDM with the energy equivalent of 12 massless neutrino species, ΛCDM with $h = 0.65$ and $\Omega_\Lambda = 0.6$, νCDM with $\Omega_\nu = 0.2$, and CDM with $n = 0.7$ (unspecified parameters have their standard CDM values).

measurements are made (e.g., the small anticipated deviation from scale invariance and the level of gravitational waves).

Cold dark matter, which is an important means of testing inflation, is a ten-parameter theory, h, $\Omega_B h^2$, S, n, $dn/d\ln k$, T/S, n_T, Ω_ν, g_*, and Ω_Λ. While this is a daunting number of parameters, especially for a cosmological theory, there is good reason to believe that within ten years the data will overdetermine these parameters. Crucial to achieving this goal are the high-precision, high-resolution measurements of CBR anisotropy that will be made over the next decade by earth-based, balloon-borne and satellite-borne instruments (see Fig. 11). As a reminder of the power of high-quality data, the standard model of particle physics has nineteen parameters; precision measurements at Fermilab, SLAC, CERN and other accelerator laboratories, as well as nonaccelerator experiments, have both sharply tested the theory as well as accurately determining the parameters.

Within five years we should be well on our way to precisely testing inflation and cold dark matter. In the next few years ground-based and balloon-borne anisotropy experiments (e.g., VSA, VCA, Boomerang, TOPHAT, QMAX, and others) should be able to determine the approximate position of the Doppler peak and thereby Ω_0 to an accuracy of around 10%. Because flatness is a fundamental prediction of inflation, perhaps the most fundamental, this is a landmark test. On the same timescale, the Supernova Cosmology Project and the High-Redshift Supernova Team will use the SNeIa magnitude – redshift diagram to determine the deceleration of the Universe. While the Doppler peak determines $\Omega_M + \Omega_\Lambda$, the SNeIa measurement determines an almost orthogonal quantity, $\Omega_M - \Omega_\Lambda$; together, they can determine Ω_M and Ω_Λ.

In the same time frame measurements of the Hubble constant will play an important role. Since the Universe is at least 10 Gyr old, a definitive determination that the Hubble constant is $65 \, \mathrm{km \, s^{-1} \, Mpc^{-1}}$ or greater would rule out all models but ΛCDM; on the other hand, a determination that the Hubble constant is below $55 \, \mathrm{km \, s^{-1} \, Mpc^{-1}}$ would undermine much of the motivation for ΛCDM. The Hubble Space Telescope calibration of secondary distance indicators such as SNeIa with Cepheids in nearby galaxies and the maturation of physics-based methods such as gravitational time delay and Zel'dovich – Sunyaev are helping to pin down H_0 more accurately.[105]

There are other important tests that will be made on a longer time scale. The level of inhomogeneity in the Universe today is determined mainly from redshift surveys, the largest of which contains of order 30, 000 galaxies. Two larger surveys are underway. The Sloan Digital Sky Survey will cover 25% of the sky and obtain positions for two hundred million galaxies and redshifts for a million galaxies.[106] The Anglo – Australian Two-degree Field is targeting hundreds of two-degree patches on the sky and will obtain 250,000 redshifts.[107] These two projects will determine the power spectrum much more precisely and on scales large enough ($500h^{-1}$ Mpc) to connect with measurements from CBR anisotropy on angular scales of up to five degrees, allowing bias to be probed.

The most fundamental element of cold dark matter – the existence of the CDM particles – will be tested over the next decade. Experiments with sufficient sensitivity to detect the CDM particles that hold our own galaxy together if they are in the form of axions of mass $10^{-6} \, \mathrm{eV} - 10^{-4} \, \mathrm{eV}$[81] or neutralinos of mass tens of GeV[82] are now underway. Evidence for the existence of the neutralino could also come from particle accelerators searching for other supersymmetric particles.[108] In addition, a variety of experiments sensitive to neutrino masses are operating or are planned: accelerator-based neutrino oscillation experiments at Fermilab, CERN, and Los Alamos; solar-neutrino detectors in Japan, Canada, Germany, Russia and Italy; and experiments at

$e^+ e^-$ colliders (LEP at CERN and CESR at Cornell) to the study the tau neutrino.

The most telling test of inflation and cold dark matter will come with the two new space missions that have recently been approved – MAP to be launched in 2000 by NASA and Planck to launched by ESA in 2005. Each will map CBR anisotropy over the entire sky with more than thirty times the angular resolution of COBE (resolution of 0.2° for MAP and 0.1° for Planck). MAP should determine the angular power spectrum out to multipole number 1000, and Planck out to multipole number 2000, each to a precision close to that dictated by sampling variance alone. Theoretical studies[109] indicate that the results of Planck should be able to determine n to a precision of less than one percent, Ω_Λ to a few percent, Ω_0 to less than one percent, Ω_ν to enough precision to test νCDM,[110] the baryon density to less than ten percent, and even the Hubble constant to one percent.

Inflation and cold dark matter are a bold attempt to extend our knowledge of the Universe to within 10^{-32} sec of the bang. The number of observations that are testing the cold dark matter theory is growing fast, and prospects for not only testing the theory, but also discriminating among the different CDM models and models of inflation are excellent. If cold dark matter is shown to be correct, an important aspect of the standard cosmology – the origin and evolution of structure – will have been resolved and a window to the early moments of the Universe and physics at very high energies will have been opened.

While the window has not been opened yet, I would like to end with one example of what one could hope to learn. As discussed earlier, S, $n-1$, T/S and n_T are related to the inflationary potential and its first two derivatives. If one can measure the power-law index of the scalar perturbations and the amplitudes of the scalar and tensor perturbations, one can recover the value of the potential and its first two derivatives around the point on the potential where inflation took place:[62]

$$V_* = 1.65 T \, m_{\mathrm{Pl}}{}^4, \tag{20}$$

$$V_*' = \pm \sqrt{\frac{8\pi}{7} \frac{T}{S}} \, V_*/m_{\mathrm{Pl}}, \tag{21}$$

$$V_*'' = 4\pi \left[(n-1) + \frac{3}{7} \frac{T}{S} \right] V_*/m_{\mathrm{Pl}}{}^2, \tag{22}$$

where the sign of V' is indeterminate (under $\phi \leftrightarrow -\phi$ the sign changes). If the tensor spectral index can also be measured the relation, $T/S = -7 n_T$, can be used to test the consistency of inflation. Reconstruction of the inflationary scalar potential would shed light both on inflation as well as physics at energies of the order of 10^{14} GeV.

Acknowledgements

Much of the material in these lectures derives from collaborative work with Scott Dodelson, Evalyn Gates, Lloyd Knox, Andrew Liddle, and Martin White. I thank my collaborators for teaching me so much. This work was supported by the DoE (at Chicago and Fermilab) and by the NASA (at Fermilab by grant NAG 5-2788).

References

1. See e.g., P.J.E. Peebles, *Physical Cosmology* (Princeton University Press, Princeton, 1993).
2. A.H. Guth, *Phys. Rev. D* **23**, 347 (1981).
3. A. H. Guth and S.-Y. Pi, *Phys. Rev. Lett.* **49**, 1110 (1982); S. W. Hawking, *Phys. Lett. B* **115**, 295 (1982); A. A. Starobinskii, *ibid* **117**, 175 (1982); J. M. Bardeen, P. J. Steinhardt, and M. S. Turner, *Phys. Rev. D* **28**, 697 (1983).
4. M.S. Turner and L.M. Widrow, *Phys. Rev. Lett.* **57**, 2237 (1986); L. Jensen and J. Stein-Schabes, *Phys. Rev. D* **35**, 1146 (1987); A.A. Starobinskii, *JETP Lett.* **37**, 66 (1983).
5. See e.g., A.D. Linde, *Inflation and Quantum Cosmology* (Academic Press, San Diego, CA, 1990); or A. Vilenkin, *Phys. Rev. D* **52**, 3365 (1995).
6. R. Penrose, in *General Relativity: An Einstein Centenary Survey*, edited by S.W. Hawking and W. Israel (Cambridge University Press, Cambridge), p. 581.
7. Y. Hu, M.S. Turner, and E.J. Weinberg, *Phys. Rev. D* **49**, 3830 (1994); A.R. Liddle, *ibid* **51**, 5347 (1995).
8. See e.g., E.W. Kolb and M.S. Turner, *The Early Universe* (Addison-Wesley, Redwood City, 1990), Ch. 8.
9. A.D. Linde, *Phys. Lett. B* **129**, 177 (1983).
10. Q. Shafi and A. Vilenkin, *Phys. Rev. Lett.* **52**, 691 (1984); S.-Y. Pi, *ibid* **52**, 1725 (1984).
11. R. Holman, P. Ramond, and G.G. Ross, *Phys. Lett. B* **137**, 343 (1984).
12. K. Olive, *Phys. Repts.* **190**, 309 (1990).
13. H. Murayama et al., *Phys. Rev. D(RC)* **50**, R2356 (1994).
14. M. Cvetic, T. Hubsch, J. Pati, and H. Stremnitzer, *Phys. Rev. D* **40**, 1311 (1990).

15. E.J. Copeland et al., *Phys. Rev. D* **49**, 6410 (1994).

16. See e.g., M. Gasperini and G. Veneziano, *Phys. Rev. D* **50**, 2519 (1994); R. Brustein and G. Veneziano, *Phys. Lett. B* **329**, 429 (1994); T. Banks et al., *Phys. Rev. D* **52**, 3548 (1995).

17. K. Freese, J.A. Frieman, and A. Olinto, *Phys. Rev. Lett.* **65**, 3233 (1990); L. Randall, M. Soljacic, and A. Guth, *Nucl. Phys. B* **472**, 377 (1996).

18. L. Knox and M.S. Turner, *Phys. Rev. Lett.* **70**, 371 (1993).

19. J. Silk and M.S. Turner, *Phys. Rev. D* **35**, 419 (1986); L.A. Kofman, A.D. Linde, and J. Einsato, *Nature* **326**, 48 (1987).

20. D. La and P.J. Steinhardt, *Phys. Rev. Lett.* **62**, 376 (1991).

21. E.W. Kolb, *Physica Scripta* **T36**, 199 (1991).

22. A. Albrecht et al, *Phys. Rev. Lett.* **48**, 1437 (1982); L. Abbott and M. Wise, *Phys. Lett. B* **117**, 29 (1992); A.D. Linde and A.D. Dolgov, *ibid* **116**, 329 (1982).

23. See e.g., L. Kofman et al., *Phys. Rev. Lett.* **73**, 3195 (1994); **76**, 1011 (1996); E.W. Kolb et al., *ibid* **77**, 4290 (1996); S. Khebnikov and I. Tkachev, *ibid* **77**, 219 (1996); D. Boyanovsky et al, hep-ph/9701304.

24. V.A. Rubakov, M. Sazhin, and A. Veryaskin, *Phys. Lett. B* **115**, 189 (1982); R. Fabbri and M. Pollock, *ibid* **125**, 445 (1983); A.A. Starobinskii *Sov. Astron. Lett.* **9**, 302 (1983); L. Abbott and M. Wise, *Nucl. Phys. B* **244**, 541 (1984).

25. M.S. Turner and F. Wilczek, *Phys. Rev. Lett.* **65**, 3080 (1990); A. Kosowsky, M.S. Turner, and R. Watkins, *ibid* **69**, 2026 (1992).

26. M.S. Turner and L.M. Widrow, *Phys. Rev. D* **37**, 2743 (1988); B. Ratra, *Astrophys. J.* **391**, L1 (1992).

27. See e.g., A. Vilenkin, astro-ph/9610125.

28. See e.g., A.D. Linde, *Phys. Lett. B* **158**, 375 (1985); D. Seckel and M.S. Turner, *Phys. Rev. D* **32**, 3178 (1985).

29. M.S. Turner, A. Cohen, and D. Kaplan, *Phys. Lett. B* **216**, 20 (1989).

30. P.J.E. Peebles, astro-ph/9704xxx (*Astrophys. J.*, in press).

31. M.S. Turner, *Phys. Rev. D* **44**, 3737 (1991).

32. M. Bucher A.S. Goldhaber, and N. Turok, *Phys. Rev. D* **52**, 3314 (1995); A.D. Linde and A. Mezhlumian, *ibid* **52**, 6789 (1995); J.R. Gott, *Nature* **295**, 304 (1992); A. Green and A. Liddle, *Phys. Rev. D* **55**, 609 (1997)

33. J.R. Gott, *Nature* **295**, 304 (1992); A.H. Guth and E.J. Weinberg, *Nucl. Phys. B* **212**, 321 (1983).

34. A. Vilenkin and S. Winitzki, *Phys. Rev. D* **55**, 548 (1997).

35. C. Copi, D.N. Schramm, and M.S. Turner, *Science* **267**, 192 (1995).

36. D. Tytler, X.-M. Fan and S. Burles, *Nature* **381**, 207 (1996); D. Tytler, S. Burles and D. Kirkman, astro-ph/9612121.

37. Y. Sigad et al, in preparation (1997); J. Willick et al, astro-ph/96 A. Dekel, D. Burstein, and S. White, astro-ph/9611108. M. Strauss and J. Willick, *Phys. Repts.* **261**, 271 (1995); A. Dekel, *Ann. Rev. Astron. Astrophys.* **32**, 319 (1994); N. Bahcall, L.M. Lubin and V. Dorman, *Astrophys. J.* **447**, L81 (1995); A. Dekel and M.J. Rees, *Astrophys. J.* **422**, L1 (1994); R.G. Carlberg, H.K.C. Yee, and E. Ellingson, astro-ph/9512087.

38. See e.g., M.S. Turner, *Phys. Rept.* **197**, 67 (1990); G.G. Raffelt, *ibid* **198**, 1 (1990).

39. See e.g., G. Jungman, M. Kamionkowski, and K. Griest, *Phys. Rept.* **267**, 195 (1996).

40. See e.g., G.G. Ross, *Grand Unified Theories* (Addison-Wesley, Redwood City, 1985), or B. Kayser, F. Gibrat-Debu, and F. Perrier, *The Physics of Massive Neutrinos* (World Scientific, Singapore, 1989).

41. A. Vilenkin, *Phys. Rev. Lett.* **53**, 1016 (1984); M.S. Turner, G. Steigman, and L. Krauss, *Phys. Rev. Lett.* **52**, 2090 (1984); R.L. Davis, *Phys. Rev. D* **35**, 3705 (1987); M. Kamionkowski and N. Toumbas, *Phys. Rev. Lett.* **77**, 587 (1996); M. Ozer and M.O. Taha, *Nucl. Phys.* **B287** 776 (1987); K. Freese et al., *Nucl. Phys.* **B287** 797 (1987); B. Ratra and P.J.E. Peebles, *Phys. Rev. D* **37**, 3406 (1988); L.F. Bloomfield-Torres and I. Waga, astro-ph/9504101; K. Coble, S. Dodelson, and J. Frieman, *Phys. Rev. D* **55**, 1851 (1996).

42. G.F. Smoot et al., *Astrophys. J.* **396**, L1 (1992).

43. J.M. Bardeen, J.R. Bond, N. Kaiser, and A.S. Szalay, *Astrophys. J.*, **304**, 15 (1986).

44. A.R. Liddle and M.S. Turner, *Phys. Rev. D* **54**, 2980 (1996).

45. M.S. Turner and M. White, *Phys. Rev. D* **53**, 6822 (1996).

46. P.J. Steinhardt and M.S. Turner, *Phys. Rev. D* **29**, 2162 (1984).

47. M.S. Turner, *Phys. Rev. D* **48**, 3502 (1993).

48. D.H. Lyth and A.R. Liddle, *Phys. Lett. B* **291**, 391 (1992).

49. A.D. Linde, *Phys. Lett. B* **108**, 389 (1982); A. Albrecht and P.J. Steinhardt, *Phys. Rev. Lett.* **48**, 1220 (1982).

50. D.H. Lyth and E. Stewart, *Phys. Rev. D* **54**, 7186 (1996); A. Linde, *ibid* **49**, 748 (1994).

51. A. Kosowsky and M.S. Turner, *Phys. Rev. D* **52**, R1739 (1995).

52. J. Bartlett et al., *Science* **267**, 980 (1995); P.T.P. Viana et al., *Mon. Not. R. astron. Soc.* **283**, 107 (1996).

53. See e.g., W. Hu and N. Sugiyama, *Phys. Rev. D* **51**, 2599 (1991). For a very fast and accurate numerical routine, see U. Seljak and M. Zaldarriaga, astro-ph/9603033.

54. R.K. Sachs and A.M. Wolfe, *Astrophys. J.* **147**, 73 (1967).

55. C.L. Bennett et al, *Astrophys. J.* **464**, L1 (1996).

56. E.F. Bunn et al., *Phys. Rev. D* **54**, 5917 (1997); also see e.g., K. Gorski et al, astro-ph/9608054.

57. R.H. Dicke and P.J.E. Peebles, in *General Relativity: An Einstein Centenary Survey*, edited by S.W. Hawking and W. Israel (Cambridge University Press, Cambridge), p. 504.

58. E.R. Harrison, *Phys. Rev. D* **1**, 2726 (1970); Ya.B. Zel'dovich, *Mon. Not. R. astr. Soc.* **160**, 1p (1972).

59. See e.g., M.S. Turner, J.E. Lidsey and M. White, *Phys. Rev. D* **48**, 4613 (1993).

60. M.S. Turner and Y. Wang, *Phys. Rev. D* **53**, 5727 (1996); B. Allen and S. Koranda, *ibid* **52**, 1902 (1995); S. Dodelson et al, *Phys. Rev. Lett.* **72**, 3444 (1994); R. Crittenden et al, *ibid* **71**, 324 (1993).

61. R. Davis et al., *Phys. Rev. Lett.* **69**, 1856 (1992); F. Lucchin, S. Mattarese, and S. Mollerach, *Astrophys. J.* **401**, L49 (1992); D. Salopek, *Phys. Rev. Lett.* **69**, 3602 (1992); A. Liddle and D. Lyth, *Phys. Lett. B* **291**, 391 (1992); J.E. Lidsey and P. Coles, *Mon. Not. R. astron. Soc.* **258**, 57p (1992); T. Souradeep and V. Sahni, *Mod. Phys. Lett. A* **7**, 3541 (1992).

62. See e.g., E.J. Copeland, E.W. Kolb, A.R. Liddle, and J.E. Lidsey, *Phys. Rev. Lett.* **71**, 219 (1993); *Phys. Rev. D* **48**, 2529 (1993); M.S. Turner, *ibid*, 3502 (1993); *ibid* **48**, 5539 (1993); S. Dodelson, W.H. Kinney and E.W. Kolb, astro-ph/9702166.

63. L. Knox and M.S. Turner, *Phys. Rev. Lett.* **73**, 3347 (1994).

64. U. Seljak, astro-ph/9608131; M. Kamionkowski, A. Kosowsky, and A. Stebbins, astro-ph/9609132.

65. M.S. Turner, *Phys. Rev. D* **55**, R435 (1997).

66. See e.g., A. Abramovici et al., *Science* **256**, 325 (1992) and in *Particle and Nuclear Astrophysics and Cosmology in the Next Millennium*, eds. E.W. Kolb and R.D. Peccei (World Scientific, Singapore, 1995), p. 398; N. Christensen, *Phys. Rev. D* **46**, 5250 (1992); E. Flanagan, *ibid* **48**, 2389 (1993).

67. B. Allen, qr-gc/9604033.

68. Pre-phase A Design Study for LISA.

69. K.S. Thorne, in *300 Years of Gravitation*, eds. S.W. Hawking and W. Israel (Cambridge Univ. Press, Cambridge, 1987), p.330.

70. S.D.M. White, C. Frenk and M. Davis, *Astrophys. J.* **274**, L1 (1983).
71. S. Colombi, S. Dodelson and L. Widrow, *Astrophys. J.* **458**, 1 (1996).
72. G. Blumenthal et al., *Nature* **311**, 517 (1984).
73. See e.g., J.P. Ostriker and R. Cen, astro-ph/9601021; Y. Zhang et al, astro-ph/9711224; T. Abel et al, astro-ph/9608040;
74. S. Dodelson, E. Gates and M.S. Turner, *Science* **274**, 69 (1996).
75. For example, S. Parke, *Phys. Rev. Lett.* **74**, 839 (1995); C. Athanassopoulos et al, *ibid* **75**, 2650 (1995); J.E. Hill, *ibid*, 2654 (1995); K.S. Hirata et al, *Phys. Lett. B* **280**, 146 (1992); Y. Fukuda et al, *ibid* **335**, 237 (1994); R. Becker-Szendy et al, *Phys. Rev. D* **46**, 3720 (1992).
76. S. Weinberg, *Rev. Mod. Phys.* **61**, 1 (1989).
77. K. Fisher et al., *Mon. Not. R. astron. Soc.* **266**, 50 (1994).
78. S. Hancock and G. Rocha, astro-ph/9612016.
79. S.J. Perlmutter et al, astro-ph/9608192.
80. M.S. Turner and M. White, astro-ph/9701138.
81. C. Hagmann et al, *Nucl. Phys. B (Proc. Suppl.)* **51**, 209 (1996).
82. T. Shutt et al, in *Proceedings of the XVIIIth Texas Symposium on Relativistic Astrophysics (Chicago, IL, 1996)*, edited by A. Olinto (World Scientific, Singapore, 1997).
83. J. Peacock and S. Dodds, *Mon. Not. Roy. astron. Soc.* **267**, 1020 (1994).
84. A.R. Liddle et al., *Mon. Not. R. astron. Soc.* **282**, 281 (1996); *ibid* **281**, 531 (1995); J.S. Bagla, T. Padmanabhan and J.V. Narlikar, astro-ph/9511102; S. Cole et al, astro-ph/9702082.
85. S.D.M. White, G. Efstathiou, and C.S. Frenk, *Mon. Not. Roy. astron. Soc.* **262**, 1023 (1993).
86. R. Carlberg et al., *J. R. Astron. Soc. Canada* **88**, 39 (1994); P.T.P. Viana and A. Liddle, *Mon. Not. R. astron. Soc.* **281**, 323 (1996); J.R. Bond and S. Myers, *Astrophys. J. Supp.* **103**, 63. (1996); V.R. Eke, S. Cole and C.S. Frenk, astro-ph/9601088; U.-L.Pen, astro-ph/9610147.
87. K. Lanzetta, A.M. Wolfe, and D.A. Turnshek, *Astrophys. J.* **440**, 435 (1995); L.J. Storrie-Lombardi, R.G. McMahon, M.J. Irwin, and C. Hazard, astro-ph/9503089 (1996).
88. J.P. Ostriker, *Ann. Rev. Astron. Astrophys.* **31**, 689 (1993).
89. A. Liddle and D. Lyth, *Phys. Repts.* **231**, 1 (1993).
90. M. White and J. Silk, *Phys. Rev. Lett.* **77** 4704 (1996).
91. See e.g., W. Freedman, astro-ph/9612024; A. Riess, R.P. Krishner, and W. Press, *Astrophys. J.* **438**, L17 (1995); R. Giovanelli et al, astro-ph/9612072.
92. M. Bolte and C.J. Hogan, *Nature* **376**, 399 (1995).

93. B. Chaboyer et al, *Science* **271**, 957 (1996).

94. J. Cowan, F. Thieleman, and J. Truran, *Ann. Rev. Astron. Astrophys.* **29**, 447 (1991).

95. S.D.M. White et al, *Nature* **366**, 429 (1993).

96. U.G. Briel et al, *Astron. Astrophys.* **259**, L31 (1992).

97. D.A. White and A.C. Fabian, *Mon. Not. R. astron. Soc.* **273**, 72 (1995).

98. A. Evrard, astro-ph/9701148.

99. E. Turner, *Astrophys. J.* **365**, L43 (1990).

100. C.S. Kochanek, *Astrophys. J.* **466**, 638 (1996).

101. L. Krauss and M.S. Turner, *Gen. Rel. Grav.* **27**, 1137 (1995); J.P. Ostriker and P.J. Steinhardt, *Nature* **377**, 600 (1995); J.S. Bagla, T. Padmanabhan and J.V. Narlikar, astro-ph/9511102; A.R. Liddle et al., *Mon. Not. R. astron. Soc.* **282**, 281 (1996).

102. J. Bartlett et al., *Science* **267**, 980 (1995);

103. C.H. Lineweaver, D. Barbosa, A. Blanchard, and J. Bartlett, astro-ph/9612146.

104. See e.g., E.L. Turner, R. Cen, and J.P. Ostriker, *Astron. J.* **103**, 1427 (1992); X. Shi, L.M. Widrow, and L.J. Dursi, *Mon. Not. R. astron. Soc.* **281**, 565 (1996); X. Shi and M.S. Turner, astro-ph/9705xxx.

105. T. Kundic et al, astro-ph/9610162; P.L. Schechter et al, astro-ph/9611051; W.L. Holzapfel et al, astro-ph/9702224; S.T. Meyers et al, astro-ph/9703123.

106. See e.g., http://www-sdss.fnal.gov:8000/.

107. See e.g., http://www.ast.cam.ac.uk/AAO/2df.

108. See e.g., J.E. Ellis, *Physics World*, July 1994, p. 31.

109. See e.g., L. Knox, *Phys. Rev.* **D 52**, 4307 (1995); G. Jungman, M. Kamionkowski, A. Kosowsky, and D. Spergel, *Phys. Rev.* **D 54**, 1332 (1996); J.R. Bond, G. Efstathiou, and M. Tegmark, astro-ph/0702100; M. Zaldarriaga, D.Spergel and U. Seljak, astro-ph/9702157.

110. S. Dodelson, E. Gates, and A.S. Stebbins, *Astrophys. J.* **467**, 10 (1996).

LARGE–SCALE STRUCTURE, PECULIAR VELOCITIES AND THE DENSITY FIELD IN THE LOCAL UNIVERSE

RICCARDO GIOVANELLI

Department of Astronomy
Cornell University
Ithaca, NY 14853, USA

1. Lecture 1: Large–Scale Structure and the Relationship between Density Fluctuations and the Peculiar Velocity Field

Peculiar velocities of galaxies, i.e. devitions from smooth Hubble flow, are thought to arise from inhomogeneities in the matter distribution. The amplitude of those velocities is regulated by the spatial scale and the amplitude of the matter density irregularities as well as by the value of the mean matter density. The direct measurement of peculiar velocities thus provides a means to infer the characteristics of the mass distribution, luminous and dark, in the universe and cosmological parameters such as Ω_0. Conversely, the distribution of luminous matter (i.e. galaxies) and fair assumptions on the nature of biasing can be used to obtain an estimate of the peculiar velocity field. In the first lecture of this series of three, the relationships between the matter density and peculiar velocity fields will be surveyed. In doing so we heavily borrow from Peebles (1993) and Strauss and Willick (1995). We will also explore the estimates of the peculiar velocity field resulting from photometric and redshift catalogs of galaxies and clusters of galaxies. In the second lecture, we will review the methods most commonly used for the direct measurement of peculiar velocities, with emphasis on the Tully–Fisher (1977) relation (hereafter TF), and in lecture 3 we will concentrate on the results of those measurements and their implications for the determination of the density field and parameters of cosmological interest.

D. N. Schramm and P. Galeotti (eds.), Generation of Cosmological Large-Scale Structure, 193–222.

1.1. AUTOCORRELATION FUNCTION OF THE DENSITY FIELD

Given a mass density field $\rho(\mathbf{r})$ of average value $< \rho >$, let

$$\delta(\mathbf{r}) = \frac{\rho(\mathbf{r}) - < \rho >}{< \rho >} \tag{1}$$

We define its autocorrelation function as

$$\xi(r) = < \delta(\mathbf{x})\delta(\mathbf{x} + \mathbf{r}) > \tag{2}$$

where the average is over $\mathbf{x}$; if the universe is homogeneous, $\xi(\mathbf{r})$ is independent on the direction of $\mathbf{r}$. A random observer measuring the mass within a spherical shell of radius r and thickness dr will obtain

$$dM(r) = 4\pi r^2 dr < \rho > [1 + \xi(r)] \tag{3}$$

The galaxy distribution is a discrete process. The function $\xi(r)$ can thus be understood as the joint probability that two galaxies be found, respectively within volumes dV_1 and dV_2, separated by r:

$$dP = < n >^2 dV_1 dV_2 [1 + \xi(r)] \tag{4}$$

or operationally defines as the excess number, over random, of galaxy pairs of separation r. Given a sample of galaxies of mean density $< n >$, distributed within a volume V,

$$\xi(r) = \frac{\Delta N(r)}{4\pi r^2 \Delta r < n >^2 V} - 1 \tag{5}$$

where $\Delta N(r)$ is the number of pairs found in V with separation between r and $r + \Delta r$, while $4\pi r^2 \Delta r < n >^2 V$ is the number expected for a random distribution.

In real life, V is sampled in a biased manner. That bias is described by means of a selection function $\phi(r, \ldots)$ which, to the first order, may be considered purely a function of distance. In a strictly flux–limited catalog,

$$\phi(r) \propto \int_{L_{min}(r)}^{\infty} \Phi(L) dL \tag{6}$$

where $\Phi(L)$ is the galaxy luminosity function (LF) and $L_{min}(r)$ is the smallest L observable at distance r, at the given flux limit. More generally, however, catalogs are not likely to be strictly flux–limited, and ϕ can be a complex and non–analytical function of distance, surface brightness, morphological type, galactic extinction, velocity width and other quirks of the survey. To make things even murkier, the LF $\Phi(L)$ may **not** be universal: if $\Phi_t(L)$ is the LF of galaxies of morphological type t, $\Phi(L) = \Sigma_t \Phi_t(L)$; since in general $\Phi_t \neq \Phi_{t'}$ and the space distribution of galaxies of type t and type t' are different, there is no reason to expect that Φ will be space invariant. Fortunately, departures of Φ from space invariance appear to be relatively mild, and perhaps limited to cluster cores. Intrinsic biases in the catalogs from which galaxy sample are extracted may play a more important role, as underscored by Disney, Davis, Phillipps and collaborators at Cardiff (see Davies et al. 1995 and refs. therein)

The most frequently used estimator of $\xi(r)$ relies on the construction of a simulated catalog of galaxies, randomly spread over the volume V and sampled according to the survey's selection function. Then

$$\xi(r) = \frac{(\Delta N_{rr}(r)/ < n_r >)}{(\Delta N_{ss}(r)/ < n_s >)} - 1 \tag{7}$$

where $\Delta N_{rr}(r)$ is the number of galaxy pairs of separation between r and $r + \Delta r)$ for the real data, $\Delta N_{rr}(r)$ is the number of galaxy pairs for the simulated data, $< n_r(r) >$ and $< n_s(r) >$ are the mean density of respectively real and simulated galaxies in V. Distances r are usually obtained directly from redshifts, on the assumption of smooth Hubble flow, which has limited validity. Much work has gone in gauging accuracy and limitations of $\xi(r)$ estimators (e.g. Landy & Szalay 1993; cf Strauss & Willick 1995). The main problems in such estimates arise from shot noise, redshift space distortions, the question of whether V encloses a fair sample of the universe, galactic extinction and other catalog biases.

If redshifts are unavailable, large catalogs listing galaxy coordinates and fluxes can be used to estimate the *angular* correlation function $w(\theta)$, defined as the joint probability that two galaxies be found within solid angles $d\omega_1$ and $d\omega_2$, separated by angle θ:

$$dP = < N >^2 d\omega_1 d\omega_2 [1 + w(\theta)] \tag{8}$$

where $< N >$ is the sky mean surface density of galaxies in the catalog, and operationally obtained via a procedure similar to that in eqn. (7). The *Limber equation* relates $\xi(r)$ and $w(\theta)$. If the selection function can be expressed as a function of distance alone, then

$$w(\theta) = S^{-1} \int \phi(r_1) \, \phi(r_2) \, \xi(r_{12}) \, d^3\mathbf{r_1} d^3\mathbf{r_2} \tag{9}$$

where $S = [\int \phi(r) \, d^3\mathbf{r}]^2$ and $r_{12} = |\mathbf{r_2} - \mathbf{r_1}|$. The Limber equation can be used to show that if

$$\xi(r) \propto r^{-\gamma} \tag{10}$$

then

$$w(\theta) \propto \theta^{-(\gamma-1)} \tag{11}$$

Measurements of $\xi(r)$ and $w(\theta)$ have been carried out for virtually all galaxy surveys. The angular correlation function has most notably been determined by Groth and Peebles (1986) using the Lick counts, and by Collins *et al.* (1992) and Maddox *et al.* (1990) using the photometrically more accurate surveys obtained with the *Cosmos* and APM machines. For scales smaller than $20h^{-1}$ Mpc, both functions are well fitted by power laws

$$\xi(r) = (r/r_0)^{-\gamma} \tag{12}$$

with $\gamma \simeq 1.8$ and $r_0 \simeq 5.5h^{-1}$ Mpc, for optical galaxies. These parameters imply a density field where the r.m.s. galaxy count fluctuations, among randomly chosen spheres of radius $8h^{-1}$ Mpc is ~ 1, i.e. $\sigma_8 \simeq 1$. So on scales $r < 8h^{-1}$ Mpc, clustering becomes nonlinear. Deviations in the power law behavior of (13) have been noted by several (e.g. Guzzo et al. 1991), prompting discussion on the true slope of $\xi(r)$ in the linear regime (Maddox *et al.* 1990). Some have suggested that the power law behavior of $\xi(r)$ indicates that the galaxy distribution is a fractal or a multi–fractal (Pietronero 1997 and refs. therein). In such a model, there is no definable mean density of the universe: the mean density depends on the volume a survey samples.

Figure 1 shows $1 + \xi(r)$, as obtained by Guzzo *et al.* (1991) from the Perseus–Pisces redshift survey. Note that $1 + \xi(r)$ is the autocorrelation function of the density ρ, while $\xi(r)$ is the autocorrelation function of the density fluctuation δ. Guzzo *et al.* identify a change of slope in $1+\xi(r)$ near a scale of a few Mpc, presumably related to nonlinear effects associated with the onset of gravitational clustering. The lack of correlation at scales in excess of 20 Mpc is of doubtful reality in this case, as the volume sampled is relatively small.

Figure 2 shows $w(\theta)$ as obtained by Maddox *et al.* from the APM survey, a large catalog of $\sim 2 \times 10^6$ galaxies sampling a volume of approximately $2 \times 10^8 h^{-3}$ Mpc^3. The spectrum is roughly a power law up to a scale of about $3°$, after which it steepens significantly. In this case, the authors interpret the presence of the "knee" at large separations as a real feature, rather than an artifact associated with sampling bias. The angular scale of $3°$ corresponds to a linear scale of few tens of Mpc.

1.2. POWER SPECTRUM OF THE DENSITY FIELD

The Fourier Transform of the density field is $\tilde{\delta}$, so:

$$\delta(\mathbf{r}) = (2\pi)^{-3} \int \tilde{\delta}(\mathbf{k}) e^{-i\mathbf{k}\cdot\mathbf{r}} d^3\mathbf{k} \tag{13}$$

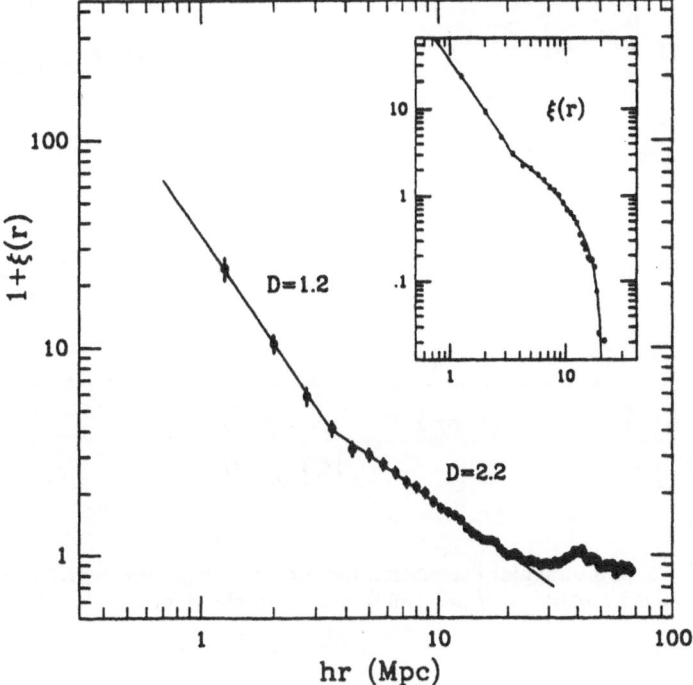

Figure 1. Two–point correlation function of the galaxy distribution in the Perseus–Pisces supercluster region, obtained by Guzzo *et al.* (1991). Both $\xi(r)$ and $\xi(r) + 1$ are shown.

and the function

$$P(k) = (2\pi)^{-3} < |\tilde{\delta}(\mathbf{k})|^2 >$$ (14)

is referred to as the power spectrum. Power spectrum and autocorrelation function are a FT pair:

$$\xi(r) = (2\pi)^{-3} \int P(k)e^{-i\mathbf{k}\cdot\mathbf{r}} d^3\mathbf{k}$$ (15)

$P(k)$ or $\xi(r)$ yield a complete description of the density field, as long as the phases of the Fourier components δ are random (i.e. the density field is Gaussian). When non–linear effects develop, that is no longer true, and additional information is necessary in gauging the density field, e.g. higher order statistics such as the 3–point, 4–point...correlation functions.

Given a redshift catalog, rather than deriving $P(k)$ from $\xi(r)$ by inverting eqn. (15), it is preferred to obtain it via the density field:

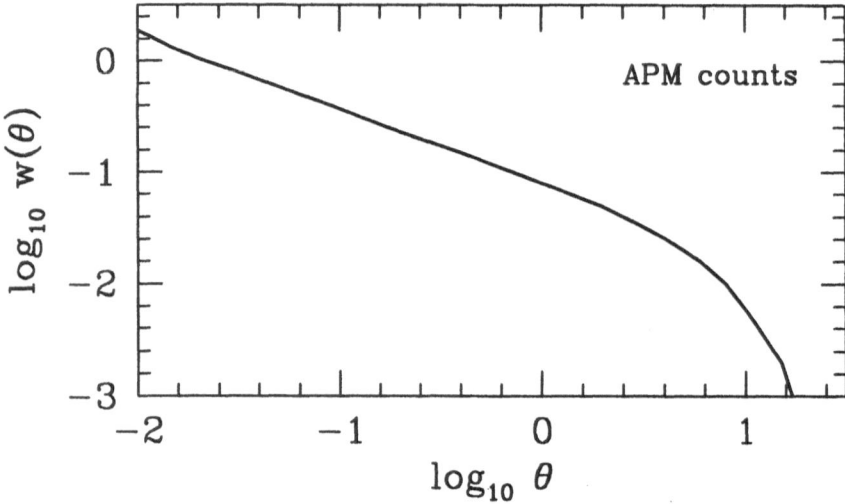

Figure 2. Two–point angular correlation function of the galaxy distribution of the APM survey, obtained by Maddox *et al.* (1990), scaled to the depth of the Lick survey.

$$\delta(\mathbf{r}) = \frac{1}{<n> V} \Sigma_i \frac{\delta_{Dirac}(\mathbf{r}_i)}{\phi(r_i)} - 1 \tag{16}$$

whose FT

$$\tilde{\delta}_{\mathbf{k}} = \frac{1}{<n> V} \Sigma_i \frac{e^{i\mathbf{k}_i \cdot \mathbf{r}_i}}{\phi(r_i)} - \frac{1}{V} \int e^{i\mathbf{k}\cdot\mathbf{r}} d^3\mathbf{r} \tag{17}$$

yields $P(k) \propto |\tilde{\delta}_{\mathbf{k}}|^2$ (see Fisher *et al.* 1993; Feldman *et al.* 1994; Peacock & Dodds 1994).

Galaxy surveys sample the power spectrum up to scales on the order of a few tens of Mpc. The "knee" in $w(\theta)$ shown in figure 2 suggests a turnover in the power spectrum at the corresponding spatial frequency. Figure 3 shows a derivation of $P(k)$, obtained by combining the information garnered from galaxy surveys, which samples the high spatial frequencies k, and the Cosmic Microwave Background (CMB) fluctuation spectrum as obtained with the COBE satellite, reconstructed by Peacock & Dodds (1994). The CMB observations sample the lower spatial frequencies. The smooth lines are standard CDM models of $P(k)$, parametrized via $\Gamma = \Omega_o h$.

In the early universe, the power spectrum $P(k)$ is thought to have been of the form $P(k) \propto k$ (see Peebles 1993). The observed $P(k)$ does however have a peak near scales $2\pi/k \sim 10^2 - 10^3$ Mpc, and drops steeply at larger k's. That is because, during the radiation era, the growth of perturbations on scales smaller than the horizon scale was inhibited by the pressure of relativistic particles. The scale of the $P(k)$ turnover thus reflects the size of the horizon scale at the time of equality between radiation and matter energy density. $P(k)$ of perturbations at $k > t_{turnover}$ is depressed by a factor approximately $\propto k^4$ if dark matter is cold, hence the expectation would be that

$$P(k > k_{turnover}) \sim k^{-3} \tag{18}$$

Efstathiou et al. (1992) give the following useful analytical fit to the expected CDM $P(k)$:

$$P(k) = \frac{Ak^n}{\left[1 + \left[ak + (bk)^{3/2} + (ck)^2\right]^{1.13}\right]^{2/1.13}} \tag{19}$$

where $a = 6.4/\Gamma$, $b = 3.0/\Gamma$, $c = 1.7/\Gamma$, $\Gamma = \Omega_o h$ and k is in $h\mathrm{Mpc}^{-1}$. For standard CDM models with $\Omega_o = 1$, $h = 0.5$ and $n = 1$, the only free parameter is the amplitude A, which can be set using CMB fluctuation measurements:

$$A = 1.45 \times 10^{16} \frac{<Q>}{T_{cmb,o}^2} \tag{20}$$

where $<Q>$ is the CMB quadrupole anisotropy and $T_{cmb,o}$ its current temperature. As shown in figure 3, standard CDM fails to match agreeably the combined observational evidence.

1.3. DENSITY FLUCTUATIONS AND PECULIAR VELOCITY FIELD

In the linear approximation of small perturbations, inhomogeneities in the mass density field produce a peculiar velocity

$$\mathbf{v}(\mathbf{r}) = \frac{H_o f}{4\pi} \int \delta(\mathbf{r}') \frac{(\mathbf{r}' - \mathbf{r})}{|\mathbf{r}' - \mathbf{r}|^3} d^3\mathbf{r}' \tag{21}$$

where

$$f \simeq \Omega_o^{0.6} + \frac{\Omega_\Lambda}{70}\left(1 + \frac{\Omega_o}{2}\right) \tag{22}$$

The influence of a nonzero Λ on local dynamics is small, and we can replace f with $\Omega_o^{0.6}$.

   In the same approximation of small fluctuations, the divergence of the peculiar velocity field is equal to $-f\delta(\mathbf{r})$ (Peebles 1993), so that the power spectrum of the peculiar velocity field is related to $P(k)$ via

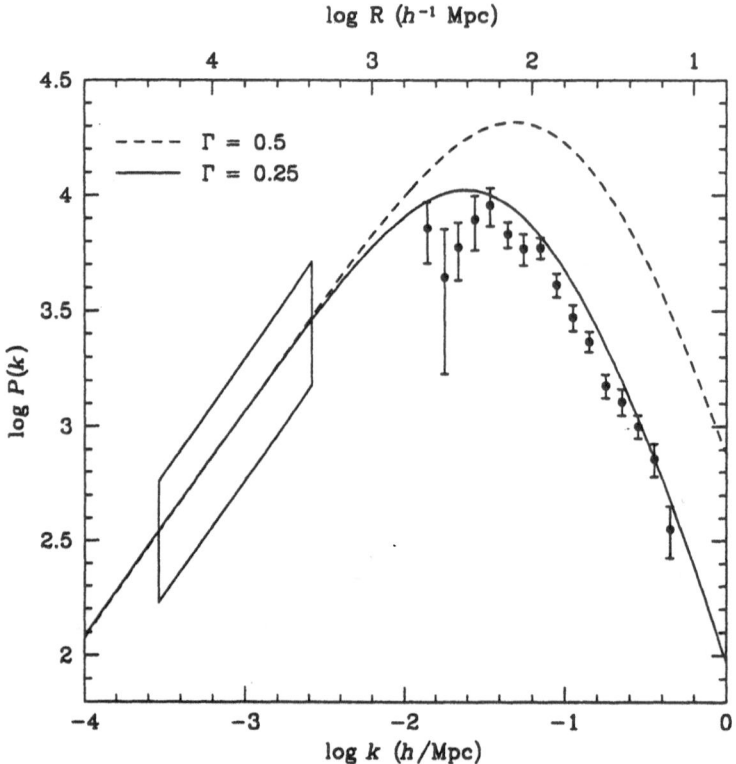

Figure 3. Power spectrum as reconstructed by Peacock & Dodds (1994). Filled symbols correspond to data obtained from galaxy catalogs, the large parallelogram is the error box allowed by the COBE CMB measurement. Smooth lines correspond to CDM models for the indicated values of Γ.

$$P_v(k) = \Omega^{1.2} k^{-2} P(k). \tag{23}$$

If it is assumed that *light traces mass*, i.e. in the simplest approximation (linear biasing)

$$\delta_{gal}(\mathbf{r}) = b\ \delta(\mathbf{r}), \tag{24}$$

then

$$\mathbf{v}(\mathbf{r}) = \frac{H_o \beta}{4\pi} \int \delta_{gal}(\mathbf{r}') \frac{(\mathbf{r}' - \mathbf{r})}{|\mathbf{r}' - \mathbf{r}|^3} d^3 \mathbf{r}' \tag{25}$$

with

$$\beta = \frac{\Omega_o^{0.6}}{b} \tag{26}$$

Note that from the assumption of linear biasing it follows that

$$\xi_{gal}(r) = b^2 \xi_{mass}(r). \tag{27}$$

1.4. THE MOTION OF THE LOCAL GROUP. CONVERGENCE DEPTH

Eqns. (21) or (25) can be modified to yield the peculiar velocity induced on an observer, e.g. the Local Group (LG), by the inhomogeneities present within a sphere of radius R:

$$\mathbf{V}_{pec,LG}(R) = \frac{H_0\beta}{4\pi} \int \delta_{gal}(\mathbf{r}) \frac{\mathbf{r}}{r^3} W(r,R) d^3\mathbf{r} \tag{28}$$

where $W(r, R)$ is a window function, e.g. Gaussian or top–hat of width R.

The CMB dipole of 3.365 ± 0.027 mK (Kogut *et al.* 1993) has been interpreted as a Doppler shift arising from the motion of the LG with respect to the comoving reference frame, with a velocity of 627 ± 22 km s^{-1} in the direction $(l_{cmb}, b_{cmb}) = (276° \pm 3°, 30° \pm 3°)$, which is about 45° away from the Virgo cluster. The unknown parameters in eqn. (28) can then be constrained by matching the CMB dipole to the asymptotic value $\mathbf{V}_{pec,LG}(R \to \infty)$.

The concept of *convergence depth*, often referred to in the literature, is defined by the distance R at which $\mathbf{V}_{pec,LG}(R)$ reaches some large fraction, e.g. 90%, of its asymptoric value $\mathbf{V}_{pec,LG}(R \to \infty)$. If the universe approaches homogeneity on large scales, such convergence occurs logarithmically according to eqn. (28).

Note that the gravitational pull of a galaxy is $\propto M_{gal} r^{-2}$; so, if $M_{gal} \propto L_{gal}$, its contribution to $\mathbf{V}_{pec,LG}(\infty)$ is proportional to its flux. It is then possible to infer *from an all–sky catalog of positions and fluxes* $\mathbf{V}_{pec,LG}(\infty)/\beta$ (see Plionis 1988; Lynden–Bell *et al.* 1989; Rowan–Robinson *et al.* 1991; Kaiser & Lahav 1989; Scharf *et al.* 1992).

The first attempt to measuring the large scale motion of the LG was carried out by Rubin *et al.* (1976). For an all–sky sample of 96 Sc I galaxies, enclosed in the redshift shell bound by 3500 and 6500 km s^{-1}, Rubin *et al.* measured the dipole of the quantity $HM = \log(cz) - 0.2m$, where z is the redshift, m the galaxy's apparent magnitude and c the speed of light. Such dipole suggested a LG motion of 454 ± 125 km s^{-1} towards $(l_{Sc}, b_{Sc}) = (163°, -11°)$, significantly discrepant with that indicated by the CMB dipole.

1.5. PECULIAR VELOCITY FIELD FROM REDSHIFT SURVEYS

In an all–sky catalog, let the number of galaxies with distance $< R$ be $N(R)$, and $\mathbf{r}_i$ the distance vector to an individual galaxy. Since the galaxian distribution is a point process, eqn. (28) can be rewritten as

$$\mathbf{V}_{pec,LG}(R) = \frac{H_\circ \beta}{4\pi <n>} \sum_{i=1}^{N(R)} w_i \frac{\mathbf{r}_i}{|\mathbf{r}_i|^3} \tag{29}$$

We define the *monopole* and the *dipole* of the distribution respectively as

$$\mathcal{M}(R) = \frac{1}{4\pi} \sum_{i=1}^{N(R)} \frac{w_i}{r_i^2} \qquad \mathbf{D}(R) = \frac{3}{4\pi} \sum_{i=1}^{N(R)} w_i \frac{\mathbf{r}_i}{|\mathbf{r}_i|^3} \tag{30}$$

where the w_i's are the statistical weights of each cataloged galaxy. In a well–behaved catalog, the weights are related to the selection function: $w_i \propto 1/\phi(r_i)$. If it is assumed that $M_i \propto L_i$, then $w_i \sim L_i/\phi(r_i)$.

In a uniform distribution: $\mathcal{M}(R) = 4\pi <n> R$, thus $\mathcal{M}(R)/4\pi R$ can be used as an estimator of the mean density, and eqn. (29) can be written as

$$\mathbf{V}_{pec,LG}(R) = \frac{4\pi}{3} \beta H_\circ R \frac{\mathbf{D}(R)}{\mathcal{M}(R)} \tag{31}$$

More properly, however, $\mathcal{M}(R_{large})/4\pi R$ should be used as a mean density estimator, where R_{large} is large enough (a *fair sample* of the universe) to yield a meaningful $<n>$.

This approach has been adopted in many studies, utilizing different kinds of luminous sources as mass tracers. For example, Hudson (1993), Scaramella *et al.* 1994 and Pellegrini *et al.* (1990) have done so using redshift catalogs and optical fluxes of galaxies; Strauss *et al.* (1992) and Rowan–Robinson *et al.* (1990) used IRAS fluxes; Scaramella *et al.* (1991), Plionis & Valdarnini (1991), Tini–Brunozzi *et al.* (1995), Lahav *et al.* (1989), Branchini *et al.* (1996) used clusters of galaxies (see also Juskiewicz *et al.* 1990; Freudling *et al.* 1995; Strauss & Willick 1995; Kolokotronis *et al.* 1996).

Figure 4 shows the quantity $|\mathbf{V}_{pec,LG}(R)|/\beta$ plotted versus R, both expressed in km s^{-1}, as obtained in several studies, adapted from Hudson (1993). In Table 1, the asymptotic value of $|\mathbf{V}_{pec,LG}(R)|/\beta$, as well as the direction of the apex of motion and its angular distance from the CMB dipole apex, are shown for the various samples plotted in fig. 4. Note that while the agreement of the amplitude of the velocity with that indicated by the CMB can be forced by choosing an appropriate value of β, of between

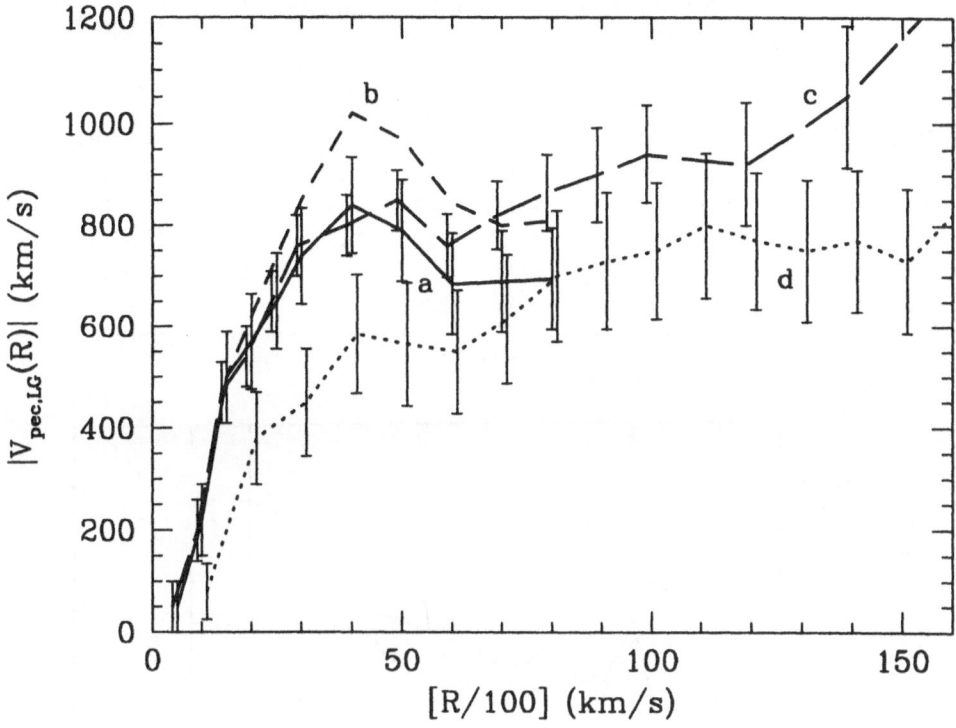

Figure 4. Amplitude of the peculiar velocity of the Local Group, scaled by the inverse of the bias parameter, produced by fluctuations in the galaxy distribution within distance R, for different samples of galaxies. Adapted after Hudson (1993). Labels by each curve identify the samples listed in Table 1.

0.5 and 0.9, the discrepancy in the apex directions from that of the CMB cannot as easily be reconciled. An important source of uncertainty in all these studies is however introduced by the fact that a large solid angle of sky is not sampled by galaxy catalogs, that corresponding to the zone of avoidance, which for optical galaxies can amount to $\sim 40\%$ of the sky. Note how the manner in which the region of the zone of avoidance is treated, by comparing the "avg" and "cloned" approaches used by Hudson for the same, leads to substantial differences in the final results.

Figure 5 illustrates the results of two studies (Branchini *et al.* 1996 and Tini–Brunozzi *et al.* 1995) which use the Abell/Abell–Corwin–Olowin catalogs of galaxy clusters, adapted from the data shown in those sources. Note that the β value estimated from the Tini–Brunozzi *et al.* sample assumes that the Abel/ACO cluster set produces an asymptotic velocity on the LG of 500 km s^{-1}, as the sample does not include Virgo, to which the remainder of the LG velocity is ascribed.

TABLE 1. Local Group Reflex Motion

Sample	Type	R_{max}	l_{apex}	b_{apex}	$\Delta\theta_{cmb}$	V_{LG}/β	β
a. Hudson "avg"	Opt. gal.	8000	242	49	28	696 ± 105	0.87
b. Hudson "cloned"	Opt. gal.	8000	231	40	32	816 ± 102	0.74
c. Strauss et al. 92	IRAS	20000	249	40	20	1100 ± 150	0.55
d. RR90	IRAS	15000	237	43	30	736 ± 137	0.82
e. Tini–Brunozzi 95	Abell/ACO	24000			20	2050 ± 500	0.25*
f. Branchini 96	Abell/ACO	18000	278	20	11	3020 ± 333	0.21

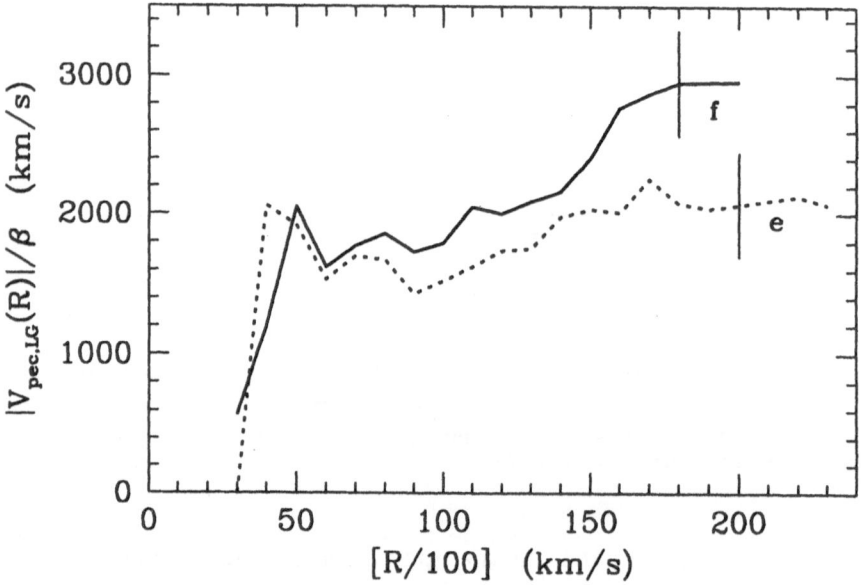

Figure 5. Amplitude of the peculiar velocity of the Local Group, scaled by the inverse of the bias parameter, produced by fluctuations in the distribution of Abell/ACO clusters of galaxies within distance R, as estimated by Branchini *et al.* (1996) and Tini–Brunozzi *et al.* (1995).

The large–scale distribution of luminous matter suggests that a large fraction of the LG motion is produced by mass fluctuations within $cz \sim$ 5000 km s^{-1}, perhaps as much as 80% or more of the value indicated by the CMB dipole. However, the cluster distribution suggests that more distant features also play a role, and convergence does not appear to be reached until distances as large as 15,000 km s^{-1}. For clusters, match with the CMB dipole requires a lower value of β, between 0.20 and 0.30, than for the galaxian population. The apex direction of the LG motion suggested by the Abell/ACO clusters, listed in Table 1, is discrepant from that indicated by galaxy samples, but within the errors is in general agreement with that of the CMB dipole. The effect of incompleteness in the sky coverage and other catalog biases obviously have an important effect on the results.

1.6. DENSITY FIELD FROM A REDSHIFT SURVEY

The density field can be reconstructed from a redshift survey, using eqn. (16). However, that approach assumes that galaxies are located at the distances implied by their redshifts, under the assumption of smooth Hubble flow. However, the Doppler shift of a galaxy at distance d measures

$$cz = H_o d + [\mathbf{V}_{pec}(\mathbf{d}) - \mathbf{V}_{pec}(0)] \cdot \frac{\mathbf{d}}{d} \qquad (32)$$

$\mathbf{V}_{pec}(0)$, the peculiar velocity of the LG, we obtain from the CMB dipole; $\mathbf{V}_{pec}(\mathbf{d}$ can be obtained from eqn. (25), if the density field is known, provided that density fluctuations are small. Peculiar velocities, especially in the neighborhood of large mass fluctuations, can be quite large and, at low redshifts they amount to a sizable fraction of the total redshift. The circularity in the problem is solved by an iterative approach:
   – Place galaxies at $H_o d = cz \quad \rightarrow \quad$ infer $\delta_{gal,o}(\mathbf{r})$
   – Smooth $\delta_{gal,o}(\mathbf{r})$
   – Fill ZoA by interpolation, cloning, averaging, etc
      [Note: ZoA accounts for up to 40% of sky]
   – Apply eqn. (25) to obtain $\quad \rightarrow \quad \mathbf{V}_{pec,o}(\mathbf{r})$
   – Use $\mathbf{V}_{pec,o}(\mathbf{r})$ to assign $V_{pec,los}$ to each galaxy in sample.
   – Recompute $H_o d$, $\delta_{gal}(\mathbf{r})$ and repeat process.
After a few iterations, the process usually converges, although special measures may be necessary in the neighborhood of the densest regions (see Yahil *et al.* 1991; Strauss *et al.* 1992).

2. Lecture 2: The Direct Measurement of Peculiar Velocities

The second lecture of this series deals mainly with the technical aspects in the measurement of galaxy distances. We provide a general survey of the techniques that are being used, and concentrate on the more detailed description of the treatment of bias and the production of template relations in the case of the TF relation, used as a case study.

The measurement of redshift–independent distances of galaxies relies on the use of a "standard candle", i.e. a class of objects of constant intrinsic properties which can be inferred through a set of easily repeatable observations. Ideally, the method should be applicable up to large distances and to a population of objects easily obervable in different density regimes and environments, have a well–understood physical basis, rely on calibrating relations of low dispersion, and be observationally economic. The methods commonly used meet these conditions to a varying degree. The observation of Cepheids for example, extended by HST to galaxies in nearby clusters such as Virgo and Fornax, can yield relatively accurate distances, with errors in the distance modulus (DM) on the order of 0.15 mag; it cannot however be applied at large distances and it is observationally quite expensive. Similar in their distance limitations are other techniques that rely on the detection of stellar systems within galaxies, such as those that use the LF of planetary nebulae, the LF of globular clusters, the detection of novae or of giant stars near the tip of the HR diagram. An exception in this class are SN of type Ia, which can be seen at large distances and whose light curve can be used to estimate DM's with accuracies better than 0.2 mag; however, in spite of the fact that the number of detections is rising rapidly — currently hovering near 3 dozens — thanks principally to the efforts of the Cerro Calán/CTIO collaboration, they still sample the local universe too sparsely. The most commonly used distance determination techniques, for the purpose of reconstructing the peculiar velocity and density field, are then somewhat less precise but more readily applicable to the large volumes of interest. We survey them in the next section.

2.1. TECHNIQUES

The main techniques for the direct measurement of the peculiar velocity field are:
– the TF relation, which applies to spiral galaxies;
– the Fundamental Plane (FP) or D_n–σ relation which applies to spheroidal systems;
– the Surface Brightness Fluctuations (SBF) technique, also applicable to spheroidals;
– the Hoessel relation for Brightest Cluster Ellipticals (BCG).

All these techniques are based on the combination of photometric data and at least one spectroscopic datum (the redshift); in the case of the first two, a second spectroscopic parameter — the velocity width of one or more spectral lines — is necessary as well.

The TF relation has been applied most extensively. It expresses proportionality between the luminosity and the rotational velocity of disk systems:

$$L \propto V_{rot}^{\alpha} \tag{33}$$

At this time, approximately 3000 spiral galaxy distances $H_o d$ are available. The main advantages of this technique, which will be discussed in detail in the next few sections, are that it is easily applicable to a widely spread population of galaxies, found over all density regimes, and it is characterized by fair accuracy, averaging about 0.3 mag on the DM of individual objects.

The FP relation, which is equivalent to D_n-σ, has also been applied to a large sample of elliptical galaxies as well as to bulges of lenticulars. This relationship between effective radius R_e, effective surface brightness Σ_e and stellar velocity dispersion σ,

$$R_e \propto \Sigma_e^\eta \sigma^\theta \qquad (34)$$

was independently proposed by Dressler *et al.* (1987) and Djorgovski & Davis (1987); R_e and the Σ_e isophote enclose half of the light emitted by a system with a surface brightness profile $\Sigma(r) = \Sigma_e exp\{-7.67[(r/R_e)^{1/4} - 1]\}$. It is particularly effective in the determination of the peculiar velocity of clusters of galaxies, whose cores are rich in spheroidals and the high galaxy density allows multiplexing with multiobject spectrographs. The typical accuracy on the DM for single objects with this technique is on the order of 0.4 mag (Scodeggio 1997; Joergense 1997).

The SBF technique, applied by Tonry and collaborators (see Tonry & Schneider 1988) exploits the Poisson variations, pixel–to–pixel in an image, of the surface brightness produced by the brightest stars in a galaxy, which are related to the galaxy's distance. This technique works best with dust–free elliptical bulges, and under the best seeing conditions it has been applied to distances of a few thousand km s^{-1}. The reported accuracy of the method under good conditions is high, on the order of 0.1 to 0.15 mag, albeit the superior observing conditions necessary for its application and lingering doubts about its dependence on the vagaries of the bright stellar population, that may result from for example metallicity variations, have so far limited its scope.

BCG's have long been used as "standard candles"; they are the most luminous galaxies, generally located in the central region of clusters, and can thus be easily identified to large distances (Sandage & Hardy 1973). By definition, however, their space density is limited to one per cluster, and their most likely unusual formation histories set them apart from other spheroidal galaxies. In 1980, Hoessel, noted a correlation between the slope of the BCG's photometric profile at a radius of $10h^{-1}$ kpc and the luminosity interior to that radius. This relation improves the "standard candle" quality of these objects, and it has been used by Lauer & Postman (1994) to inspect the global motion of a sample of 119 clusters to a redshift of $\sim 15,000$ km s^{-1}. The accuracy of this technique is reported to be on the order of 0.3 mag, comparable to TF for a single object. The peculiar formation histories of BCG's, which may be related to the overall evolutionary state of the cluster, and the large scatter of the Hoessel relation are the main reasons for caution in the application of this technique.

The ultimate limit to the application of these techniques for the measurement of peculiar velocities is related to evolutionary changes in the structure of galaxies, that may lead them to depart from the adopted template relations. Studies of the evolution of M/L with z indicate that no significant changes have occurred between lookback times corresponding to $z \sim 0.1$ and now, for both spirals and spheroidals. Such changes may start to be detectable at $z \sim 0.2$ to 0.3, although the case, especially for spirals, is still quite unsettled (Bershady 1996; Joergensen 1997). It should however be kept in mind that lognormal errors on the distance imply that the distance determination quality of all the methods discussed deteriorates rapidly well before evolutionary concerns may arise. For example, the typical uncertainty on the DM of a single galaxy obtained using the TF relation, of 0.3 mag, corresponds to a distance error of 15%. At $z \sim 0.05$, this translates in an uncertainty on the peculiar velocity larger than 2000 km s^{-1}. Typical peculiar velocities of clusters of galaxies may not generally exceed 500 km s^{-1}, so in order to obtain a significant estimate of the peculiar velocity of a system, the signals of a large number N of galaxies need to be averaged together, on the assumption that the statistical error behaves as $N^{-1/2}$. The practical distance limit of meaningful measures of peculiar velocity by the techniques discussed here is thus $z \sim 0.05$ or so.

2.2. CASE STUDY: THE TULLY–FISHER RELATION

The understanding in physical terms of the TF relation is based on general scaling laws and numerical simulations of structure formation. The parameters relating the luminosity and rotational velocity of a galaxy can therefore only be quantitatively obtained through empirical inspection, rather than analytical inference, a common limitation to other of the techniques discussed in the preceding section.

The scaling arguments for TF are simple. Consider a pure exponential disk of central disk surface brightness $I(0)$ and scale length r_d ; its total luminosity is then $L_d \propto r_d^2 I(0)$. The mass internal to radius R is $M(R) \propto RV^2$, so if the rotation curve flattens in the outer regions of the disk, as is usually the case for spiral galaxies, the total mass is $M_{tot} \propto r_d V_{max}^2$. We can then write $L_d \propto (M_{tot}/L_d)^{-2} V_{max}^4 / I(0)$. If a dark matter halo is present so that the disk mass is $M_d = \Gamma M_{tot}$, then

$$L_d \propto (M_d/L_d)^{-2} \Gamma^2 V_{max}^4 / I(0) \tag{35}$$

When a number of "standard" assumptions are made, i.e. that $\Gamma \sim$ const, $M_d/L_d \sim$ const and $I(0) \sim$ const (Freeman's law, 1970), $L_d \propto V_{max}^4$, which resembles eqn. (33). In practice, none of the assumptions of constancy for M_d/L_d, Γ and $I(0)$ apply; all those parameters exhibit mild dependencies on V_{max} (or L_d), reducing the exponent to values $\alpha < 4$, in a measure that depends on the adopted photometric band.

Empirical calibrations of the TF relation yield power law behavior of the type shown by eqn. (33), with values of $\alpha \sim 3$ in the I band. Some workers (e.g. Aaronson *et al.* 1986; Willick 1990) have found significant departures from a single power law behavior, and quadratic or bilinear TF fits have been adopted. It has been argued that inappropriate extinction corrections (Giovanelli *et al.* 1995) or samples including a mixture of morphological types (Giovanelli *et al.* 1997) may result in TF departures from linearity. There is however no *a priori* reason to expect that the TF relation be strictly linear. Recent N–body simulations (Navarro *et al.* 1996) provide further insights in the understanding of the TF relation.

The main parameters necessary to carry out the TF exercise are:
– the galaxy flux;
– the disk inclination to the line of sight;
– the rotational width of one or more spectral lines;
– the galaxy systemic radial velocity;
– the galaxy type;
– a template TF relation;
– and the associated uncertainties with each of the items listed above.

It has become common practice to obtain the photometric datum for TF work using I or R band CCD images. The sky background at those wavelengths is still relatively low (as compared to H and K bands), detectors have high efficiency and large fields of view, and data acquisition is relatively fast even with small aperture telescopes. The population dominating the light at I band is comprised of stars that are several Gyr old. Thus, disks are well outlined but of smoother appearance than as seen in blue portions of the visible spectrum, and their apparent inclinations to the line of sight can be more reliably determined. In addition, the processes operating in clusters that may alter the star formation rate in galaxies will have a very retarded effect on the red and infrared light of disks; thus, smaller — if any — systematic differences are expected between the I and R TF relation of cluster and field galaxies.

Rotational widths are observationally more demanding. The two most favored techniques use either the line of HI in the radio spectrum at 21cm, with large single dish telescopes, or single–slit spectra of optical emission lines in, such as H_α at 6563 Å. Although synthesis radio telescopes and

Fabry–Perot imaging spectrographs are sometimes also used, the bulk of the work has been done in the two former modes. While optical and synthesis spectra have the advantage of a detailed rendition of the shape a galaxy's rotation curve, 21cm single–dish spectra sample sensitively the outer regions of a disk, where the maximum rotational speed is more likely to have been achieved, and are not vulnerable to uncertainties in the position angle of the major axis of the galaxy.

An accurate determination of the inclination of the disk to the line of sight is necessary for two important reasons: (1) it is used to estimate a correction factor to the observed velocity width, which increases in amplitude and uncertainty as the disk appears more face–on, in order to obtain a true rotational width, and (2) it helps to estimate the amount of extinction suffered by the stellar light in propagating through the interstellar medium of the disk, which increases as the disk appears more edge–on; the amplitude of this correction is not well agreed on (see for example Giovanelli *et al.* 1995 and the proceedings edited by Davies & Burstein 1995). Observational errors on the measurement of velocity widths are generally larger, in relative terms, than those affecting the flux. Thus, the amplification of those errors by the width correction factor associated with the inclination of the disk poses a practical limit of 30° to 40° for the lower value of the inclination of a galaxy (with respect to face–on), below which TF work becomes unreliably noisy.

Low rotational velocity systems are of limited use for TF work for several reasons. On one hand, the amplitude of the relative errors on the measured widths increase substantially, to dominate the error budget, as the measurement error does not decrease with the measured width. On the other, the contribution to the observed width by purely rotational motions becomes more uncertain: turbulent motions in the interstellar medium broaden the 21cm linewidth by an increasingly important amount as the rotational width decreases. Since an accurate correction for that effect is difficult to estimate, in practice, additional noise is introduced when it is applied. It also appears that the "cosmic" sources of scatter become increasingly important in the TF error budget as the velocity width decreases. Thus, faint spirals, with rotational widths (i.e. twice the maximun rotational velocity) smaller than 100–150 km s^{-1}are very poor candidates for the measure of peculiar velocities by means of the TF relation (for more details, see Giovanelli *et al.* 1997 and refs. therein).

2.3. THE CONSTRUCTION OF A TEMPLATE RELATION

Clusters of galaxies are favorite targets for applications of the TF technique of determining redshift–independent distances $H_o d$ for two important reasons. First, a cluster provides a large number of objects located at a common distance, thereby allowing the determination of a TF relation slope which will be exempt from the vagaries that are introduced by an *a priori* unknown peculiar velocity field in a field galaxy sample. Clusters are thus well suited for the generation of a TF template relation. Second, the combination of independent distance estimates of several galaxies in a cluster provides a more accurate determination of the cluster distance: to the extent that N galaxies in a cluster can be considered to be at the same distance, the cluster distance can be found $\sim N^{1/2}$ times more accurately than as determined for a single galaxy. Well–sampled clusters and groups can thus provide "hard points" in a map of the large scale peculiar velocity field.

One commonly–adopted approach to the determination of a TF template relation is to select a single cluster of galaxies as a reference, thereby equating the universal template with the TF relation defined by its member galaxies. In this instance, all cluster members are assumed to be at approximately the same distance from the observer as predicted by the systemic velocity of the cluster, and their relative locations in the TF diagram are assumed to be unaffected by peculiar motions. This strategy requires that the cluster be endowed with several special qualities: (a) a substantial number of individual galaxy distances must be measurable for that cluster, either by virtue of its large spiral population or of its proximity; (b) there must be a reasonable expectation that the cluster is at rest in the comoving reference frame; (c) there must be strong reasons to believe that the cluster's TF relation is a particularly good approximation to the true one. The practical advantages of having a TF relation defined by a large sample of objects, as required by (a), are obvious; the likelihood that a reliable template can be obtained from a *single* cluster is however questionable on statistical grounds.

Consider the case of a nearby cluster first. We assume that a cluster TF relation is a random realization of some universal relation, defined (in a logarithmic version of eqn. 33) by a zero offset and a slope. Uncertainty is contributed by the limited accuracy of measurements, by an intrinsic amount of scatter in galaxies' properties and by the *a priori* unknown amplitude of the cluster motion with respect to the comoving reference frame. If the number of observed galaxies is large and the dynamic range in the observed parameters high, then the slope of the universal relation should be increasingly well approximated by that of the chosen cluster. These requirements favor the selection of a nearby cluster in which the fainter galaxies are more easily observable. The uncertainty on the offset or zero point of the TF relation can also be initially reduced by increasing the size of the sample. However, the limit of this process depends on the amplitude of the cluster motion V_{pec} itself, the uncertainty on the zero point increasing with cluster proximity: $\Delta m \simeq -2.17 V_{pec}/cz$ mag. In other words, if the chosen cluster is nearby, the TF zero point uncertainty remains high, independent of the size of the sample.

Consider now the case of a distant cluster, where the average single–point scatter about a best fit linear relation is on the order of a third of a magnitude. A typical TF cluster sample may include two dozen galaxies, so that the cluster zero point can be determined to better than 0.1 mag. A cluster motion of 300 km s^{-1}— about half the velocity of the LG with respect to the CMB reference frame — translates into a displacement of the cluster TF relation smaller than 0.1 mag only if the cluster is at $cz > 6000$ km s^{-1}. There is no a $priori$ validity to expecting that the chosen cluster be at rest. If we were to use a single cluster as reference, we would then be forced to use a distant one, which would limit both the number and the dynamic range in luminosity of the sampled galaxies.

A more attractive approach is that of building a template by combining data of many clusters. It is reasonable to expect that the mean peculiar motion of a "basket of clusters", if well distributed over all the sky, approximates null velocity better than any single cluster. The TF relation parameters can be statistically better determined, as the larger number of objects sampled helps to reduce statistical uncertainties. Nearby clusters in the set play an important role in the definition of the template slope, while the more distant clusters allow a fairer definition of the kinematical zero point. In order for the TF relation of each cluster to be adequately "spliced" to that of the others, it is necessary to ascertain its peculiar velocity with respect to the reference frame defined by the cluster set. Thus, the construction of a TF template relation and the determination of the peculiar motions of the cluster set need to be carried out at the same time or, rather, iteratively.

Figure 6 shows a set of TF relations for a sample of 24 clusters of galaxies with systemic velocities ranging between 1000 and 9000 km s^{-1}, after Giovanelli et $al.$ (1997). They can be combined to obtain a template relation, after the peculiar velocities are estimated via a maximum likelihood technique. While relative peculiar velocities among clusters in the set are easily determined, their absolute value depends ultimately on the degree to which the cluster set can be believed to yield a mean peculiar velocity of zero, as measured from an Earth observer.

2.4. BIASES

Samples extracted from just about any galaxy catalog are affected by a variety of biases. The very process of sample extraction is also tainted.

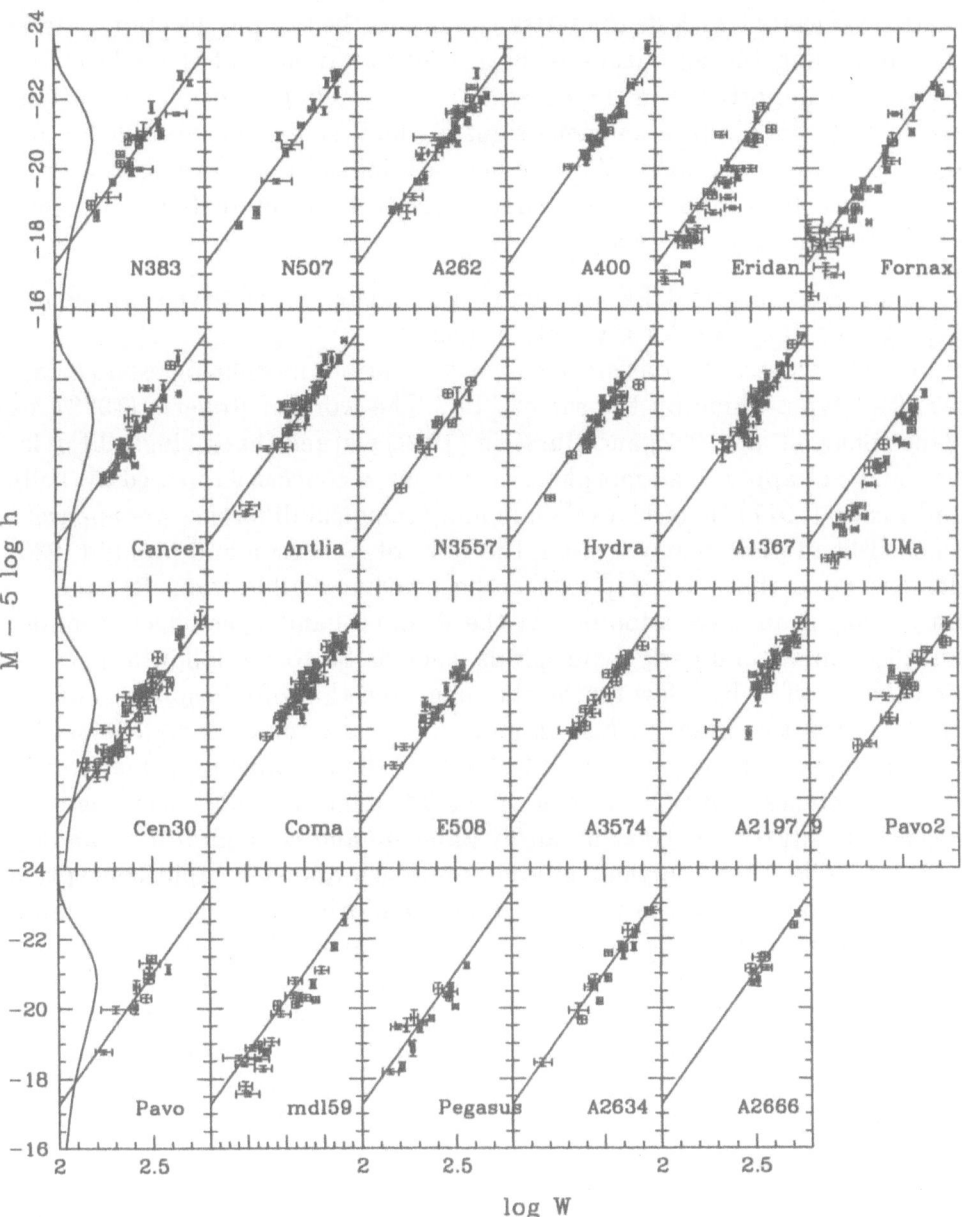

Figure 6. TF diagrams for 24 clusters of galaxies, after Giovanelli *et al.* (1997). Clusters A2197 and A2199 are included in the same panel.

Because the TF relation, as any empirical template, has a certain measure of scatter associated with it, the latter couples to the sample selection criteria in determining the amplitude of bias. Clear awareness of this effect is of paramount importance in the determination of reliable peculiar velocities. Its neglect can produce spurious signals which can easily overwhelm any real signature in the data. We give here a brief summary of bias concerns, and address the reader to other sources for further details (e.g. Giovanelli *et al.* 1997 and refs. therein).

Suppose a universal TF relation (UTF) exists. An observed sample can depart from the UTF for a variety of reasons:
(a) If a cluster sample consists of galaxies of various morphologies: do galaxies of all types abide by the same UTF? The work of Roberts (1978), de Vaucouleurs et al. (1982) and Burstein (1982) warned that a single TF relation may not apply to all morphological types, a concern expressed by Tully and Fisher (1977) themselves. While morphological differencs are apparent when B band photometry is used, the work of Aaronson and Mould (1983) demonstrated that such dependence becomes negligible when H–band infrared magnitudes are adopted. At the R and I bands, the effect is milder but discernible, and early type spirals, such as Sa to Sb, follow a TF relation that is offset by a few tenths of a mag from that of late spirals, making an adjustment necessary when an morphologically mixed sample is used.
(b) An important source of bias results from the coupling of a cluster sample completeness with the scatter in the TF relation. This effect has been recently illustrated in great detail by Sandage and collaborators (Sandage 1994 and refs. therein); their graphic approach allows a rapid conceptual grasp of the problem. In essence, the problem arises because, as the luminosity decreases and the sample becomes progressively more incomplete, the regions on the faint side of the UTF become depopulated faster thant those on the bright side, thus falsifying the slope, offset and scatter of the relation. Many studies have obtained analytical corrections for the problem, albeit some (including this writer) believe that a numerical simulation is best suited to appropriately correct for the effect. Differences in the application of this correction are very commonly invoked as the principal reason for discrepancy in the results of cluster TF samples.
(c) Samples extracted from size– or flux–limited catalogs will in varying degree have built–in biases, which depend on the accuracy of the cataloged photometric quantities, later used to select the sample, rather than on that of the successively measured parameters. In particular, biases can arise from the proximity of sample objects to the catalog limit, and they may be strongly coupled to the amplitude of the scatter in the TF relation.
(d) When the size of a cluster is not negligible in comparison to its distance, the estimated mean distance to a cluster can be in error. The effect

is however generally small.

(e) The so called "Malmquist bias" is important in the correction of the estimated TF distances of individual field galaxies, and depends on the characteristics if the distribution of galaxies in space. This bias, which couples with the scatter in the TF diagram, is small in the case of clusters but it can be very important in the case of individual field galaxies. An accurate knowledge of the characteristics of the scatter in the TF relation is extremely important in the correction for this effect.

(f) It has been feared that the cluster environment may affect both the photometric and kinematical characteristics of galaxies, depending on their location in the cluster gravitational potential well. As a result, the location of individual objects in the TF diagram may depend on distance from the cluster center or on the cluster richness. While such an effect may be discernible when blue photometry is involved (Schroeder 1996), it appears that in the red and infrared bands the effect is negligible (Giovanelli *et al.* 1997).

3. Lecture 3: Measurements of the Peculiar Velocity Field

3.1. BACKGROUND

De Vaucouleurs (1958) pioneered the study of large scale structure in the local universe, promoting the idea of the existence of a Local Supercluster (LS) and of perturbations in the local Hubble flow produced by the density excess associated with the LS. These predictions were not comforted by reliable distance determination techniques. After the introduction of the TF relation in 1977, Aaronson *et al.* (1982) used it to obtain an estimate of the Virgocentric infall velocity of the LG of $V_{r_0} = 480 \pm 75$ km s^{-1}, later revised to 250 ± 64 km s^{-1}. A re–analysis of the Aaronson *et al.* data by Kraan–Korteweg (1985) favored even lower values of the Virgocentric infall, of about 200 km s^{-1}. Using the relative distance moduli of Virgo and nearby clusters, Jerjen & Tammann (1993) obtained a Hubble velocity of the LG relative to the center of the Virgo cluster of 1179 ± 17 km s^{-1}, and a Virgocentric infall velocity of 233 km s^{-1}.

Virgocentric infall and the apex of the CMB dipole do however point in directions that are $\sim 45°$ apart, making clear that Virgo alone could not account for the CMB dipole. As the complexity of the large–scale structure of the galaxian distribution started to emerge, Shaya (1984) and Tammann and Sandage (1985) suggested that the Hydra–Centaurus supercluster at a redshift of about 3000 km s^{-1} or supercluster structures hidden in the "zone of avoidance" in the general direction of (l_{cmb}, b_{cmb}) could play an important role in determining the peculiar velocity of the LG.

Using the D_n–σ relation on an all–sky sample of about 400 elliptical galaxies, Dressler *et al.* (1987) and Lynden–Bell *et al.* (1988) found that Hydra–Centaurus could not be the gravitational attractor mainly responsible for the LG motion, because the supercluster itself appeared to participate in the flow. They postulated the existence of a large mas concentration at a distance of 4350 km s^{-1}, towards which both the LG and Hydra–Centaurus appeared to be falling, which became known as "the Great Attractor" (GA). The study of Lynden–Bell et al. (1988) suggested a much reduced velocity of Virgo infall for the LG, with respect to previous estimates. Faber and Burstein (1988) corroborated that result, suggesting that the magnitude of the Virgo infall at the LG might be as small as 85 km s^{-1}. On the other hand, re–analyzing the Aaronson et al. cluster sample within the framework of a bi–infall model (Virgo plus GA), Han and Mould (1990) found optimal solutions with Virgo infall velocities on the order of 200 km s^{-1}, and GA infall solutions of the order of 400 km s^{-1}. The results of Lynden–Bell et al. placed the GA near $(l_{ga}, b_{ga}) = (309°, 18°)$, at $R_m = 4350$ km s^{-1}, and recent applications of the *Potent* method yield $(l_{ga}, b_{ga}) = (320°, 0°)$, at $R_m = 4000$ km s^{-1}(Dekel 1994). In this region, the surface density of galaxies is greatly reduced by galactic extinction, making it difficult to find optical counterparts to the GA.

In 1989, the presence of a large concentration of clusters, near $cz \sim$ 14,000 km s^{-1}(the "Shapley Supercluster"), was noted in the general direction of the GA (Scaramella *et al.* 1989), thus prompting the idea that infall towards it may be an important component of the local velocity field. While Dressler & Faber (1990) reported back–flow into the GA beyond 4500 km s^{-1}, Mathewson *et al.* (1992) questioned their finding, as their independent data set indicated continued outflow of galaxies beyond the location of the GA, out to the limits of their survey near 7,000 km s^{-1}, a result supportive of the Scaramella *et al.* (1989) idea. Willick (1990) contributed to the debate by reporting the measurment of a peculiar velocity of $\sim$ 400 km s^{-1}for the Pisces Perseus Supercluster (PPS), at $cz \sim$ 5000 km s^{-1}and antipodal to the GA, thus reinforcing the idea that the PPS, the LS and the Hydra–Centaurus–GA region are all partaking of a *bulk flow*, roughly towards the Shapley supercluster, at a velocity of 400–600 km s^{-1}. In 1994, Lauer & Postman (LP) used the Hoessel method with a sample of 119 BCG's extending to $cz \sim$ 15000 km s^{-1}, and reported a LG motion of 561 ± 284, km s^{-1}toward $l = 220°$, $b = -28°$ ($\pm 27°$), with respect to the reference frame defined by the BCG sample. This is largely discrepant from the direction of the apex of the CMB dipole, and would indicate that, for a Doppler origin of the CMB dipole, the BCG sample as a whole is moving at 689 ± 178 km s^{-1}toward ($l = 343°$, $b = +53°$) (± 23). The dynamical implication of this result is that the LG motion and that of the BCG sample are caused *largely* by mass concentrations beyond 10000 km s^{-1}.

The comparison between cluster velocities derived from the spiral galaxy population, via the TF relation, and from the spheroidal population, via the FP technique, has not always been favorable (see, e.g. Mould *et al.* 1991, Lucey *et al.* 1991), suggesting that environmental effects, deficient template relations or data of insufficient quality were involved. The careful analysis of Scodeggio (1997) shows that when comparison is made amongst samples of sufficient size, galaxies of properly gauged cluster membership and good quality data, the agreement between the two methods is satisfactory.

Having set the stage with the preceding historical background, in the rest of this lecture we will discuss the results of several survey studies of peculiar velocities carried out by the author and his collaborators (M. Haynes, T. Herter, D. Dale, M. Scodeggio and N. Vogt at Cornell, J. Salzer at Wesleyan, G. Wegner at Dartmouth, P. Chamaraux at Paris–Meudon, L. da Costa and W. Freudling at the ESO). They are based on TF peculiar velocities of 555 cluster galaxies in 24 clusters within $cz \sim 9000$ km s^{-1}and of ~ 2000 field late spirals. These are all–sky samples including observations carried out by the above mentioned team and complementary data in the public domain.

3.2. CLUSTER PECULIAR VELOCITIES

While earlier studies reported relatively high values for cluster peculiar velocities, in some cases exceeding 3000 km s^{-1}(e.g Mould *et al.* 1991; Lucey *et al.* 1991), the results of our study suggest a significantly more quiescent cluster peculiar velocity distribution. Of the 24 clusters in our sample, none exhibit 1–d peculiar velocities in excess of 600 km s^{-1}. Bahcall & Oh (1996) and Moscardini *et al.* (1996) compared our cluster velocities with expectations of numerical simulations in the framework of different cosmological scenarios, and obtained agreement only with with models with $\Omega_o \leq 0.4$. On that basis, Bahcall & Oh estimate a 1–d r.m.s. peculiar velocity of clusters of $< v_{1d}^2 >^{1/2}= 293 \pm 28$ km s^{-1}.

The cluster set can also be used to estimate the LG reflex motion. When the subset of 14 clusters with cz between 4000 and 9000 km s^{-1}are used, the reflex motion of the LG mimics very well that of the CMB dipole:

$$V_{clu} = 577 \pm 101 \quad \text{vs.} \quad V_{cmb} = 627 \pm 22$$
$$l_{clu} = 272 \pm 20 \quad \text{vs.} \quad l_{cmb} = 276 \pm 2$$
$$b_{clu} = +31 \pm 17 \quad \text{vs.} \quad b_{cmb} = +30 \pm 2,$$

suggesting that the largest fraction of our motion with respect to the CMB arises within the shell occupied by the clusters.

TABLE 2. Dipoles of the Peculiar Velocity of Field Galaxies

V_{cmb} Window	$V_{pec}(LG)$	l_{apex}	b_{apex}	Nr.
All	417 ± 28	253 ± 09	37 ± 03	1585
$2000 - 3000$ km s^{-1}	333 ± 41	268 ± 16	41 ± 08	235
$3000 - 4000$ km s^{-1}	437 ± 57	242 ± 20	24 ± 05	303
$4000 - 5000$ km s^{-1}	551 ± 62	236 ± 24	37 ± 05	372
$5000 - 6000$ km s^{-1}	566 ± 81	281 ± 15	24 ± 15	270
CMB	627 ± 22	276 ± 03	30 ± 03	

3.3. DIPOLES OF THE PECULIAR VELOCITY FIELD

Similarly to what is observed for clusters, the peculiar velocities of field objects measured in the LG reference frame clearly reflects the motion of the LG with respect to the CMB, suggesting that the LG motion with respect to the CMB is largely a local phenomenon, not involving the majority of the volume subtended by the sample. Table 2 exhibits the dipole amplitudes and apices of the field spirals, segregated by redshift shells, and compared to the analogous quantities for the CMB dipole. It is clear that the outer shells of the sampled volume approach rest with respect to the CMB and that, according to these data, the coherence length of the LG motion would not appear to significantly exceed the radius of the sampled volume. The apex alignment of these data with the CMB is better than that obtained from the light distribution (see Section 1.5). It should however be kept in mind that these data do not sample distances $cz > 8000$ km s^{-1}, and are unable to adequately detect the convergence in $V_{pec,LG}(R)$.

3.4. BULK FLOWS

The large amplitude bulk flow reported by Lauer & Postman, as reported in section 3.1, can be tested using the data discussed in the two preceding sections. As done by Giovanelli et al. (1996), galaxies and clusters contained within two cones of 30° aperture, directed respectively toward the apex and the antapex of the LP bulk flow, can be selected, and their peculiar velocities compared with those expected from the LP flow. In fig. 7, those data are binned by shells of radial velocity. Apex and antapex cones are added in opposition, so that the sign is that of the net velocity in the LP apex direction. The median peculiar velocity within $cz = 5000$ km s^{-1} hovers between 0 and 350 km s^{-1}, while beyond that distance the component of

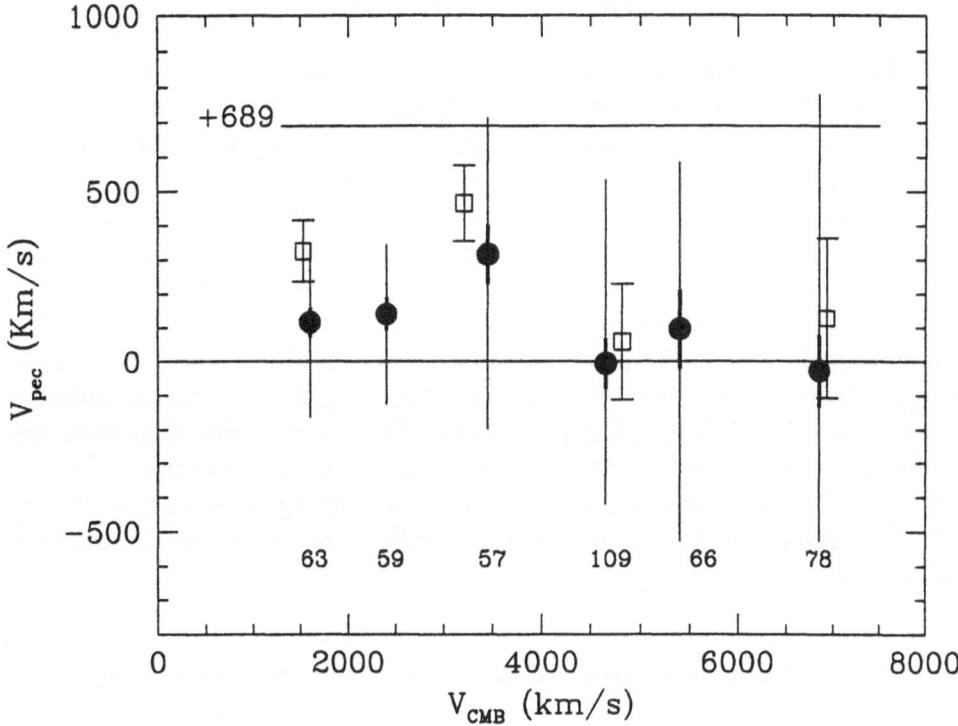

Figure 7. Median peculiar velocities of field galaxies (filled symbols) and of four clusters (unfilled) in the apex–antapex cones of the LP bulk flow. Numbers at bottom identify the number of galaxies used in each bin. The +689 km s^{-1}horizontal line refers to the LP bulk flow amplitude. Thin error bars outline the range spanned by data within the two inner quartiles, and thich error bars approximate a standard error on the median (after Giovanelli *et al.* 1996).

the flow in the LP apex direction subsides to values indistinguishable from zero. An average bulk flow with respect to the CMB as large as 689 km s^{-1}in that direction can be excluded for galaxies within the volume subtended by these data. The depth of the data is however limited, and it cannot exclude that the LP flow may entirely take place outside the depth of our survey but within the outer limits of the LP sample, i.e. between 8000 and 13000 km s^{-1}. Ongoing work by Dale *et al.* (1997) aims to extend the application of the TF technique to clusters to $cz \sim 18,000$ km s^{-1}, thus providing a suitable reference frame to test the BCG result. At the same time, Lauer, Postman and Strauss are extending the BCG sample to include approximately 500 objects.

3.5. RECONSTRUCTION OF THE DENSITY FIELD

Our data was also used to obtain a reconstruction of the 3–d peculiar velocity and density fields, using the *Potent* algorithm developed by Bertschinger and Dekel (1989). The result is shown in fig. 8 (see also previous reconstructions in Dekel 1994), and its details are described in da Costa *et al.* (1996). Reconstructions such as this rely on severe smoothing of the grid of values derived from the observed peculiar velocities; thus, only the coarsest details of the matter density distribution are recovered. Details of the reconstruction near the edges of the map are of poor reliability, due to the effects of smoothing. However, the principal characteristics of the mass distribution match relatively well the large conglomerates in the distribution of luminous matter, such as the Local, Coma, Perseus–Pisces and Hydra–Centaurus superclusters, as well as the vast voids between them. A comparison of a map such as this with one obtained from redshift surveys can be used to obtain an estimate of the parameter $\beta = \Omega_\circ^{0.6}/b$. This comparison is currently under way.

This work was supported by the National Science Foundation grant AST94–20505.

References

1. Aaronson, M. & Mould, J. 1983, *ApJ* **265**, 1
2. Aaronson, M. *et al.* 1982, *ApJ* **258**, 64
3. Aaronson, M. *et al.* 1986, *ApJ* **302**, 536
4. Bahcall. N.A., & Oh, S.P. 1996, *ApJ* **462**, L49
5. Bershady, M. 1996, preprint
6. Bertschinger, E. & Dekel, A. 1989, *ApJ* **335**, L5
7. Branchini, E. *et al.* 1996, *ApJ* **461**, L17
8. Burstein, D. 1982, *ApJ* **253**, 539
9. Collins, C.A., Nichols, R.C. & Lumsden, S.L. 1992, *MNRAS* **254**, 295
10. Dale, D.A., Giovanelli, R., Haynes, M.P., Scodeggio, M., Hardy, E. & Campusano, L. 1997, *AJ*, submitted
11. da Costa, L. *et al.* 1996, *ApJ* **468**, L5
12. Davies, J., Jones, H. & Trewhella, M. 1995, *MNRAS* **273**, 699
13. Davies, J. & Burstein, D. 1995, *Opacity of Spiral Disks*, Kluwer Pub., Dordrecht.
14. Dekel, A. 1994, *ARAA* **32**, 371
15. Djorhovski, S. & Davis, M. 1987, *ApJ* **313**, 59
16. Dressler, A. *et al.* 1987, *ApJ* **313**, L37
17. Dressler, A. *et al.* 1987, *ApJ* **313**, 43
18. Dressler, A. and Faber, S.M. 1990, *ApJ* **354**, L45
19. Efstathiou, G.P., Bond, J.R. & White, S.D. 1992, *MNRAS* **258**, 1P
20. Feldman, H.. Kaiser, N. & Peacock, J. 1994, *ApJ*, **426**, 23
21. Fisher, K.B. al 1993, *ApJ*, **402**, 42
22. Freudling, W. *et al.* 1995, *MNRAS* **268**, 943
23. Giovanelli, R. *et al.* 1995, *AJ*, **110**, 1059
24. Giovanelli, R. *et al.* 1996, *ApJ*, **464**, L99

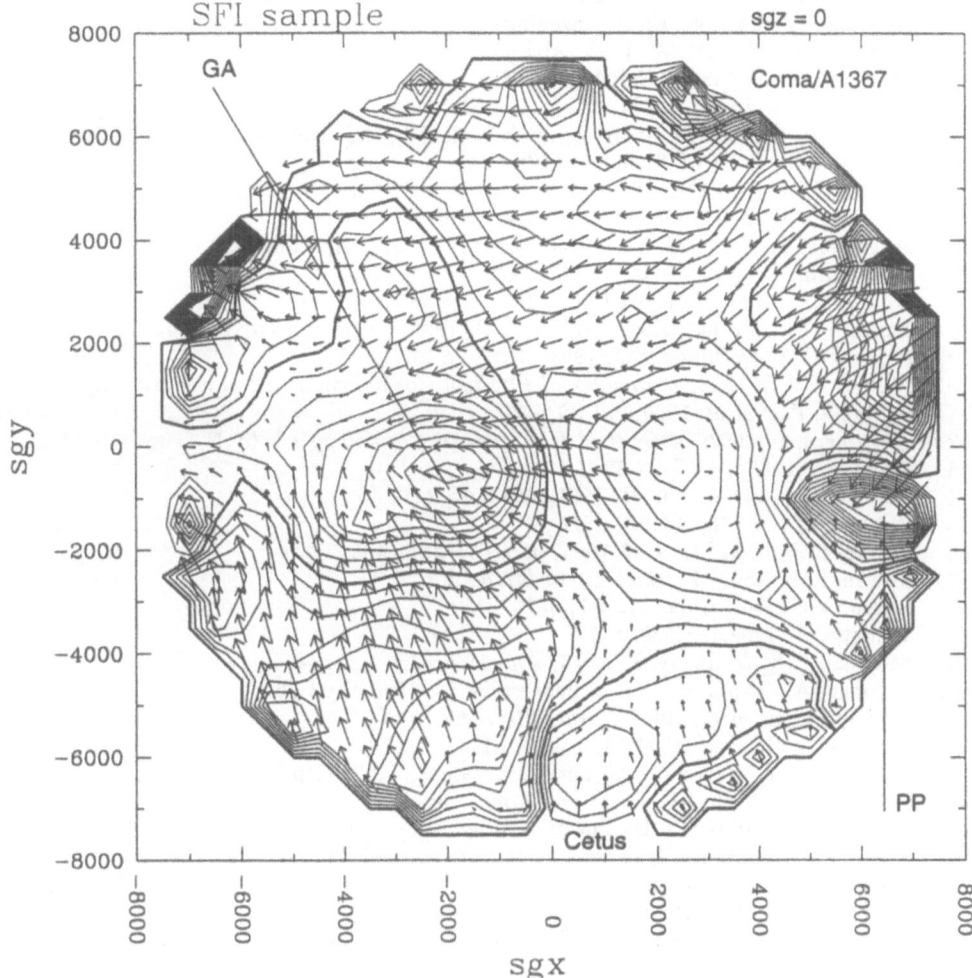

Figure 8. Potent reconstruction of the density and peculiar velocity field using our field spiral galaxy sample. Cut of the 3–d density field across the supergalactic plane (After da Costa *et al.* 1996).

25. Giovanelli, R. *et al.* 1997, *AJ*, in press
26. Groth, E.J. & Peebles, P.J.E. 1986, *ApJ* **310**, 499
27. Guzzo, L. *et al.* 1991, *ApJ* **382**, L5
28. Han, M, & Mould, J.R. 1990, *ApJ* **360**, 448
29. Hoessel, J.G. 1980, *ApJ* **241**, 493
30. Hudson, M. 1993, *MNRAS* **265**, 72
31. Jerjen, H. and Tammann 1993, *AA* **273**, 354
32. Joergensen, I. 1997, in *Galaxy Scaling Relations*, ed. by L. da Costa, in press
33. Juskiewicz, R. *et al.* 1990, *ApJ* **349**, 408
34. Kaiser, N. & Lahav, O. 1989, *MNRAS* **237**, 129
35. Kogut, A. *et al.* 1993. *ApJ* **419**, 1
36. Kolokotronis, A. *et al.* 1996, *MNRAS* **280**, 186

37. Kraan–Korteweg, R. 1985, in *ESO Worskshop on the Virgo Cluster*, ed. by O.–G. Richter and B. Binggeli, ESO, Garching
38. Landy, S.D. & Szalay, A. 1993, *ApJ* **412**, 64
39. Lahav, O. *et al.* 1989, *MNRAS* **238**, 881
40. Lauer, T.R. & Postman, M. 1994, *ApJ* **425**, 418
41. Lucey, J.R., Gray, P.M., Carter, D. & Terlevich, R.J. 1991, *MNRAS* **248**, 804
42. Lynden–Bell, D. *et al.* 1988, *ApJ* **326**, 19
43. Lynden–Bell, D., Lahav, O. & Burstein, D. 1989, *MNRAS* **241**, 325
44. Maddox, S.J. *et al.* 1990, *MNRAS* **242**, 43P
45. Mathewson, D.S., Ford, V. & Buckhorn, M. 1992a, *ApJ* **389**, L5
46. Moscardini, L. *et al.* 1996, preprint astro-ph/9511066
47. Mould, J.R. *et al.* 1991, *ApJ* **383**, 467
48. Navarro, J.F., Frenk, C.S. & White, S.D. 1996, preprint astro–ph/9611107
49. Peacock, J.A. & Dodds, S.J. 1994, *MNRAS* **267**, 1020
50. Peebles, P.J.E. 1993, *Principles of Physical Cosmology*, Princeton University Press
51. Pellegrin, P.S. *et al.* 1990, *AJ* **99**, 751
52. Pietronero, L. 1997, in *Critical Dialogues in Cosmology*, ed. by N. Turok, in press
53. Plionis M. 1988, *MNRAS* **234**, 401
54. Plionis M. & Valdarnini, R. 1991, *MNRAS* **249**, 46
55. Roberts, M.S. 1978, *AJ* **83**, 1026
56. Rowan–Robinson, M. *et al.* 1991, *MNRAS* **253**, 485
57. Rowan–Robinson, M., *et al.* 1990, *MNRAS* **247**, 1
58. Rubin, V.C. *et al.* 1976, *AJ* **81**, 687
59. Sandage, A. 1994, *ApJ* **430**, 1
60. Sandage, A.R. & Hardy, E. 1973, *ApJ* **183**, 743
61. Sandage, A., Tammann, G. & Federspiel, M. 1995, *ApJ* **452**, 1
62. Scaramella, R. *et al.* 1989, *Nature* **338**, 562
63. Scaramella, R. *et al.* 1991, *ApJ* **376**, L1
64. Scaramella, R., Vettolani, G. and Zamorani, G. 1994, *ApJ* **422**, 1
65. Schroeder, A. 1996, Ph. D. Thesis, Univ. Basel
66. Scodeggio, M. 1997, Ph.D. Thesis, Cornell University.
67. Scharf, C., Hoffman, Y., Lahav, O. & Lynden–Bell, D. 1992, *MNRAS* **256**, 229
68. Shaya, E.J. 1984, *ApJ* **280**, 470
69. Strauss, M.A. *et al.* 1992, *ApJ* **385**, 421
70. Strauss, M.A. & Willick, J.A. 1995, *Phys. Rep.* **261**, 271
71. Tammann, G.A. and Sandage, A. 1985, *ApJ* **294**, 81
72. Tini–Brunozzi, P. *et al.* 1995, *MNRAS* **277**, 1210
73. Tonry, J.L. & Schneider, D.P. 1988, *AJ* **96**, 807
74. Tully, R.B. & Fisher, J.R. 1977, *AA* **54**, 661
75. de Vaucouleurs, G. 1958, *AJ* **63**, 252
76. Vaucouleurs, G. de *et al.* 1982, *ApJ* **254**, 8
77. Willick, J.A. 1990, *ApJ* **351**, L5
78. Yahil, A., Strauss, M.A., Davis, M. & Huchra, J.P 1991, *ApJ* **372**, 380

HIGHER ORDER STATISTICS FROM THE EDSGC CATALOG

ISTVÁN SZAPUDI

Fermi National Accelerator Laboratory
Theoretical Astrophysics Group
Batavia, IL 60510

1. Introduction

The strongly non-Gaussian distribution of galaxies must be described with the full hierarchy of the N-point correlation functions, since none of the phase information is retained by the two-point function and its Fourier transform, the power spectrum. Higher-order statistics, however, are inherently more complex than two-point statistics, primarily because of the combinatorial explosion of terms and the large number of parameters involved. For this reason special techniques are necessary for the extraction of cosmologically important information.

An efficient method based on factorial moments of cell counts is used to analyse the higher order functions in the EDSGC catalog, an angular survey overing approximately 1000 square degrees ([14, 5]). The infinite sampling of the catalog [23] eliminates the measurement errors arising from the use of a finite number of sampling cells [4, 24]. In the next section EDSGC catalog is described, followed by an account of the measurement technique in §3. The results of the analysis are presented in §4, their relation to theoretical expectations is discussed in §5. More details can be found in [25].

2. The Edinburgh/Durham Southern Galaxy Catalogue

The Edinburgh/Durham Southern Galaxy Catalogue (EDSGC) is a catalogue of 1.5 million galaxies covering $\simeq$ 1000 square degrees centered on the South Galactic Pole (SGP). The database was constructed from COSMOS scans (a microdensitometer) of 60 adjacent UK IIIa–J Schmidt photographic plates and reaches a limiting magnitude of $b_j = 20.5$. The entire catalogue has < 10% stellar contamination and is $\gtrsim$ 95% complete for

D. N. Schramm and P. Galeotti (eds.), Generation of Cosmological Large-Scale Structure, 223–230.
© 1997 *Kluwer Academic Publishers.*

galaxies brighter than $b_j = 19.5$ [14]. The two–point galaxy angular correlation function measured from the EDSGC has been presented by [5, 21].

A rectangular area of the catalog between $\alpha = 22^h$, passing through
0^h to 3^h, and declination $-42 \leq \delta \leq -23$, was suitable for our purposes.
The original coordinates were converted to physical ones using an equal
area projection: $x = (\alpha - \alpha_{\min}) \cos \delta$, $y = \delta - \delta_{\min}$. This simple formula is
suitable for the small angular scales considered in this paper. The projection
did not affect the declination range, but to obtain a rectangular area the
physical coordinates corresponding to right ascension were restricted to
values less than 55°. This resulted in a sample of 2.9×10^5 galaxies, and a
total effective survey area of 1045 square degrees, or $\simeq 997$ square degrees
after accounting for the cut-out regions.

Magnitude cuts were determined by practical considerations. The catalog is complete to about 20.3 magnitude. We adopt a limit half a magnitude
brighter for our analysis to be conservative. For consistancy with the APM
survey [16, 17, 18, 12], the magnitude cut $16.98 \leq m_{\text{EDS}} \leq 19.8$ was used.
There is an offset in the magnitude scales of the two catalogs [20]. Based
on matching the surface densities listed in [26], the magnitude range we
have adopted corresponds approximately to the APM magnitude range
$17 \leq m_{\text{APM}} \leq 20$.

3. The Method of Analysis

The calculation of the higher order correlation functions consists of a sequence of three consecutive steps: estimation of the probability distribution,
calculation of the factorial moments, and extraction of the normalized, averaged amplitudes of the N-point correlation functions. We present the
relevant definitions and theory below.

Let P_N denote the probability that a cell contains N galaxies, with
implicit dependence on the cell size ℓ. The best estimator for P_N from
the catalog is the probability that a randomly thrown cell *in the catalog*
contains N galaxies (excluding edge effects, which are negligible for the
scales in the present study, except perhaps on the largest scales as a result
of the holes cut out around bright stars). This may either be calculated
from the configuration of the points using a computer algorithm (see [23]),
or estimated by actually throwing cells at random,

$$\tilde{P}_N = \sum_{i=1}^{C} \delta(N_i = N), \tag{1}$$

where C is the number of cells thrown and N_i is the number of galaxies in
cell i. It is desirable to use as many cells as possible, since for large C, the

errors behave as [24]

$$E^{C,V} = (1 - \frac{1}{C})E^{\infty,V} + E^{C,\infty}, \qquad (2)$$

where the $E^{C,V}$ is the total theoretical error (not including the sytematic errors of the catalog), $E^{\infty,V}$ is the 'cosmic' error associated with the finiteness of the catalog, and $E^{C,\infty}$ is the error associated with the finite number of cells used for the estimator. Since $E^{C,\infty} \propto C^{-1}$ [24], the lowest possible error is obtained for $C \longrightarrow \infty$. We employed such a code on scales up to $2°$.

The factorial moments (see e.g. [27]), may be obtained from the probability distribution using

$$F_k = \sum P_N(N)_k, \qquad (3)$$

where $(N)_k = N(N-1)..(N-k+1)$ is the k-th falling factorial of N. The F_k's directly estimate the moments of the underlying continuum random field which is Poisson sampled by the galaxies.

The average of the N-point angular correlation function on a scale ℓ, $\bar{\omega}_N$, is related to the cumulants of the distribution $s_N = \bar{\omega}_N/\bar{\omega}_2^{N-1}$. They a are related to the factorial moments through the recursion relation [27],

$$s_k = \frac{F_k\bar{\omega}_2}{N_c^k} - \frac{1}{k}\sum_{q=1}^{k-1}\frac{(k-q)s_{k-q}F_q\binom{k}{q}}{N_c^q}, \qquad (4)$$

where $N_c = \langle N \rangle \bar{\omega}_2$. Note that although the notation indicates a projected catalog, there are corresponding expressions for three dimensions.

The deprojection of the s_N's to their 3D counterparts has an intrinsic limitation due to the finite sizes of the cells. While the deprojection of any individual tree-structure is well-defined, care must be taken in interpreting the deprojected values of the cell-count determined s_N's, since these implicitly contain an averaging over trees within each cell (see [26] for a discussion). On small scales, where clustering is strongly nonlinear, the coefficients deproject to the 3D coefficients S_N defined by $S_N = \bar{\xi}_N/\bar{\xi}_2^{N-1}$, where the hierarchical assumption may be presumed valid. In this case,

$$s_N = R_N S_N, \qquad (5)$$

where S_N is the corresponding three dimensional value for the spherically averaged $\bar{\xi}_N$'s. The projection coefficients R_N's are fairly insensitive to slight variations of the magnitude cut (see Table 2 in [26]), and the shape dependence is neglected according to the findings of [3]. We adopt the R_N's of [26] with a Hubble constant of $H_0 = 100 \text{km s}^{-1} \text{Mpc}^{-1}$.

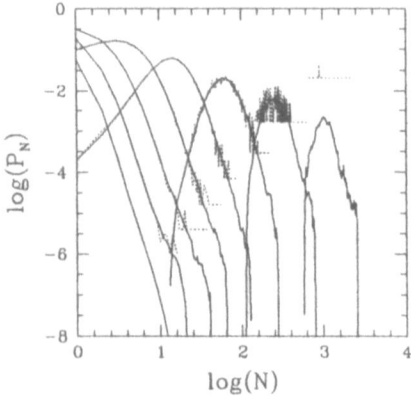
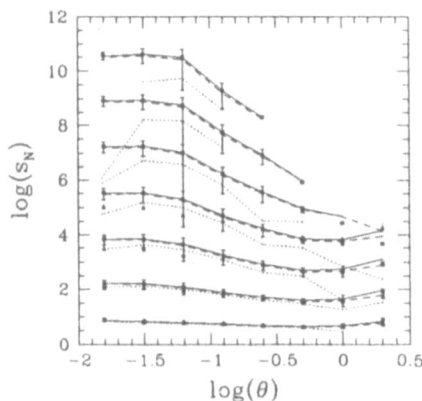

Figure 1. Figure 1. a. Shows the distributions P_N of counts in cells measured in the EDSGC catalog. The solid line corresponds to infinite sampling, while the dotted line to severe undersampling. The curves from left to right correspond to cell sizes from $0.015125°$ doubling up to $2°$. Exhaustive sampling is essential on all scales. b. The solid line is the measurement of the s_N's over the entire survey area with infinite sampling, the dotted line is the same with low sampling. Undersampling results in a systematic underestimate of the coefficients. The squares show the mean of the measurements (see text for details) in four equal parts of the survey, and the errors are calculated from the dispersion. The misalignment of the squares and the solid line at the largest scales may be a result of edge effects. The triangles show the s_N's corresponding to the best fitting formal n_{eff} (see text).

4. Results

We measured counts in cells by calculating the results corresponding to an infinite number of square cells, placed according to the algorithm of [23], with sizes in the range $0.015125° - 2°$ (corresponding to $0.1 - 13h^{-1}$Mpc with $D \simeq 370h^{-1}$Mpc, the approximate depth of the catalog). The largest scale is limited by the geometry induced by the cutout holes: the number of available cells would be severely limited for a measurement on significantly larger scales, since cells intersecting with the cutout holes were rejected. The smallest scale approaches that of galaxy halos for the typical depth of the catalog.

The results of both infinite and low sampling measurements for P_N are displayed in Figure 1a. The low sampling corresponds to covering the area with cells once only, i.e. the number of sampling cells is $C_V = V/v$, where V is the volume of the survey and v is the volume of the sampling cell at the given scale. As proved in ([24]), the 'number of statistically independent cells', C^*, depends strongly both on scales and on the aims of the measurement, but for higher order statistics it is generally much larger than C_V. Therefore to minimize the errors as much as possible we used infinite oversampling for all measurements in this paper. A comparison of the two curves shows the substantial improvement in accuracy achieved through oversampling. Note that covering the area fully with C_V number of cells is not fully equivalent to throwing the same number of random cells, because these might overlap, thus more effectively sampling clusters, and in principle decreasing the bias toward low values visible in Figure 1b.

Figure 1b shows the scale dependence of the s_N's determined from the counts in cells. The solid line corresponds to the measurements of the entire survey area with high oversampling. The dotted line is the same measurement with undersampling. For the error determination we divided the survey into four equal parts, similar to the approach of [12]; this procedure can overestimate the cosmic error, because subcatalogs have smaller area, but it could also lead to underestimation, because the subcatalogs are not independent volumes [24]. The squares show the mean of the measurements taken in the following way: to avoid the bias introduced by the fact that the s_N's are nonlinear functions of the factorial moments, the mean of the moments was taken first, and then the cumulants were calculated. The error bars, estimated by a determination of the dispersion of the (possibly biased) s_N's calculated from the factorial moments from each zone, are shown only for those points for which there was sufficient physically valid data permitting a determination.

Figure 1 exhibits two plateaus, one at small scales ($< 0.03°$) and a second at large ($> 0.5°$). The large scale plateau is approaching the width of the survey, and so may merely reflect edge effects. The plateau at small separations, however, may indicate that the strongly nonlinear clustering limit has been reached, in which case the hierarchical form for the angular correlations should apply, for which the coefficients appear to converge.

5. Comparison with Theory

At small (non-linear) scales, the hierarchical tree model [22] is believed to be a good approximation to the clustering. At larger (weakly non-linear) scales, perturbation theory of gravitating matter starting from Gaussian initial conditions provides a basis for comparison. In a projected catalog, the transition between scales is somewhat uncertain, since a lengthscale is assigned to angular scales using the depth of the catalog. This procedure is physically correct although somewhat arbitrary, and there could be effects associated with mixing of different scales in the selection function weighted cone corresponding to a cell in an angular catalog. While no existing measurement has clearly demonstrated the validity of either of the above models, the results based on moment correlators seem to support the hierarchical model, at least on small scales [28, 19, 26], as does the present work. In what follows, a direct comparison is made without taking into account the possibility of biasing: the data are consistent with the galaxies' acting as faithful tracers of the underlying mass distribution.

A plateau at small separations may be expected when the clustering becomes strongly nonlinear. The effect is found, for example, in the N-body experiments for scale-free clustering by [7]. If the clustering we measure is strongly nonlinear on the smallest scales, then we are permitted to identify $S_N = s_N/R_N$ in Table 2 at small separations. We may then in turn derive the 3D clustering amplitudes.

For $r = 0.1$ Mpc we find for $N = 3-9$, $Q_N = (2.02, 7.3, 30, 108, 320, 745, 1298)$. The values for Q_3 and Q_4 somewhat exceed those found for the Lick-Zwicky sample [13, 10, 28]), and greatly exceed the values found for the CfA1 and SSRS surveys [11]. The discrepancy between the larger angular samples and the smaller samples used for the redshift surveys has been previously noted by [9]. Our results suggest the discrepancy at small scales may be even larger. The reason for the difference in the values is unknown, but may be a result of cosmic variance [8]. It appears not to be a result of the added redshift information, since [12] found that suppressing the redshifts in the CfA1 and SSRS surveys and treating the samples as angular catalogs had little effect on the values.

Since our measurements are most uncertain at the largest scales, we omit here the comparision with the theory of weakly nonlinear clustering, nevertheless, details can be found in [25]. For the highly non-linear regime [7] finds from N-body experiments for scale-free initial conditions that the values for S_N vary only slightly with scale, increasing for small separations where the clustering becomes strongly nonlinear. They find, independent of n,

$$S_N \simeq [D(\bar{\xi}_2)]^{N-2} \tilde{S}_N, \qquad (6)$$

for $N = 3, 4$, and 5, where $D(\bar{\xi}_2) = (\bar{\xi}_2/100)^{0.045}$ and $\tilde{S}_N$ is the value of S_N for $\bar{\xi}_2 = 100$. The relation implies a weak departure from the hierarchical clustering behavior, since the S_N depend on scale. The dependence is so weak, however, that the departure is slight. We compare the clustering amplitudes found in the EDSGC with this relation in Figure 2a. The agreement in the strong clustering limit is remarkably good, particularly for $N = 3$ and 4. Because we have only angular information, it is not possible to determine whether the deviation from the scaling relation for $\log \bar{\xi}_2 < 2$ is a real effect or a consequence of the inherent limitations of extracting 3D

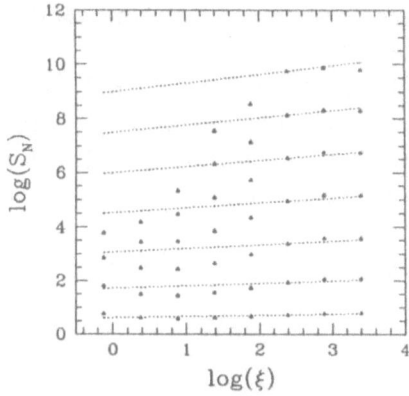
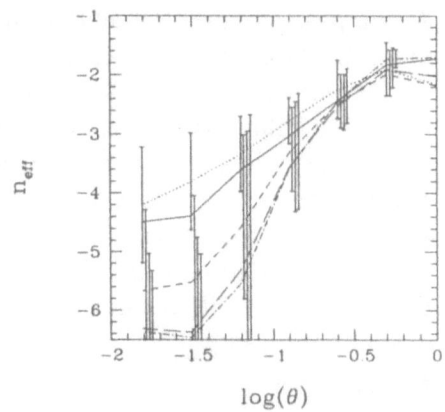

Figure 2. Figure 2. a. The clustering amplitudes s_N/R_N as a function of the average 2-point function $\bar{\xi}_2$. Also shown is the scaling relation of [6] found in the strongly nonlinear limit in N-body experiments with scale-free initial conditions. b. The best formal fits for n_{eff} (*solid*), using up to sixth order quantities. Also shown are the values determined from each N separately, including an indication of the errors based on the upper and lower quartile values for each S_N. Within the errors, the clustering may be described by a single value of n_{eff}. Shown are the values of n_{eff} for $N = 3$ (dotted), $N = 4$ (*short-dashed*), $N = 5$ (*long-dashed*), and $N = 6$ (*dot-dashed*).

information from a projected catalog. It should be noted that the agreement is particularly remarkable since hierarchical clustering is assumed for the underlying distribution in order to convert from the projected correlations to the 3D, while the relation of equation (6) violates this assumption. This suggests that the hierarchical model is a good, though perhaps not perfect, description of the clustering.

[6] finds that the clustering for $N = 3$, 4, and 5 may be described by a single effective spectral index n_{eff}, found from the expressions for weakly nonlinear clustering [15, 1, 2]. Although the relations between the S_N and n from weakly nonlinear theory do not apply for strong clustering, and even less so in an angular catalog, we may adopt them to obtain a formal value for n_{eff} as done by [7, 6]. We do so by fitting s_N/R_N to the expressions for S_N in the limit of weakly nonlinear clustering for $N = 3 \ldots 6$ using least squares, for $\theta \le 1°$. The results are shown in Figure 2b, including the values derived for each N individually. Within the error estimates, a single value of n_{eff} appears to provide an adequate description of the

clustering amplitudes, although the errors are large for small separations. A comparison with N-body results for scale-dependent clustering models, like a CDM-dominated cosmology, could be very illuminating.

The author would like to thank A. Meiksin and R.C. Nichol for an enjoyable collaboration, S. Colombi, J. Frieman, and A. Szalay for stimulating discussions. We are indebted to C. Collins, S. Lumsden, N. Heydon-Dumbleton, and H. MacGillivray for full use of the EDSGC. The author would like to thank the support and hospitality of the International School of Particle Astrophysics in Erice, and support by DOE and NASA through grant NAG-5-2788 at Fermilab.

References

1. Bernardeau, F. 1994a, *ApJ*, **433**, 1
2. Bernardeau, F. 1994b, *A&A*, **291**, 697
3. Boschán, P., Szapudi, I., & Szalay, A. 1994, *ApJS*, **93**, 65
4. Bouchet, F.R., Schaeffer, R., & Davis, M. 1991, ApJ, **383**, 19
5. Collins, C. A. Nichol, R. C., & Lumsden, S. L. 1992, *MNRAS*, **254**, 295
6. Colombi, S., Bernardeau, F., Bouchet, F.R., & Hernquist, L. 1996, (in preparation)
7. Colombi, S., Bouchet, F.R., & Hernquist, L. 1995, *A&A*, **281**, 301
8. Colombi, S., Bouchet, F.R., & Schaeffer, R. 1994, *A&A*, **281**, 301
9. Fry, J. N. & Gastañaga, E. 1994, *ApJ*, **425**, 1
10. Fry, J., & Peebles, P.J.E. 1978, *ApJ*, **221**, 19
11. Gastañaga, E. 1992, *ApJ*, **319**, L17
12. Gastañaga, E. 1994, *MNRAS*, **268**, 913
13. Groth, E.J., & Peebles, P.J.E. 1977, *ApJ*, **217**, 385
14. Heydon-Dumbleton, N. H., Collins, C. A., & MacGillivray, H. T. 1989, *MNRAS*, **238**, 379
15. Juszkiewicz, R., Bouchet, F. R., & Colombi, S. 1993, *ApJ*, **412**, L9
16. Maddox, S. J., Efstathiou, G., Sutherland, W. J., & Loveday, L. 1990a, *MNRAS*, **242**, 43P
17. Maddox, S. J., Sutherland, W. J., Efstathiou, G., & Loveday, L. 1990b, *MNRAS*, **243**, 692
18. Maddox, S. J., Sutherland, W. J., Efstathiou, G., & Loveday, L. 1990b, *MNRAS*, **246**, 433
19. Meiksin, A., Szapudi, I., & Szalay, A., 1992, *ApJ*, **394**, 87
20. Nichol, R. C. 1992, PhD thesis, University of Edinburgh
21. Nichol, R. C., & Collins, C. A. 1994, *MNRAS*, **265**, 867
22. Peebles, P.J.E. 1980, The Large Scale Structure of the Universe (Princeton: Princeton University Press)
23. Szapudi, I. 1997, (in preparation)
24. Szapudi, I., & Colombi, S. 1996, *ApJ* **470**, 131
25. Szapudi, I., Meiksin A., & Nichol, R.C. 1996, *ApJ* **473**, 15
26. Szapudi, I., Dalton, G., Efstathiou, G.P., & Szalay, A. 1995, *ApJ*, **444**, 520
27. Szapudi, I. & Szalay, A. 1993, *ApJ*, **408**, 43
28. Szapudi, I., Szalay, A., & Boschán, P. 1992, *ApJ*, **390**, 350

ON THE COSMOLOGICAL MASS FUNCTION

PIERLUIGI MONACO
SISSA & Dipartimento di Astronomia,
via Beirut 4, 34014 Trieste

Abstract. The theoretical determination of the mass function is performed by means of Lagrangian perturbation theory and ellipsoidal collapse. The statistical framework necessary to determine the mass function is briefly described. The main results and future applications are outlined.

1. Introduction

The cosmological mass function gives the number density of cosmic structures (galaxies, galaxy groups and clusters etc.) of given mass, per unit mass interval. It can be compared to direct mass estimates (see, e.g., the lectures by Kaiser in this volume for galaxy clusters), as well as to the distribution of observational quantities connected to the mass of the structures, as optical or X-ray luminosities or X-ray temperatures.

Press & Schechter (1974, hereafter PS) were the first to propose an approximate theoretical determination of the mass function: they assumed that every point in a smoothed density field enters a collapsed structure if its linearly evolved density contrast overcomes a threshold value δ_c; according to spherical collapse, $\delta_c \simeq 1.69$. After PS, many authors extended the validity of their procedure, mainly considering the statistical aspects of the problem, or proposed alternative formulations. All these works are reviewed in Monaco (1997). I present here a new formulation of the mass function problem, based on Lagrangian perturbation theory. Full details can be found in Monaco (1996a,b).

2. Dynamics

As already noted in Monaco (1995), the Lagrangian formulation of fluidodynamics is suitable to face the mass function problem. In the Lagrangian for-

D. N. Schramm and P. Galeotti (eds.), Generation of Cosmological Large-Scale Structure, 231–234.

mulation, every mass particle is followed in its motion from the original (Lagrangian) position $\mathbf{q}$ to the final (Eulerian) position $\mathbf{x}$: $\mathbf{x}(\mathbf{q}, t) = \mathbf{q} + \mathbf{S}(\mathbf{q}, t)$, where $\mathbf{S}(\mathbf{q}, t)$ is the displacement field, and its derivatives constitute the deformation tensor. The whole Newtonian problem can be recast in terms of the field $\mathbf{S}$, and the equations can be solved for small displacements (see, e.g., Buchert 1994, Bouchet et al. 1995, Catelan 1995).

Lagrangian perturbations can be used to estimate the collapse time, defined as the instant at which all the quantities relative to a mass element, namely density contrast, expansion and shear, diverge. This instant can be found by solving the equation:

$$\det(\delta_{ij} + S_{i,j}(\mathbf{q}, t)) \equiv J(\mathbf{q}, t) = 0 \tag{1}$$

i.e. when $J(\mathbf{q}, t)$, the Jacobian determinant of the Lagrangian – Eulerian transformation, vanishes. This instant corresponds to the well-known orbit crossing (or shell crossing) event. Noteworthy, it is possible to parameterize out, with great accuracy, the dependence of dynamics on the background cosmology by using the growing mode $b(t)$ as time variable. Then, the collapse "time" will be called b_c.

An important point is to verify whether the Lagrangian series converges at collapse; this is not assured by construction, as the various perturbative terms are of the same order at collapse. It has been verified (Monaco 1996a) that, in the easy case of a homogeneous ellipsoid, the Lagrangian series converges to the exact solution; second-order perturbations show a bad behavior for initial underdensities. In the more general case of Gaussian fields with scale-free spectra, the Lagrangian series shows a convergence toward a solution if the point collapses fast enough ($b_c < 1$ with unit mass variance); the convergence ceases to hold even at a qualitative level at $b_c > 2$. In the range of convergence, the Lagrangian series tends to the same solution as the homogeneous ellipsoidal model.

Such an analysis has revealed a number of interesting connections between ellipsoidal collapse and Lagrangian perturbations. In particular, the ellipsoidal model can be considered as a particular truncation of the Lagrangian series.

3. Statistics

The statistics of inverse collapse times determines the statistics of collapsed regions, and then the mass function. The most important quantity is the one-point probability distribution function (hereafter PDF) of collapse times. It is convenient to deal with inverse collapse times, $F \equiv (b_c)^{-1}$ as they are better behaved: they take the value 0 in the delicate passage

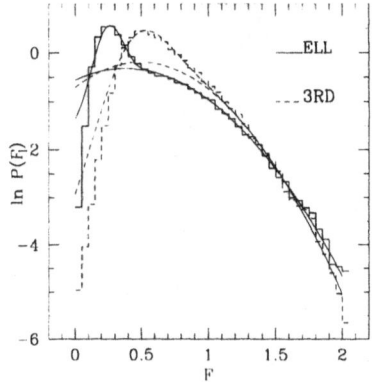

Figure 1. PDFs of inverse collapse times

from collapse to non-collapse; if spherical collapse is assumed, F is simply proportional to the density contrast.

Fig. 1 shows the PDF of inverse collapse times, for 3rd-order Lagrangian perturbations (3RD) and ellipsoidal collapse (ELL). The two curves are very similar for $F > 1$, and show the same qualitative behavior: they are Gaussian at intermediate and large F values (fitting Gaussian curves are also shown), and show a peak at small F values.

In Monaco (1996b) it is shown that it is possible to extend the diffusion formalism, developed by Bond et al. (1991), to the non-Gaussian process F, to calculate the mass function. Such formalism corrects for the so-called cloud-in-cloud problem, and gives a correctly normalized mass function; however, it is valid only if the initial field is smoothed with a filter which is a top-hat in the Fourier space (sharp k-space filter). For Gaussian filters, the problem can be approximately solved by using a procedure proposed by Peacock & Heavens (1990).

Fig. 2 shows, as an example, the outcoming mass function (with Gaussian smoothing) for a power-law spectrum with spectral index $n = -2$. It can be noted that both the ELL and 3RD mass functions are very similar to a PS one with $\delta_c=1.5$.

4. Conclusions and Further Work

The main conclusions about the mass function are that more large-mass objects are predicted to form with respect to the usual PS prediction with the "spherical" $\delta_c=1.69$ (which corresponds to an effective lowering of the δ_c parameter; this had already been found by Monaco 1995), and that the small-mass part cannot be robustly determined, as it relies on the uncertain dynamics of slowly-collapsing regions.

234

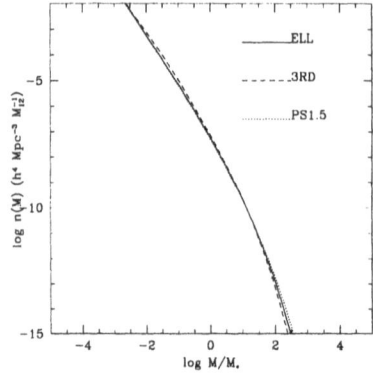

Figure 2.   Mass functions for a power-law spectrum

This dynamical framework provides a useful tool for describing the formation history of cosmological structures; because of the realistic dynamical treatment, plenty of dynamical information is available for the collapsing structures. Future work in this direction will be a more accurate treatment of the geometry of collapsed regions in Lagrangian space, an accurate comparison to numerical simulations, and applications to the formation of cosmologically relevant objects, from high-redshift objects to galaxy clusters.

References

1.   Bond, J.R., Cole, S., Efstathiou, G., and Kaiser, N. (1991) Excursion Set Mass Functions for Hierarchical Gaussian Fluctuations, *The Astrophysical Journal*, 379, 440
2.   Bouchet, F.R., Colombi, S., Hivon, E., and Juszkiewicz, R. (1995) Perturbative Lagrangian Approach to Gravitational Instability, *Astronomy & Astrophysics*, 296, 575
3.   Buchert, T. (1994) Lagrangian Theory of Gravitational Instability, *Monthly Notices of the Royal Astronomical Society*, 267, 811
4.   Catelan, P. (1995) Lagrangian Dynamics in Non-flat Universes and Non-linear Gravitational Evolution, *Monthly Notices of the Royal Astronomical Society*, 276, 115
5.   Monaco, P. (1995) The Cosmological Mass Function with Non-spherical Collapse, *The Astrophysical Journal*, 447, 23
6.   Monaco, P. (1996a) A Lagrangian Dynamical Theory for the Cosmological Mass Function; I Dynamics, *Monthly Notices of the Royal Astronomical Society*, in press
7.   Monaco, P. (1996b) A Lagrangian Dynamical Theory for the Cosmological Mass Function; II Statistics, *Monthly Notices of the Royal Astronomical Society*, in press
8.   Monaco, P. (1997) The Cosmological Mass Function, *Fundamentals of Cosmic Physics*, submitted
9.   Peacock, J.A., and Heavens, A.F. (1990) Alternatives to the Press-Schechter Cosmological Mass Function, *Monthly Notices of the Royal Astronomical Society*, 243, 133
10.   Press, W.H., and Schechter, P. (1974) Formation of Galaxies and Clusters of Galaxies by Self-similar Gravitational Condensation, *The Astrophysical Journal*, 187, 425

A FEW POINTS CONCERNING GALAXY LUMINOSITY FUNCTIONS

MARK U. SUBBARAO
The Johns Hopkins University
Department of Physics and Astronomy
Baltimore, MD 21218, USA

1. Introduction

The luminosity function provides, arguably, the two most basic descriptions of the galaxy population: their spatial density and their distribution in luminosity. In this chapter I review some topics which I considered in my thesis. In the presence of large scale structure sophisticated techniques are needed for luminosity function. I will discuss one of these and its extension to a photometric data set. All luminosity function estimation techniques which are insensitive to large scale structure require that only the normalization, not the shape, of the luminosity function vary with radial position. In the final two sections of this chapter we will see that this is not strictly true.

2. Luminosity Function Estimation

2.1. C- METHOD

The C-method [8] is a powerful estimator for the luminosity function. It is a non-parametric technique which is insensitive to density inhomogeneities and makes full use of all the data. The quantity $C(M)$ is defined by the following equation: $d\Psi/\Psi = dX/C$, where $\Psi(M)$ is the cumulative luminosity function. In the absence of differential K–corrections $C(M)$ is the number of galaxies in the volume limited sample defined by the absolute magnitude M. This equation has the solution,

$$\Psi(M) = A \exp\left[\int_{-\infty}^{M} \frac{dX}{C}\right].$$

(1)

D. N. Schramm and P. Galeotti (eds.), Generation of Cosmological Large-Scale Structure, 235–238.
© 1997 *Kluwer Academic Publishers.*

2.2. A SURVIVAL ANALYSIS DERIVATION OF LYNDEN-BELL'S FORMULA

For discrete galaxies the function $dX(M)$ is a series of Dirac delta functions and the evaluation of equation of 1 is non–trivial. A simple solution comes from an analogy from biostatistics. We construct the hazard function, which in this astronomical context is better termed the intensity function (Grace Yang, priv. comm.). If p_i is the probability that a galaxy with absolute magnitude greater than or equal to M_i has an absolute magnitude M_i, then $\Psi_k = \prod_{i<k} p_i$. We can relate p_i to C as follows, $p_i = (1 + C^-(M_i))^{-1}$, where we observe a galaxy and $p_i = 1$ where we do not, here $C^-(M_i)$ does not include the ith galaxy. This leads to Lynden–Bell's solution,

$$\Psi(M) = A \prod_{M_i < M} \left(\frac{C(M_i)}{1 + C(M_i)} \right). \tag{2}$$

2.3. FUZZY DATA

The photometric redshift technique [3] provides a quantum leap in the number of galaxies with redshift information. If we consider the intrinsic scatter in the photometric redshift relation then we have a probability distribution for the estimate of the absolute magnitude. Since we are now dealing with continuous quantities the luminosity function can be directly calculated from equation 1 [11].

3. Evolution of the Luminosity Function at Moderate Redshift

A recent investigation of the evolution of the galaxy luminosity function from the Canada France Redshift Survey [7] has shown that when divided by rest frame color the blue galaxies show significant evolution, while the red galaxies show little change in their distribution. Using a photometric sample, the greater number of galaxies available will allow us to further sub–divide the sample by color, allowing us to further isolate the evolving population of galaxies. We construct a sample of galaxies with photometric redshifts from the photometric survey data of the Kitt Peak Galaxy Redshift Survey (KPGRS) ([5], [4]). Details of the data can be found in [1],[9]. We construct a sample with limiting magnitude $B_J = 23$ in the fields SA68 and SA57 which consists of nearly 4000 galaxies. This is significantly more than the 600 galaxies in the [7] data set. However their magnitude limit of $I_{AB} = 22.5$ is deeper (roughly 1 magnitude for a Sbc galaxy at $z = 0.5$). We limit our analysis to galaxies within the redshift range $0.1 < z < 0.6$.

Luminosity functions are calculated using the C-method modified to handle photometric redshift data as described in [11]. We divide the sam-

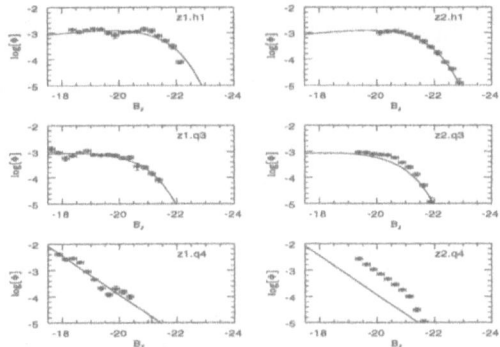

Figure 1. The luminosity functions of the subsamples ($H_0 = 50$, $q_0 = 0.5$. The best fit Schecter function of the lower redshift bin is plotted as a baseline for comparison. z1 corresponds to $0.1 < z < 0.3$ and z2 corresponds to $0.3 < z < 0.6$. The red half of the galaxy population is denoted h1, the third bluest quartile q3, and the bluest quartile q4.

ple into quartiles based on their restframe color. Since the reddest galaxies are the most difficult to observe at high redshift in a B_J limited sample, and since the red half of the galaxy population in the CFRS survey shows little evidence of evolution; we combine the two reddest quartiles, resulting in a sample split three ways in color. We split the sample six ways into two redshift bins, $0.1 < z < 0.3$ and $0.3 < z < 0.6$, and three color bins. The luminosity functions are plotted along with a Schecter function fit to the low redshift bin, which provides a baseline for assessing evolution. Only the bluest quartile shows evidence of evolution. As steep as the luminosity functions for the bluest quartile are; when integrated over the redshift range for which they have been calculated they do not overpredict the blue number counts.

4. LF vs. Environment

Regions of high and low galaxy density presumably correspond to fluctuations in the initial density field. In this sense these regions are equivalent to different laboratories in which to study galaxy formation. The galaxy density could also have an effect on their subsequent evolution, merging depleting the number of small galaxies and creating larger ones [2], and interactions possibly triggering dwarf galaxy formation [6].

Using the Canada France Redshift Survey data set [7] the sample is divided into regions of high and low galaxy density according to the one–dimensional redshift distribution. The redshift distribution of the entire sample is smoothed, multiplied by $\frac{2}{5}$, and the result is used to threshold

238

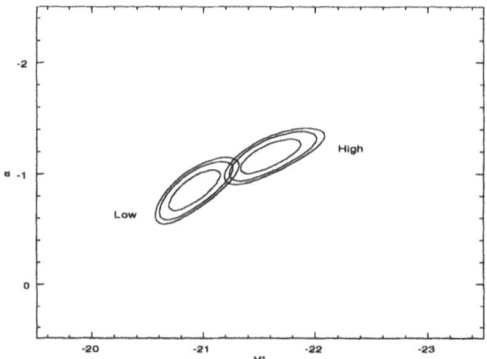

Figure 2. STY 68%, 90% and 95% likelihood contours are plotted for the high and low density samples.

the redshift distributions of the five separate fields which make up the CFRS. This splitting technique is model independent and provides roughly equal numbers of galaxies in the high and low density samples.

Once the sample is divided according to density, the luminosity functions of both the high density and low density sample are calculated using the parametric maximum likelihood STY [10] technique. Contours in the M^*, α space are calculated by finding the iso-likelihood contour interior to which is a certain percentage of the integrated likelihood.

The luminosity functions of the different regions are shown to be distinct. The region of high galaxy density has brighter galaxies as well as a steeper faint end slope. This is consistent with the effects of merging and tidally triggered star formation. This is different than the expectation that regions of high galaxy would have the luminosity function of early type galaxies and the low density regions that of the late types. Although in a magnitude limited sample this would be the impression.

References

1. Bershady, M.A. et al., 1994, AJ, 108, 870.
2. Carlberg, R.G., & Charlot, S.,1992, ApJ, 397, 5.
3. Connolly, A.J., Csabai, I., Szalay, A.S., Koo, D.C., Kron, R.C., Munn, J.A., 1995, AJ, 110, 2655.
4. Koo, D.C., 1986, ApJ, 311, 651.
5. Kron, R.G., 1980, ApJS, 43, 305.
6. Lacey, C., Silk, J., 1991, ApJ, 381, 14.
7. Lilly, S.J., Tresse, L., Hammer, F., Crampton, D., LeFevre,O., 1995, ApJ, 455, 108.
8. Lynden–Bell, D., 1971, MNRAS, 155, 95.
9. Munn, J., 1997.
10. Sandage, A., Tammann, G.A., & Yahil, A., 1979, ApJ, 232, 352.
11. SubbaRao, M.U., Connolly, A.J., Szalay, A.S., Koo, D.C., 1996, AJ, 112, 929.

SPECTROSCOPY OF SOLAR NEUTRINOS WITH BOREXINO

TH. P. GOLDBRUNNER AND M. E. NEFF

Department of Physics, Technische Universität München,
James Franck Str., 85748 Garching, Germany
E-mail: goldbrunner(neff)@e15.physik.tu-muenchen.de

1. What will we learn from BOREXINO ?

The results of all solar neutrino experiments carried out so far are shown in Tab.1 (the gallium experiments GALLEX and SAGE are taken together). For comparison the range of predictions of several Standard Solar Models (SSM) [1] [2] [3] is listed for each type of experiment together with the calculated contributions of the strongest solar neutrino branches. A basic analysis shows that the result of Kamiokande [4], which is only sensitive to the high energetic and less accurately predicted ^{8}B-neutrinos, is in the range of predicted values. That means, depending on the underlying SSM, there is a deficit of neutrinos or not. Combining the Kamiokande result with the data of the radiochemical Homestake-experiment [5] the derived flux for the ^{7}Be-neutrinos seems to be strongly suppressed for all SSM's. Using a minimal SSM the strong suppression of ^{7}Be-neutrinos is indepen-

	Gallium	Homestake	Kamiokande
pp	51 - 61%	-	-
^{7}Be	24 - 28%	13 - 25%	-
^{8}B	5 - 12%	70 - 83%	100%
data	69 ± 8 SNU	2.56 ± 0.22 SNU	$(2.8 \pm 0.38) \cdot 10^6 \mathrm{cm}^{-2}\mathrm{s}^{-1}$
theory	115 - 137 SNU	4.2 - 9.3 SNU	$(2.5 - 6.6) \cdot 10^6 \mathrm{cm}^{-2}\mathrm{s}^{-1}$

TABLE 1. Comparison between the results and the range of predictions for the solar neutrino experiments.

D. N. Schramm and P. Galeotti (eds.), Generation of Cosmological Large-Scale Structure, 239–244.
© *1997 Kluwer Academic Publishers.*

dently obtained by the radiochemical gallium experiments [6] [7], which in contrast to the Homestake-experiment are calibrated with neutrino sources of known activity and therefore can exclude specific systematic errors of radiochemical experiments. Assuming that the sun is in equilibrium and that the pp-cycle is entirely responsible for the energy released in the solar interior, the pp-neutrinos contribute with about 70 SNU to the signal in the gallium experiments. The measured value around 70 SNU leaves no space for a contribution from the ^{7}Be-neutrinos. This deficit of ^{7}Be-neutrinos is the essence of the so called solar neutrino problem in its present stage. What is the origin of this deficit? Certainly it can't be ruled out, that our picture of the processes inside the sun is partly wrong or incomplete or that the systematic errors of the data are partly underestimated, but these explanations for the solar neutrino problem are based on ad-hoc assumptions.

By contrast ν-oscillations can explain the data of all solar neutrino experiments without further ado [8]. To explain in addition the LSND-result and the problem of atmospheric neutrinos it is proposed, that the neutrinos from the sun are converted to sterile neutrinos. The further discussion considers both possibilities, active neutrinos and sterile neutrinos. Using the two neutrino picture there are three parameter sets in the Δm^2-$\sin^2 2\Theta$-plot, which solve the solar neutrino problem. First there are two solutions for matter enhanced oscillations (MSW-effect), the "small mixing angle" solution ($\sin^2 2\Theta$ between $10^{-2} - 10^{-3}$ depending on the underlying SSM) and the "large mixing angle" solution ($\sin^2 2\Theta \approx 0.6$) both with $\Delta m^2 \approx 10^{-5} eV^2$. For SSM's with a low ^{8}B-ν flux or for sterile neutrinos the large mixing angle solution gets much more unlikely. But also vacuum oscillations ('Just so') with $\Delta m^2 \approx 6 \cdot 10^{-11}$ and $\sin^2 2\Theta \approx 0.8$ can explain the solar neutrino problem for active neutrinos, whereas for sterile neutrinos this solution is much more unlikely. In Fig. 1 [9] the MSW-suppression curve for the small mixing angle solution is shown for different mixing angles, which correspond to different SSM's. One can easily see how the experimental situation is "naturally" obtained by neutrino oscillations.

Future solar neutrino experiments have to clarify whether neutrino oscillations are responsible for the measured deficit of neutrinos from the sun and if so, which of the presently allowed parameter sets applies. The next generation experiments Superkamiokande and SNO are only sensitive to ^{8}B-neutrinos. They look for a distortion of the ^{8}B-neutrino spectrum and in addition SNO will be able to measure a deviation from the charged current to neutral current signal ratio. The significance of the effects depends on the real ^{8}B-ν flux (Fig. 1), on the achieved background level of the experiments and the tested parameter set.

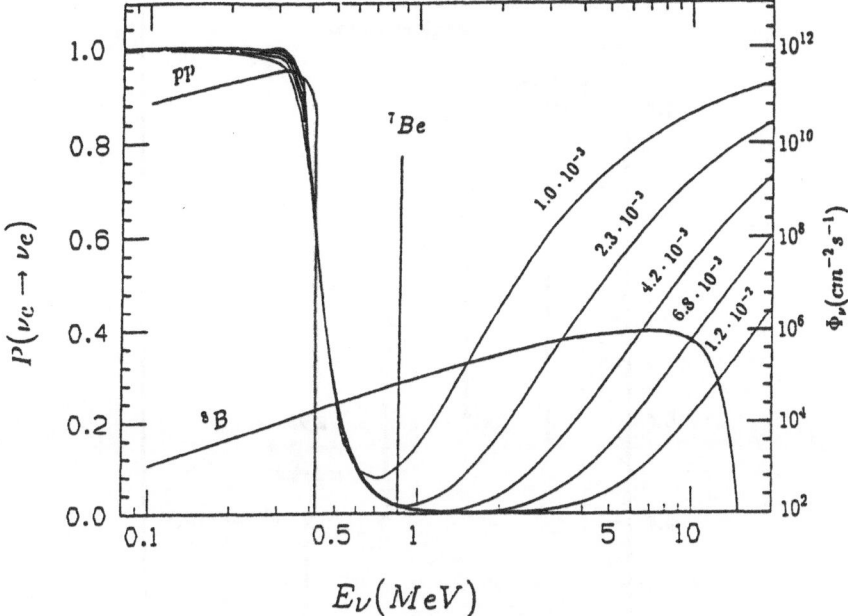

Figure 1. Solar neutrino spectrum and the MSW-suppression curve for the small mixing angle solution depending on the underlying SSM.

As discussed above the solar neutrino problem indicates a strong suppression of the ^{7}Be-neutrinos independent of the underlying solar model, if the neutrinos are standard. For the small mixing angle solution shown in Fig. 1 the intermediate energy range with the ^{7}Be-neutrino line is affected most strongly. A first direct measurement of the ^{7}Be-neutrinos by BOREXINO, the second strongest neutrino branch of the sun, is therefore important for a better understanding of the solar neutrino problem (see Fig. 2 [8]). Depending on the scenario the expected signal in BOREXINO varies strongly. It is important to note that in BOREXINO charged and neutral current events contribute to the signal as explained in the next section. Looking at seasonal variations a discrimination between vacuum oscillations and MSW-solutions is obtained very likely.

2. Concept of BOREXINO

BOREXINO will detect neutrinos via ν-e^--scattering in approx. 100 t of ultrapure organic liquid scintillator [10]. This concept implies:

- the cross section for the pure leptonic detection reaction is well predicted by the theory of electroweak interaction. This is an advantage over radiochemical detection reactions where systematic errors of the cross sections are often not negligible and difficult to estimate.

242

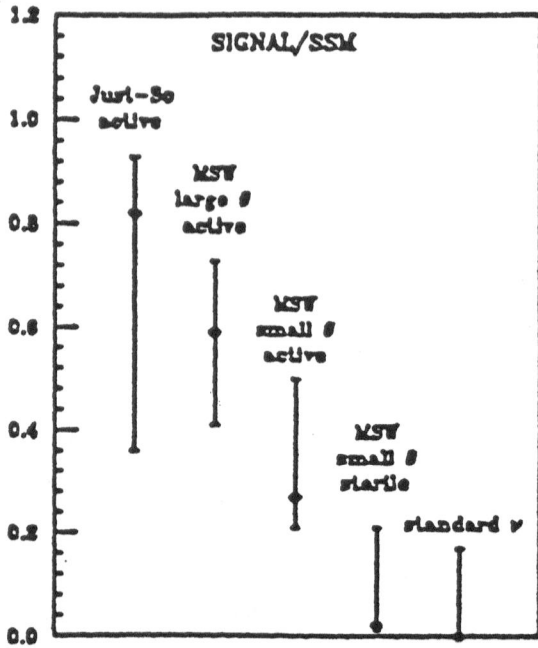

Figure 2. The Beryllium (CC+NC) signal, in units of the SSM prediction. Diamonds indicate the best fit points, bars correspond to 90% C.L.

- the energy threshold for the recoil electrons can be set below 100 keV. For background reasons the threshold will be at 250 keV corresponding to an energy of the neutrinos of 450 keV. In contrast to radiochemical experiments spectroscopy of neutrinos is possible, because different branches of the solar neutrino spectrum result in recoil electron spectra with different energies. Monoenergetic neutrino lines cause a "compton like" edge and therefore a good signature. For the ^{7}Be-neutrinos at 862 keV this edge lies at 660 keV (Fig. 3).
- BOREXINO is sensitive to all neutrino flavours, because $\nu_{\mu,\tau}$ can interact via neutral current reactions. The ratio of the cross-sections for ν_e and $\nu_{\mu,\tau}$ is approx. 4.
- according to the SSM's the signal rate for ^{7}Be-neutrinos is about 50 c/d, for pep-neutrinos 3 c/d and for ^{8}B-neutrinos 0.5 c/d. In the case of total flavour conversion the rate for the ^{7}Be- and pep-neutrinos is approx. 4 times lower. The comparable high rate for the ^{7}Be-neutrinos allows the observation of time dependent effects as annual or seasonal variations or day/night effect.
- in contrast to Cerenkov-detectors the directional information about the neutrino source is lost and no cuts can be applied to improve the signal to background ratio as in Kamiokande or Superkamiokande. Therefore and because of the low energy threshold the requirements on

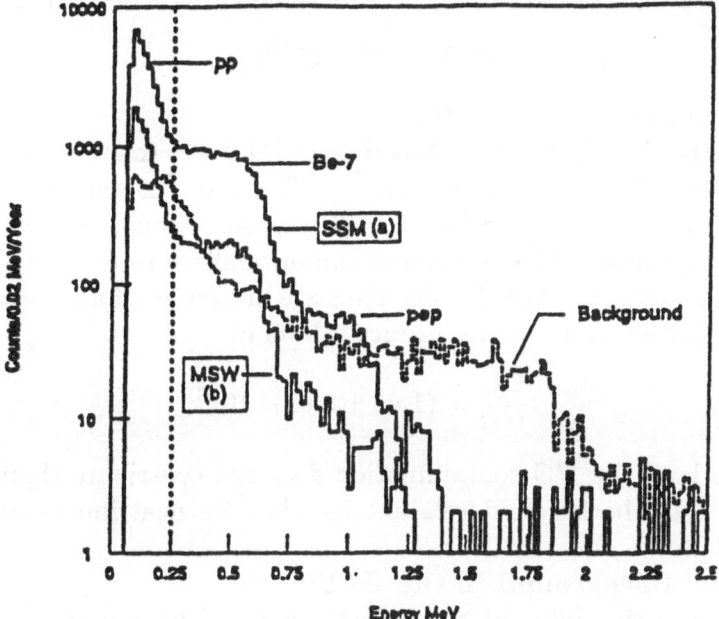

Figure 3. Expected spectrum in BOREXINO for standard and fully converted active neutrinos. For background curve see text.

the radiopurity of the detector components, especially the scintillator, are very high. In Fig. 3 the background spectrum is shown assuming an uranium and thorium content of 10^{-16} g/g and a potassium content of 10^{-14} g/g in the scintillator.

3. Results of the prototype detector CTF

To show that the above mentioned purity requirements are achievable a prototype detector CTF (Counting Test Facility) has been operated since January 1995. The results are:

– **Contamination due to the ^{232}Th family :**
   To measure the contamination due to the ^{232}Th family, the delayed α-β-coincidence: ^{212}Bi $\rightarrow$ ^{212}Po $(T_{1/2} = 300\,\mathrm{ns}) \rightarrow$ ^{208}Pb has been used. Assuming the whole decay chain in radioactive equilibrium one obtains a ^{232}Th concentration in the CTF scintillator of

$$(2.5 \pm 1.3) \cdot 10^{-16} \mathrm{g/g}$$

– **Contamination due to the ^{238}U family:**
   To measure the contamination due to the ^{238}U family, the delayed α-β-coincidence: ^{214}Bi $\rightarrow$ ^{214}Po $(T_{1/2} = 164\,\mu\mathrm{s}) \rightarrow$ ^{210}Pb has been used.

During 35 days of data taking we obtained a ^{238}U concentration of

$$(4.9 \pm 1.5) \cdot 10^{-16} \text{g/g}$$

- **Contamination due to ^{14}C:**
  One of the most important objectives of CTF was to evaluate the concentration of ^{14}C in the scintillator. Therefore the energy spectrum at low energies measured in the CTF was superimposed to the corresponding simulated ^{14}C spectrum and normalized to the same area in the region 50 keV - 100 keV. The data are in very good agreement with the simulation and one can deduce a level of

$$\frac{^{14}\text{C}}{^{12}\text{C}} = (1.45 \pm 0.01) \cdot 10^{-18}$$

  At this level the ^{14}C contamination does not contribute significantly to the total background relevant for the ^{7}Be neutrino detection in BOREXINO.

- **Internal Background in the CTF:**
  By Rn insertion into the water buffer it could be demonstrated that the external background rate in the CTF was mostly dominated by Rn contamination in the external water shielding. A comparison of the radial distribution of CTF events prior to Rn insertion with the radial distribution of the determined external background gives an 1σ upper limit for the internal background rate in the ^{7}Be neutrino energy window (250-800 keV) in 4.5 t of scintillator of 40 counts/day (1σ). This limit is consistent with a vanishing counting rate due to possible contaminations of the scintillator.

These results have shown that the detector concept of BOREXINO is feasible. The experiment is under construction and funded by Germany and Italy. The approval from the U.S. is expected soon.

References

1. J.N. Bahcall and M.H. Pinsonneault, *Rev. Mod. Phys.* **67**, 781 (1995).
2. S. Turck-Chieze and I. Lopes, Astrophys. J. **408**, 347 (1993).
3. A. Dar and G. Shaviv, *Nucl. Phys.* B *Proc. Suppl.* **48**, 335 (1996).
4. Kamiokande Coll., *Phys. Rev. Lett.* **77**, 1683 (1996).
5. R. Davis, *Nucl. Phys.* B *Proc. Suppl.* **48**, 284 (1996).
6. Gallex Coll., Gallex Solar Neutrino Observations: Results for Gallex III, to be published in Phys. Lett. B 1996.
7. Sage Coll., *Nucl. Phys.* B *Proc. Suppl.* **48**, 299 (1996).
8. E. Calabresu, *Nucl. Phys.* B *Proc. Suppl.* **48**, 343 (1996).
9. V. Berezinsky, Solar Neutrinos, Proceedings of 'Particles and Cosmology', Baksan Valley, World Scientific 1996.
10. Borexino Coll., Borexino Proposal, 1991, INFN Milan. - Addendum to the Borexino Proposal, July 1996, INFN Milan.

MEASURING Ω_0 VIA THE BIAS PARAMETER

LICIA VERDE, ALAN HEAVENS & SABINO MATARRESE
Institute for Astronomy, University of Edinburgh,
Royal Observatory, Blackford Hill, Edinburgh EH9 3HJ, U.K.
Dipartimento di fisica Galileo Galilei, Via Marzolo 8, 35131
Padova, IT

Dynamical and redshift distortion studies can give Ω_0 only in the combination $\beta = \Omega_0^{0.6}/b$. The determination of the density parameter Ω_0 by such methods is therefore impossible without a measurement of the bias parameter b.

We propose here an empirical method, based on the bispectrum, to measure the bias parameter on large scale without making any assumption about the physical mechanism for the bias.

1. Theoretical framework

As Fry [3] pointed out, the degeneracy between Ω_0 and b can be removed going to second order in perturbation theory. Under the assumption that the initial fluctuation field is Gaussian, the three point correlation function is intrinsically a second order quantity, and should be detectable in the mildly non-linear regime. We expand the fractional over-density field $\delta = \delta\rho/\rho$ to second order:

$$\delta(\vec{x}) = \delta^{(1)}(\vec{x}) + \delta^{(2)}(\vec{x}). \tag{1}$$

where $\delta^{(2)}(\vec{x})$ is $\mathcal{O}(\delta^{(1)2})$, and represents departures from gaussianity.

A general expression for the δ field to second order is given for example in [1]. Since we work in Fourier space we quote directly the expression for $\delta_{\vec{k}}^{(2)}$ (Fourier transform of the second order term of $\delta(\vec{x})$):

$$\delta_{\vec{k}}^{(2)} = \int \delta^{Dirac}(\vec{k}_a + \vec{k}_b - \vec{k}) J(\vec{k}_a, \vec{k}_b) \delta_{\vec{k}_a}^{(1)} \delta_{\vec{k}_b}^{(1)} d^3 k_a d^3 k_b \tag{2}$$

where J is a known function of the kvectors and Ω_0.

Let us assume that the biased fluctuation field (i.e. the galaxy field) is a **local** function of the unbiased field. For consistency we have to allow a second order expansion also for the biased density contrast:

$$\delta_g(\vec{x}) = f[\delta(\vec{x})]$$

245

D. N. Schramm and P. Galeotti (eds.), Generation of Cosmological Large-Scale Structure, 245–250.
© *1997 Kluwer Academic Publishers.*

$$\simeq b_1\delta(\vec{x}) + \frac{1}{2}b_2\delta^2(\vec{x}) \simeq b_1\delta^{(1)}(\vec{x}) + b_1\delta^{(2)}(\vec{x}) + \frac{1}{2}b_2\delta^{(1)2}(\vec{x}). \qquad (3)$$

The subscript g means that we are considering the galaxy over-density field which can be biased. The two coefficients b_1 and b_2 are respectively the linear and quadratic bias parameters.

Therefore the complete expression in Fourier space for the bispectrum of the biased galaxy fluctuation field to second order, in terms of observable quantities is:

$$< \delta_{g\vec{k}_1}\delta_{g\vec{k}_2}\delta_{g\vec{k}_3} >=$$

$$\left\{c_1 P_g(\vec{k}_1)P_g(\vec{k}_2)J(\vec{k}_1,\vec{k}_2)+ \text{cyc.}\right\}\delta^{Dirac}(\vec{k}_1 + \vec{k}_2 + \vec{k}_3)$$

$$+c_2\left\{P_g(\vec{k}_1)P_g(\vec{k}_2) + \text{cyc.}\right\}\delta^{Dirac}(\vec{k}_1 + \vec{k}_2 + \vec{k}_3). \qquad (4)$$

where

$$c_1 = 1/b_1, \qquad c_2 = b_2/b_1^2. \qquad (5)$$

and P_g is the spectrum of the discrete galaxy distribution, including shot noise..

The dependence of J on Ω_0 is very weak in the range of sensible values for Ω_0. This is the key point of this technique: it allows to give a valuable estimate of the bias parameter **independent of Ω_0**.

2. Key points

Equation (4) is the starting point for this work. It is worth noticing that:

- The degeneracy between Ω_0 and b has been removed since J does not significantly vary for realistic values of Ω_0.
- The two assumptions on which this is based are that the bias is local and that the primordial fluctuation field is Gaussian.
- All the quantities that appear in eq. (4), apart from c_1 and c_2, are observable: a likelihood analysis would give us a reasonable estimate of b_1 and b_2 via these two parameters.

3. Estimating the bias parameter via a likelihood analysis

In order to perform a likelihood analysis the covariance matrix needs to be calculated. Our data are the quantities $D_\mu = (\delta_{\vec{k}1}\delta_{\vec{k}2}\delta_{\vec{k}3})_\mu$, so the general element of the covariance matrix would be given by:

$$C_{\mu\nu} = \langle(\delta_{\vec{k}1}\delta_{\vec{k}2}\delta_{\vec{k}3})_\mu(\delta_{\vec{k}1}\delta_{\vec{k}2}\delta_{\vec{k}3})_\nu\rangle-$$

$$\langle(\delta_{\vec{k}1}\delta_{\vec{k}2}\delta_{\vec{k}3})_\mu\rangle\langle(\delta_{\vec{k}1}\delta_{\vec{k}2}\delta_{\vec{k}3})_\nu\rangle \qquad (6)$$

where the μ subscript means that $\vec{k}_1$, $\vec{k}_2$ and $\vec{k}_3$ are the three sides of a closed triangle. A general algorithm for the description of discrete random fields and the analysis of their statistical properties needs to be introduced (see appendix A) in order to find the expression for the six–point and three–point correlation functions for a discrete distribution in Fourier space involved in eq. (7).

In order to test the viability of the technique proposed the likelihood analysis has been carried on a numerical (N–body) simulation of an unbiased galaxy distribution.

The likelihood is a function of $b_1 \& b_2$:

$$\mathcal{L}(b_1, b_2) = \frac{1}{(2\pi)^{\frac{n}{2}}(detC)^{\frac{1}{2}}} e^{-\frac{1}{2}\sum_{\alpha\beta}(D_\alpha - \mu_\alpha)C_{\alpha\beta}^{-1}(D_\beta - \mu_\beta)} \tag{7}$$

where:

$$\mu_\alpha(b_1, b_2) = \text{ mean value of } D_\alpha$$

(obviously this is a function of b_1, b_2)

In this preliminary analysis we set $b_2 = 0$ and investigate $\mathcal{L}(b_1)$. For a start, the easiest and most sensible thing to do is to take into account only the equilateral triangle configurations since they show the greatest signal to noise. Neighbouring (closed) triangles in k-space are expected to give correlated estimates of B. This feature can be quantified by investigating the values taken by the off diagonal terms in the covariance matrix. By sconsidering triangles with no k-vectors in common, we simplify the likelihood calculation as C is diagonal. This allows us , at the expense of not including all the information, to calculate $\mathcal{L}$ readily for all linear and mildly non-linear wavevectors. The CPU time scales $\sim N$ because only the diagonal terms (the variances) need to be calculated. The breakdown of second order perturbation theory has been investigated in [4]. In this analysis it is apparent when the "true" value for the bias parameter ($b_1 = 1$) drops out of the 1-σ confidence interval.

The very encouraging result is that the likelihood function is peaked on the right value for c_1. The error on the determination of this value is still considerably big, but it is possible, in a further step, to reduce it: this quantity seems to be **independent of the size of the survey**.

Since the variance $\sigma_{c_1}^2$ for c_1 do not depend on the size of the survey (within a certain range), one can divide the survey in several subsamples, and obtain for each one of them an independent estimate of c_1. Obviously the final estimate of the bias parameter would show a better precision: if the number of subsamples is M then: $\sigma_{c_1} \propto \frac{1}{M^{1/2}}$.

A quick order of magnitude calculation shows that, if applied to the Sloan/2dF survey we can expect an error $\sigma_{c_1} \simeq 10\%$.

Before being able to apply the technique to a real redshift survey some more investigation needs to be carried on.

Since redshift distortion techniques [5] will be able to measure β with better that 10% accuracy, we would be able to know Ω_0 with about 20% accuracy.

A full report of this work is being submitted to *MNRAS*.

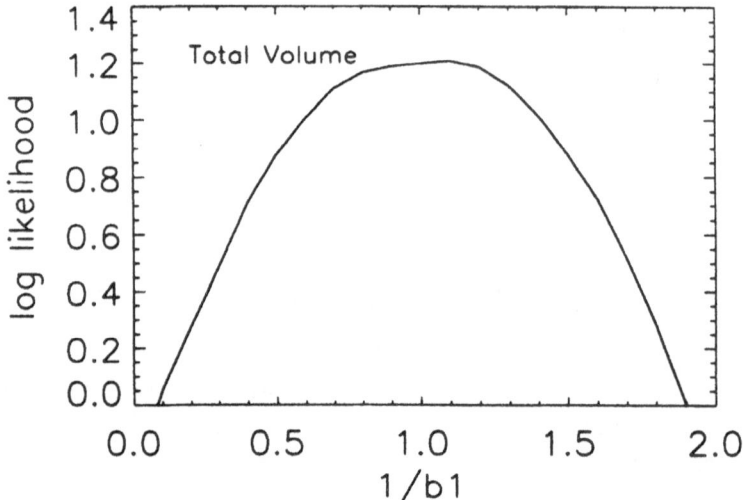

Figure 1. The likelihood function for b_1 as obtained by analizing the volume of the simulation as a whole.

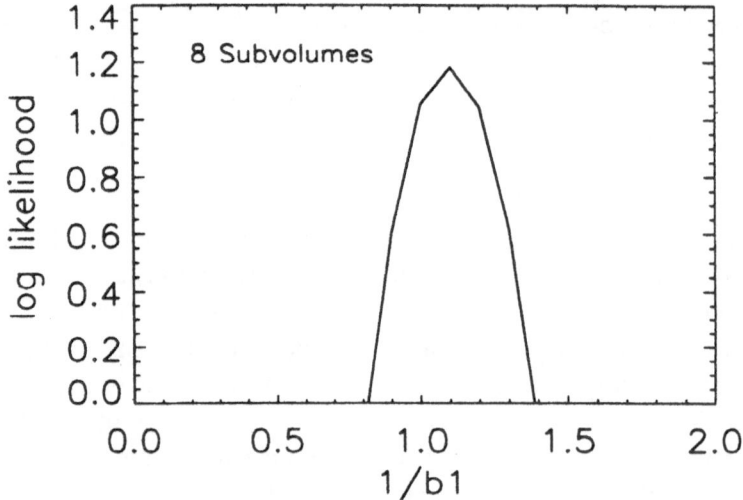

Figure2. The likelihood function as obtained by subdividing the volume in 8 subvolumes.

References 1) P.Catelan, F.Lucchin, S.Matarrese, L.Moscardini, MNRAS, $276, 39, 1995$; 2) H.A.Feldman, N.Kaiser, J.A.Peacock, ApJ 426, 23, 1994; 3) J.N.Fry, Phy.Rev.Lett. Vol 73, N 2, 1994; 4) J.N.Fry, A.L.Mellot, S.F.Shandarin ApJ 412,504,1993; 5) A.F.Heavens, A.N.Taylor, MNRAS, 275, 483, 1995.

A. Statistics of random fields and derivation of the pentaspectrum

Consider the *partition functional* or *generating functional of the correlation functions* $\mathcal{Z}$ defined as follows :

$$\mathcal{Z}[J(x)] \equiv \int \mathcal{D}[\delta]\mathcal{P}[\delta] \exp\left(i \int J(\vec{x})\delta(\vec{x})d^3x\right) =$$

$$\left\langle \exp\left[i \int J(\vec{x})\delta(\vec{x})d^3x\right]\right\rangle \tag{8}$$

where $J(\vec{x})$ is an external source perturbing the underlying statistics and $\mathcal{D}[\delta]$ is a suitable measure such that the probability distribution is correctly normalized to 1: $\int \mathcal{D}[\delta(\vec{x})\mathcal{P}[\delta(\vec{x})] = 1$. The $< \cdot >$ brackets mean the average over the ensemble.

From this function we can recover the *correlation functions* μ_N of the distribution and the *connected correlation functions* or κ_M as follows :

— From the *generating functional of the correlation functions*

$$\mu_N \equiv< \delta(\vec{x}_1)...\delta(\vec{x}_N) >= i^{-N} \frac{\delta^N \mathcal{Z}[J]}{\delta J(\vec{x}_1)...\delta J(\vec{x}_N)} \tag{9}$$

evaluated at $J = 0$

— Define the *generating functional of the connected correlation functions* as

$$\mathcal{K}[J] \equiv \ln \mathcal{Z}[J]$$

so that:

$$\kappa_M(\vec{x_1}.....\vec{x_M}) = i^{-M} \frac{\delta^M \mathcal{K}[J]}{\delta J_1...\delta J_M} \tag{10}$$

evaluated at $J = 0$.

Continuous random fields and random distributions of discrete points are cases of random processes. It is possible to relate discrete and continuous distributions by taking in account the Poisson shot noise contribution or by performing a simple substitution in the partition functional or in the generating functional for the continuous case as suggested in [3].

The quantity that we are dealing with is a complex number $Z_\triangle$ given by $\delta_{\vec{k}_1} \delta_{\vec{k}_2} \delta_{\vec{k}_3}$ where $\vec{k}_1 + \vec{k}_2 + \vec{k}_3 = \vec{0}$ (they form a closed triangle: this is the meaning of the subscript $\triangle$). The average of $Z_\triangle$ over the ensemble (of closed triangles) is the bispectrum, but the actual quantity $Z_\triangle$ fluctuates across the ensemble. In order to quantify this fluctuation it would be valuable to estimate its covariance and this involves the quantity $< Z_\triangle Z_{\triangle'} >$. This quantity is nothing but the six—point correlation function (not just the irreducible or connected one) in Fourier space (shot noise contribution included) in the particular case when the six vectors form two closed triangles. It is also possible to obtain directly the generating functional for the N—spectra choosing a suitable source function: the Fourier transform of the previous one. The expression for the six—point discrete correlation function is one of the main original results of this work.

A TWO-DIMENSIONAL PERCOLATION ANALYSIS OF THE LAS CAMPANAS REDSHIFT SURVEY

CAPP YESS
University of Kansas
Department of Physics and Astronomy
Lawrence Kansas, KS 66045

1. Introduction

Percolation analysis has been used in many branches of physics to model phase transitions. In cosmology it has become a valuable tool in detecting structure in a galaxy distribution and characterizing the topology of a distribution that is not random noise. Percolation analysis characterizes distributions as one of three topological types: connected, structureless or clumpy. Connected topologies are consistent with galaxy distributions described as filaments or sheets. Clumpy topologies are often referred to as "meatball" topologies where aggregates of galaxies are surrounded and separated by voids.

Those distributions which percolate more easily than a structureless distribution (by definition Poisson) are considered connected while those that percolate with more difficulty are considered isolated or clumpy. Determination of the topology of the local galaxy distribution is important for determining the scenario under which structure formed in our universe. In this paper we present preliminary results of a two dimensional percolation analysis of the Las Campanas Redshift Survey that indicate a connected topology out to a distance of 400 Mpc/h. For a complete description of the Las Campanas Survey see Shectman *et al.* 1996. The topology of the survey is consistent with the interpretation that the galaxies are arranged in long filaments or large sheets.

2. Percolation on a Point Set

To percolate a galaxy distribution, the positions of the galaxies are located on a grid where each grid cell containing a galaxy is labeled filled. A nearest

D. N. Schramm and P. Galeotti (eds.), Generation of Cosmological Large-Scale Structure, 251–254.
© 1997 *Kluwer Academic Publishers.*

neighbor scheme is employed to group filled cells into aggregates called clusters. The smallest cluster being a single isolated filled cell. Next the initial filling factor is determined. The filling factor is the percentage of the survey area occupied by filled grid cells. To percolate the distribution, circles are centered on each original galaxy position and increased incrementally in radius. As the circles grow each grid cell engulfed by the growing circles is labeled filled and added to the cluster. Through the process of growing circles and nearest neighbor mergers clusters grow in size and become less numerous until one cluster emerges as the largest cluster and spans the area of the survey.

At each incremental increase in the circle radii the filling factor is recalculated along with the largest cluster statistic. The largest cluster statistic is the area of the largest cluster in units of the filling factor. In a fairly sampled distribution, the largest cluster starts out as just one of many clusters and its fractional area is small. As the circle radii increase all clusters begin to grow (increasing the filling factor) and through cluster mergers the largest cluster eventually expands to dominate the entire filled area. At a filling factor particular to each distribution the largest cluster will grow in area at an accelerated rate indicating the onset of percolation. The ease of percolation is determined by the value of the filling factor when the largest cluster statistic begins to grow rapidly; the lower the filling factor value the easier the distribution percolates. For a detailed explanation of point set percolation see Klypin & Shandarin 1993.

The ease of percolation is determined by the largest cluster statistic, but a determination of the type of topology represented by a galaxy distribution is made by comparing its largest cluster statistic to the largest cluster statistic of a suitable Poisson distribution. In a two dimensional analysis of the Las Campanas Redshift Survey, suitability of a standard Poisson distribution requires a similar grid geometry, an equivalent initial filling factor, a similar selection function, and corrections for projection effects.

3. Results

Shown in figure 1 are the results of the Las Campanas percolation analysis. The left column shows the results for the three northern slices of the survey while the right column shows the results of the southern slices. In all graphs, the dashed line to the right is the result from a pure Poisson distribution with the appropriate initial filling factor and geometry. This result is shown for reference. The other dashed line is the result from an appropriate Poisson distribution corrected for selection and projection effects. In all cases the error bars represent one sigma deviations over four realizations.

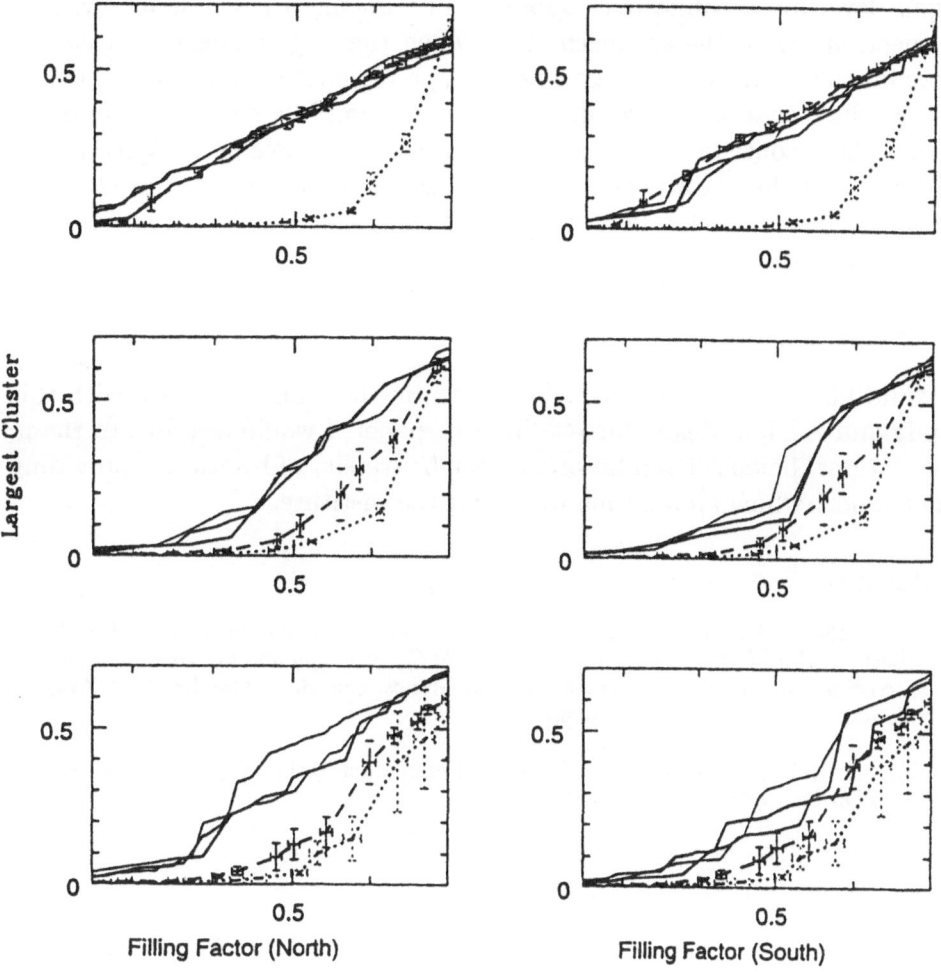

Figure 1. Two Dimensional Percolation Results of the Las Campanas Redshift Survey

In the top panels the Poisson results are indistinguishable from the survey results. An explanation for this is that the galaxy density at the outer edges of the survey is reduced to the point that the distribution is essentially random. The middle panels show the results of a magnitude limited sample in the region 60 Mpc/h < R < 400 Mpc/h. There is now a clear distinction between survey and Poisson distribution results. The slices percolate at lower filling factors than the standard in all cases. Excluding the outer extent of the survey reveals a clear and unambiguous signal for a connected topology in the region analyzed.

The bottom panels show the results in a volume limited region (60 Mpc/h < R < 400 Mpc/h). Once again there is a clear signal at the one sigma level for a connected topology. The northern slices show stronger connectivity than the southern slices when the largest cluster statistic is examined over its entire range suggesting a regional variance in the topology at these scales. The results of this study imply a connected topology for all slices consistent with a filamentary or sheet geometry. Further examination of this data is necessary, and we intend to expand our study to three dimensions with a determination of edge and orientation effects, and redshift distortions.

4. Acknowledgments

I would like to thank Sergei F. Shandarin for help and guidance with this study and Adrian Melott for reading this paper. I would also like to thank the National Science Foundation and the University of Kansas for providing the support which allowed me to attend the institute.

References

1. Shectman, S. A., Landy, S. D., Oemler, A., Tucker, D. L., Kirshner, R. P., Lin, H., & Schechter, P. L. (1996 in press), in *Wide-Field Spectroscopy and the Distant Universe, proceedings of the 35th Herstmonceux Conference*, eds. S. J. Maddox & A Aragón-Salamanca, World Scientific, Singapore

2. Kylpin A. A. and Shandarin, S. F. (1993) Percolation Technique for Galaxy Clustering, *The Astyrophysical Journal* **413**, 48-58

MINKOWSKI FUNCTIONALS IN COSMOLOGY

A unifying approach to higher–order statistics of large–scale structure

JENS SCHMALZING

Ludwig–Maximilians–Universität,
Theresienstraße 37, 80333 München, Deutschland

Max–Planck–Institut für Astrophysik,
Karl–Schwarzschild–Straße 1, 85740 Garching, Deutschland

AND

MARTIN KERSCHER

Ludwig–Maximilians–Universität,
Theresienstraße 37, 80333 München, Deutschland

1. Introduction

Ever since the first high–quality galaxy catalogues became available, statistics of large–scale structure has been an important tool to gain insight into the underlying physical processes and to allow for comparison of observations and predictions, both through theory and numerical experiment.

One of the first methods used is the two–point correlation function[11] which remains the most widely used statistics today. However, the inability to grasp important, non–Gaussian features of the large–scale structure was quickly realised. The necessity of incorporating higher–order statistics into the analysis has led to a variety of methods ranging from the void probability function[19] to genus statistics[9].

Like their predecessors, Minkowski functionals incorporate correlations of any order[8] and supply morphological, i.e. topological and geometrical information on a point distribution. In addition, they can be derived from a few simple requirements and in this sense characterise morphology in a complete and unique way.

D. N. Schramm and P. Galeotti (eds.), Generation of Cosmological Large-Scale Structure, 255–260.
© 1997 *Kluwer Academic Publishers.*

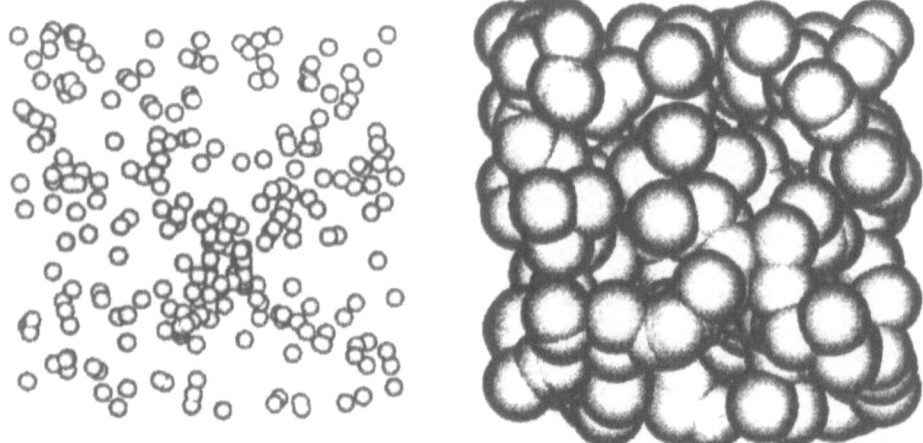

Figure 1. Boolean grain model, where the same set of points generates widely different patterns of balls. To the left, small balls form a pattern consisting of many disconnected parts, while to the right, almost all volume is covered by large balls forming an entangled network of tunnels and cavities, even though the underlying point distribution is the same.

In this talk we will give a brief outline of the mathematical background of Minkowski functionals before we present some cosmological applications.

2. Mathematical Background

Minkowski functionals were known for a long time[10] in integral geometry, a field of mathematics dealing with motions and measures of bodies in space (for an introductory overview of the field, consult Weil[18] or Schneider[16]). However, their applicability to cosmological datasets was not realised until recently[7]. While we restrict ourselves to a brief overview of the underlying concepts, more material can be found in works by the LMU cosmology group[14, 15].

Consider the set of points $\{\mathbf{x}_i\}$, $i = 1 \ldots N$ supplied, for example, by the positions of galaxies in a catalogue. A standard way to measure its morphological properties is the Boolean grain model, which is widely used in statistical physics[6]. Each point becomes the centre of a ball $B_r(\mathbf{x}_i)$ of radius r, so neighbouring points form a connected pattern of intersecting balls. The morphological properties of the union set A_r, with

$$A_r = \bigcup_{i=1}^{N} B_r(\mathbf{x}_i), \tag{1}$$

depend on the radius r (see Figure 1) which can therefore be used as a

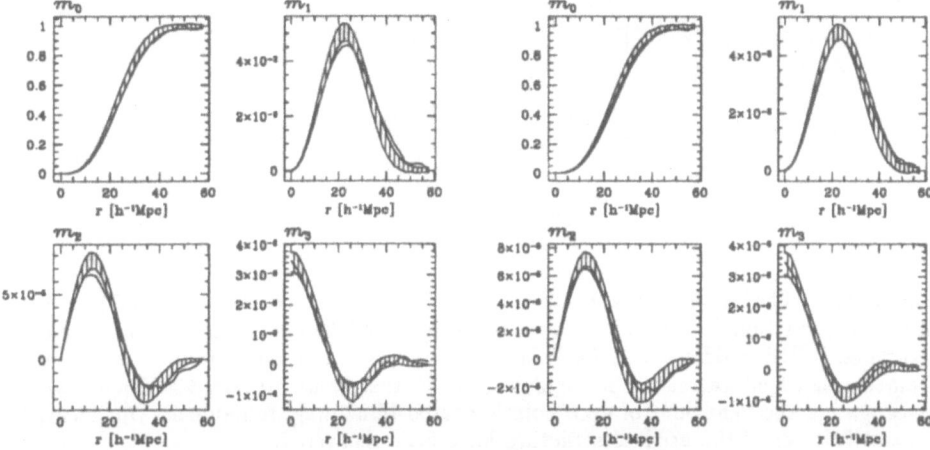

Figure 2. These panels show the Minkowski functionals of a sample of Abell/ACO clusters compared to the results of numerical simulations. The four panels on the left show the four Minkowski functionals of the Abell/ACO sample (solid line) compared to a standard cold dark matter simulation (shaded area; the shading gives an estimate of cosmic variance). Obviously, SCDM fails to reprduce the observed distribution of galaxy clusters, while adding a cosmological constant (right panels; again the solid line gives the Abell/ACO sample) yields compatible but still not perfect results.

diagnostic parameter.

In order to measure the morphological properties of this union set A_r in a quantitative fashion, we want to determine real numbers $M(A_r)$ that completely characterise the topology and geometry of A_r. Hadwiger's theorem[2] tells us that for a body in three dimensions, this can be done by evaluating only four independent quantities M_0, M_1, M_2 and M_3, the so-called *Minkowski functionals*. They can be interpreted as quantities known from differential geometry, namely the volume, the surface area, the integrated mean curvature and the Euler characteristic. Measured as functions of radius, they give a complete and unique characterisation of the point process under consideration.

3. Applications

3.1. CATALOGUES OF GALAXIES AND CLUSTERS

In analysing catalogues of galaxies or clusters, several issues arising from observational restrictions have to be adressed. We would like to mention three of them.

Redshift space distortions are an issue of ongoing research. For the samples

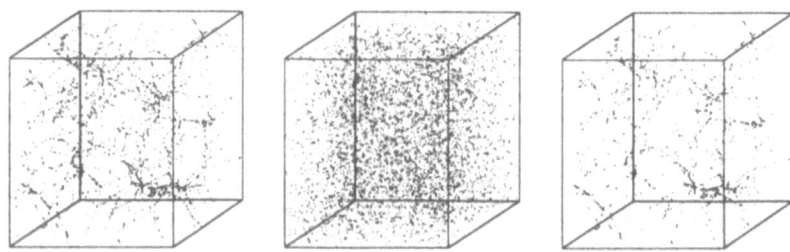

Figure 3.   The three boxes in this figure illustrate structure determination using Minkowski functionals. The left box shows a sample of 3,000 points forming a filamentary structure. In the middle box 7,000 Poisson points have been added to mock a mixture of filaments and field galaxies. The right box shows the points identified as filamentary by investigating the behaviour of each point's partial Minkowski functionals. Obviously, the general features of the original structure have been recovered.

analysed so far, comparison of real and redshift space in numerical simulations showed no influence that would alter the significance of any of our results.

Today's surveys are magnitude limited which necessarily makes them inhomogeneous. By using a volume–limited subsample homogeneity in the number density is attained at the price of throwing away a considerable proportion of the points in the survey. A way to overcome this is the construction of a series of volume–limited samples of varying depth.

The boundary effects due to limitations both in depth and in sky coverage can be completely removed by applying the principal kinematical formula[12]. The detailed prescription we have to follow can be found in [3], and in several other publications. It turns out to be equivalent to a procedure described earlier for the void probability function[5], which has thus been validated rigorously using integral geometry.

As an example Figure 2 displays some results of a comparison of the Abell/ACO cluster catalogue by Kerscher et al.[4].

3.2. STRUCTURE DETERMINATION

It is also possible to define local quantities which for each point in a sample describe the local morphology nearby. We can then compare measured values of these so–called *partial Minkowski functionals* with theoretical expectations which are known in exact form for various types of structure, such as filaments, walls and field galaxies, and determine which type of local structure fits best.

Figure 3 shows first results of structure determination in a mock sample generated from a Voronoi tesselation[17]. Knowing where filaments were placed, we are able to assess the performance of the method – almost two thirds of the points in filaments have been recovered, with a contamination over two times less than the desired signal.

Compared to other methods (see for example Sathyaprakash et al.[13] or Davé et al.[1] for a comparison of several existent methods), this is certainly a promising result. A thorough evaluation of the performance of Minkowski functionals in structure determination is subject of ongoing research.

4. Outlook

Minkowski functionals are still emerging as tools for the analysis of large–scale structure. However, we have presented a number of interesting applications. An article by Martin Kerscher in this volume explains the main results of an analysis of the IRAS 1.2Jy catalogue, while numerical studies of the evolution both of clusters of galaxies and of large–scale structure are on the way. Furthermore, the field of integral geometry itself offers a number of possibly useful prospects, such as two–dimensional Minkowski functionals and Quermaßvectors.

Another issue of interest is to establish a connection between cosmic dynamics and Minkowski functionals, which would yield direct insight into the underlying physical processes. But even without this, Minkowski functionals offer a compelling and powerful approach to higher–order statistics of large–scale structure.

Acknowledgements

We wish to thank Thomas Buchert and Herbert Wagner for enlightening discussions which inspired the work presented here. MK and JS acknowledge financial support by NATO during their stay in Erice, MK also acknowledges support by the Deutsche Forschungsgemeinschaft through "Sonderforschungsbereich SFB 375 für Astroteilchenphysik".

References

1. Romeel Davé, Doug Hellinger, Richard Nolthenius, Joel Primack & Anatoly Klypin, *Filament statistics: A quantitative comparison of cold + hot and cold dark matter cosmologies with CfA1 data*, Mon. Not. R. Astron. Soc. (1996), in press, astro-ph/9609179.
2. Hugo Hadwiger, *Vorlesungen über Inhalt, Oberfläche und Isoperimetrie*, Springer Verlag, Berlin, 1957.

3. Martin Kerscher, Jens Schmalzing & Thomas Buchert, *Analyzing galaxy catalogues with Minkowski functionals*, Mapping, measuring and modelling the universe (Valencia) (Peter Coles, Vicent Martínez & Maria Jesús Pons Bordería, eds.), Astronomical Society of the Pacific, 1996, pp. 247–252.

4. Martin Kerscher, Jens Schmalzing, Jörg Retzlaff, Stefano Borgani, Thomas Buchert, Stefan Gottlöber, Volker Müller, Manolis Plionis & Herbert Wagner, *Minkowski functionals of Abell/ACO clusters*, Mon. Not. R. Astron. Soc. **284** (1997), 73–84.

5. Sophie Maurogordato & Marc Lachièze-Rey, *Void probabilities in the galaxy distribution – Scaling and luminosity segregation*, Ap. J. **320** (1987), 13–25.

6. Klaus Mecke, *Integralgeometrie in der Statistischen Physik: Perkolation, komplexe Flüssigkeiten und die Struktur des Universums*, Harri Deutsch, Thun, Frankfurt/Main, 1994.

7. Klaus Mecke, Thomas Buchert & Herbert Wagner, *Robust morphological measures for large-scale structure in the Universe*, Astron. Astrophys. **288** (1994), 697.

8. Klaus Mecke & Herbert Wagner, *Euler characteristic and related measures for random geometric sets*, J. Stat. Phys. **64** (1991), 843.

9. Adrian L. Melott, *The topology of large-scale structure in the universe*, Physics Rep. **193** (1990), 1–39.

10. Hermann Minkowski, *Volumen und Oberfläche*, Mathematische Annalen **57** (1903), 447–495.

11. Phillip J.E. Peebles, *Principles of physical cosmology*, Princeton University Press, Princeton, New Jersey, 1993.

12. Luis A. Santaló, *Integral Geometry and Geometric Probability*, Addison-Wesley, Reading, MA, 1976.

13. B.S. Sathyaprakash, Varun Sahni & Sergei F. Shandarin, *Emergence of filamentary structure in cosmological gravitational clustering*, Ap. J. Lett. **462** (1996), 5–8.

14. Jens Schmalzing, *Morphologische Maße für kosmische Strukturen*, Diplomarbeit, Ludwig-Maximilians-Universität München, 1996, in German, English excerpts available upon request.

15. Jens Schmalzing, Martin Kerscher & Thomas Buchert, *Minkowski functionals in cosmology*, Proceedings of the international school of physics Enrico Fermi. Course CXXXII: Dark matter in the universe (Silvio Bonometto, Joel Primack & Antonello Provenzale, eds.), Societá Italiana di Fisica, 1995.

16. Rolf Schneider, *Convex bodies: the Brunn-Minkowski theory*, Cambridge University Press, Cambridge, 1993.

17. Rien van de Weygaert, *Fragmenting the universe III. The constructions and statistics of 3-D Voronoi tessellations*, Astron. Astrophys. **283** (1994), 361–406.

18. Wolfgang Weil, *Stereology: A survey for geometers*, Convexity and its applications (Peter M. Gruber & Jörg M. Wills, eds.), Birkhäuser, Basel und Boston, 1983, pp. 360–412.

19. Simon D.M. White, *The hierarchy of correlation functions and its relation to other measures of galaxy clustering*, Mon. Not. R. Astron. Soc. **186** (1979), 145–154.

A WAVELET ANALYSIS OF THE COMA CLUSTER

Statistics and morphology

M.GAMBERA, A.PAGLIARO
Istituto di Astronomia
Catania

AND

V.ANTONUCCIO-DELOGU
Osservatorio Astrofisico
Catania

Abstract. We assemble a catalog of 798 galaxy redshifts in the region of the Coma cluster and we examine the structure of the cluster at different scales. Our structure analysis method is based on the wavelet transform and on the segmentation analysis. The wavelet transform allows us to find out structures at different scales and the segmentation method allows us a quantitative statistical and morphological analysis of the sample. Our results are expressed in terms of number of structures and sphericity for different values of the threshold detection and at all the scales investigated. According to our analysis, there is strong evidence for multiple hierarchical substructure, on the scales ranging from a few hundreds of Kpc to about $4\,h^{-1}$ Mpc, and the substructure morphology is rather spherical.

1. Introduction

This paper takes up the issue of substructure in the Coma cluster (Abell 1656). In previous papers (e.g. Colless & Dunn, 1996; Biviano et al.,1996) it has been suggested that the Coma cluster may have a complex structure. The X-ray images obtained with *ROSAT* suggest clumps of emission associated with substructures. However, previous analysis were performed only on 2-D slices of the cluster. Our aim is to find out substructure at different scales through a three-dimensional analysis of the cluster, identify it and make a morphological analysis of every single structure singled out.

D. N. Schramm and P. Galeotti (eds.), Generation of Cosmological Large-Scale Structure, 261–266.

2. The data

A large body of data on the Coma cluster is available. We have used a combination of various catalogs to assemble our own. Our redshifts catalog is made of a total number of 798 galaxies with the equatorial coordinates in the range

$$
\begin{aligned}
\text{RA} &= 10^h 42^m 00^s, 15^h 28^m 00^s \\
\text{DEC} &= +27°05'00", 28°20'00"
\end{aligned} \tag{1}
$$

(B1950.0, hereafter all coordinates are referred to1950.0). The redshifts for 243 galaxies have been kindly provided us by the new redshifts survey by M. Colless & A. M. Dunn (1996). Another 305 redshifts have been taken from Biviano et al. (1995). Finally, another 320 redshifts have been taken from the catalog of 379 galaxies by I. D. Karachentsev & A. I. Kopylov (1990). The total number of galaxies collected is 868, but some objects are in common. We have considered a galaxy common when its position in a catalog is inside the mean error in the coordinate determination in another catalog. So, if for the same galaxy several redshift measurements were available, we have chosen the most accurate one. To summarize we have a heterogeneous sample of redshifts for 798 galaxies; Our coordinates of the centre of the cluster is the value derived from the survey of Godwin et al. (1983): RA = $12^h 57^m.3$ and DEC = $+28°14.4'$. The uncertainty on the redshifts measurement is about 100 Km/s and the maximum error on the positions is less than 3".

3. The method of analysis: wavelet transform and segmentation

Our method of structure detection is based on the wavelet transform evaluated at several scales and on the segmentation analysis and is very similar to the one developed by Lega(1994).

A detailed description of the implementation of the algorithms is beyond the purpose of this paper. A version for a Connection Machine *CM200* can be found in Lega(1994), a new *PVM* version will be described in Becciani et al. (1997). Our method is a threedimensional analysis. However, for the sake of simplicity, here we describe the one dimensional reduction. The generalization to the 3-D case is obvious. For a one-dimensional function $f(x)$ the wavelet transform is a linear operator that can be written as:

$$
w(s,t) = s^{-1/2} \int_{-\infty}^{+\infty} f(x)\psi^* \left(\frac{x-t}{s} \right) dx \tag{2}
$$

where $s(>0)$ is the scale on which the analysis is performed, t is the translation parameter and ψ is the analyzing wavelet function. We follow

Lega(1994) in the choice of the mother wavelet:

$$\psi(x) = \phi(x) - \frac{1}{2}\phi(\frac{x}{2})$$

(3)

where ϕ is the cubic centred B-spline function defined by:

$$\phi(x) = \frac{|x-2|^3 - 4|x-1|^3 + 6|x|^3 - 4|x+1|^3 + |x+2|^3}{12}$$

(4)

The wavelet coefficients at different scales are obtained through the à *trous* algorithm, as described by Lega(1994, pag.100). The set of scales are powers of two. The scale s in this kind of analysis may be considered as the resolution. In other words, if we perform a calculation on a scale s_0, we expect the wavelet transform to be sensitive to structures with typical size of about s_0 and to find out those structures.

The second step of our analysis is the determination of connected pixels over a fixed threshold (*segmentation*, Rosenfeld 1969), the numbering of the selected structures and their morphological analysis. The segmentation and numbering consists in the exam of the wavelet coefficients matrix; all the pixels associated with a wavelet coefficient greater than the selected threshold are labeled with a integer number. All other pixels' labels are set equal to zero. Then, the same label is associated to all the pixels connected in a single structure. Then, for each structure we calculate volume and surface and from them a morphological parameter.

The thresholding is made on the wavelet coefficient histogram. For a flat background, the wavelet transform yields coefficients equal to zero. The existence of structures at a given scale gives wavelet coefficient with large values. Obviously, a random distribution may have non zero coefficients even if there is no structure, due to the statistical fluctuations. We choose the threshold through a classical decision rule. We calculate the wavelet coefficients $W_{ran}(s)$ for each scale of our analysis, for a random distribution in the same region of space of our data. Then we calculate the probability $P[w(s) \leq W_{ran}(s)]$ and choose the value $w(s)$ so that: $P[w(s) \leq W_{ran}] \leq \epsilon$ Our threshold is the value $\nu = w(s)$. The value: $\epsilon = 0.001$ ensures a 99.9% confidence level in the structure detection. However, for the sake of completeness, we perform our analysis for several values of the threshold, calculated in terms of the standard deviation in the wavelet coefficient distribution of our data.

In order to perform a morphological analysis we have to introduce a morphological parameter that quantifies the sphericity of the structures. We choose the parameter:

$$L(s) = K(s)\frac{V^2}{S^3}$$

(5)

where V is the volume and S is the surface, as in Lega(1994), and $K(s)$ is a parameter that depends on the scale of the analysis. We want $L(s)$ to have the following behaviour: zero for very filamentary structures and 1 for spherical ones. Since we want to consider as spherical a one-pixel structure, we adopt the values:$K(s = 0, 1) = 216$ and $K(s \geq 2) = 36\pi$ Then, for every detection threshold we calculate the mean values:

$$\langle L(s) \rangle = \sum_{i=1}^{N_{obj}} \frac{L(s)}{N_{obj}} \qquad (6)$$

where N_{obj} is the number of objects detected at scale s.

4. Substructure detection in the Coma cluster

We examine two subsamples of our catalog. The first is made of 690 galaxies with redshifts between $cz = 3000$ km/s and $cz = 28000$ km/s. We call it extended Coma. We examinate four different scales: 2.02, 4.04, 8.08, $16.16\,h^{-1}$ Mpc, for six different values of the threshold: the 99.9% confidence level, 3σ, 4σ and 5σ. The second catalog is made of 485 galaxies, with redshifts between $cz = 4000$ km/s and $cz = 10000$ km/s, and we call it Coma. We examinate four different scales: 0.47, 0.94, 1.88, $3.76\,h^{-1}$ Mpc, for the same values of the threshold as before. Our results are expressed in terms of number of structures at the selected scales (see the following tables where we report: number of structures / mean morphology). Considering the 99.9% confidence level as significance level for the structure detection, we have overwhelming evidence for substructure inside Coma on the scales: 0.47 to $4.04\,h^{-1}$ Mpc. For what concerns Coma, our substructures are rather spherical on the first two scales. The value of $\langle L \rangle$ is lowered till 0.30 on the scale $1.88\,h^{-1}$ Mpc, meaning a much more filamentary morphology for those structures singled out at this resolution. The substructure of extended Coma, singled out with a greater resolution shows a spherical morphology till the scale of $2.02\,h^{-1}$ Mpc; more elongated shapes are found out at the bigger scales. The diminution of $\langle L \rangle$ and of the number of structures as a function of the scale indicates a hierarchical distribution.

Table 1: EXTENDED COMA

scale	99.9%	3σ	4σ	5σ
2.02	60 / 0.93	63 / 0.90	55 / 0.91	64 / 0.91
4.04	15 / 0.57	4 / 0.40	4 / 0.45	4 / 0.27
8.08	2 / 0.30	1 / 0.27	1 / 0.27	1 / 0.31
16.16	2 / 0.17	1 / 0.17	2 / 0.30	2 / 0.31

Table 2: COMA

scale	99.9%	3σ	4σ	5σ
0.47	54 / 0.95	205 / 0.88	200 / 0.89	75 / 0.89
0.94	21 / 0.85	42 / 0.64	35 / 0.62	25 / 0.58
1.88	3 / 0.29	3 / 0.42	2 / 0.32	3 / 0.60
3.76	2 / 0.16	1 / 0.10	1 / 0.13	2 / 0.12

5. The central region of Coma

We consider a galaxy belonging to the central region of Coma if its redshift is inside $\pm 1\sigma$ from $\langle cz \rangle$ = 7000 km/s, where $\langle cz \rangle$ and σ are the mean and the standard deviation in the redshifts distribution calculated on the 485 galaxies considered in §4. Our catalog is made of 320 galaxies, with redshifts between cz = 5850 km/s and cz = 8150 km/s. We examinate four different scales: 180, 360, 720 and 1440 h^{-1} Kpc. Our results are expressed in terms of number of structures at the selected scales. Considering the 99.9% confidence level threshold as significance level for the structure detection, we have overwhelming evidence for substructure inside the central region of the Coma cluster on the first three scales investigated: 180, 360 and 720 h^{-1} Kpc (see the table 3). Our substructure are rather spherical on all the scales but the last one, where its value is about 0.30, meaning that shapes become more elongated on a scale of typical size 720 h^{-1} Kpc, inside the central region.

Table 3: COMA CENTRE

scale	99.9%	3σ	4σ	5σ
0.18	28 / 1.00	655 / 0.90	550 / 0.93	397 / 0.94
0.36	13 / 0.61	109 / 0.68	93 / 0.60	103 / 0.66
0.72	7 / 0.35	4 / 0.32	7 / 0.35	8 / 0.45
1.44	1 / 0.28	2 / 0.50	3 / 0.41	3 / 0.42

6. Conclusions

Our results suggests that Coma can not be considered a regular cluster of galaxies, but it is filled up with substructure at scales ranging from a few hundreds of Kpc to $\sim 4h^{-1}$ Mpc. The general diminution of the mean morphological parameter, meaning more elongated shapes, and of the number of structures with the scale indicates a hierarchical distribution of the substructure. For a more complete discussion of our method and our results see Gambera et al. (1997).

Acknowledgements. A.Pa. would like to thank E. Lega for her kind and indispensable help.

References

Abell, G.O., 1958, ApJS, 3, 211

Becciani, U., Antonuccio-Delogu, V., Pagliaro, A. (in preparation)

Biviano, A., Durret, F., Gerbal ,D., Le Fevre, O., Lobo, C., Mazure, A., and Slezak, E., 1996, *A&A* Suppl. Ser. 111, 265

Colless, M., Dunn, A., M., 1996, ApJ, also astro-ph/9508070

Gambera, M., Pagliaro, A., Antonuccio-Delogu, V. (in preparation)

Godwin, J., G., Metcalfe, N., Peach, J., V., 1983, MNRAS 202, 113

Karachentsev, I., D., Kopylov, A., I., 1990, MNRAS 243, 390

Lega, E., 1994, These de Doctorat, Université de Nice

Rosenfeld, A., 1969, Picture Processing by Computer, Academic Press, New York

FLUCTUATIONS IN THE IRAS 1.2 JY CATALOGUE

MARTIN KERSCHER

Ludwig–Maximilians–Universität,
Theresienstr. 37, 80333 München, Germany

AND

JENS SCHMALZING

Ludwig–Maximilians–Universität,
and
Max–Planck–Institut für Astrophysik,
Karl-Schwarzschild-Straße 1, 85740 Garching, Germany

"It's only galaxies, nothing fundamental." (N. Kaiser)

1. Introduction

In an analysis of the IRAS 1.2 Jy redshift catalogue, with emphasis on the separate examination of northern and southern parts (in galactic coordinates), we found that the clustering properties of galaxies differ significantly in the northern and southern parts, showing fluctuations in the clustering properties at least on scales of $200h^{-1}$Mpc. Our analysis is based on Minkowski functionals, next–neighbour distributions, and the fluctuation of galaxy counts σ^2. In Section 2 we give a brief description of the way we use Minkowski functionals, in Section 3 we present the results for the Minkowski functionals obtained from the IRAS 1.2 Jy galaxy catalogue (the results for next–neighbour distributions and σ^2 will be present in Kerscher et al. [6]), and in Section 4 we discuss our results.

2. Minkowski functionals

We consider a realization $\{\mathbf{x}_i\}_{i=1}^{N}$ of N points for the process of Large Scale Structure. The realization is given by the redshift–space coordinates of the galaxies in the IRAS 1.2 Jy catalogue [3]. We characterize the properties of this point set with Minkowski functionals as they were introduced into cosmology by Mecke, Buchert and Wagner [8] (see also the article by Schmalz-

D. N. Schramm and P. Galeotti (eds.), Generation of Cosmological Large-Scale Structure, 267–272.

ing in this volume). For this purpose we decorate each point $\mathbf{x}_i$ with a ball $B_r(\mathbf{x}_i)$ of radius r. Then we look at the union set $A_r = \bigcup_{i=0}^{N} B_r(\mathbf{x}_i)$. Hadwiger [4] proved that in three dimensions the four Minkowski functionals $M_{\mu=0,1,2,3}(A_r)$ give a complete morphological characterization of a body A_r. They have a direct interpretation in terms of geometrical and topological quantities (see [8] and the article by Schmalzing in this volume). Specific, dimensionless Minkowski functionals $\Phi_\mu(A_r)$ may be constructed by dividing the Minkowski functionals $M_\mu(A_r)$ by N times the functional of a single ball $M_\mu(B_r)$ (i.e. norming the functionals to balls),

$$\Phi_\mu(A_r) := \frac{M_\mu(A_r)}{N M_\mu(B_r)}. \tag{1}$$

For a Poisson process the functionals can be calculated analytically [9]. For spheres with radius r we get for the specific Minkowski functionals:

$$\begin{aligned}
\Phi_0^P &= (1 - e^{-\eta})\,\eta^{-1}, & \Phi_1^P &= e^{-\eta}, \\
\Phi_2^P &= e^{-\eta}\,(1 - \tfrac{3\pi^2}{32}\eta), & \Phi_3^P &= e^{-\eta}\,(1 - 3\eta + \tfrac{3\pi^2}{32}\eta^2),
\end{aligned} \tag{2}$$

with the mean number of galaxies in a sphere with radius r, $\eta = \bar{n} M_0(B_r) = \bar{n}\,4\pi r^3/3$, the mean number density $\bar{n}$. As can be seen from Eqs. (2), the Minkowski functionals which are confined to the surface ($\mu = 1, 2, 3$), are damped by the exponential term $e^{-\eta}$. This is obvious for a Poisson process. It may be shown that this e^{-r^3} dependence also holds in the case of more general cluster processes for larger radii. We want to investigate the large–scale behavior of the point process, therefore, we look at the surface densities of the specific Minkowski functionals $\Phi_{\mu=1,2,3}$, defined by

$$\phi_\mu(A_r) = \frac{\Phi_\mu(A_r)}{\Phi_1^P(r)}, \tag{3}$$

and thus remove the damping term $e^{-\eta}$. Since we do not want to introduce any additional uncertainties, we divide by the analytical result for the Poisson process $\Phi_1^P(r)$.

3. Analysis of the 1.2 Jy catalogue

In this section we report on the application of the Minkowski functionals introduced above to a volume limited sample of the IRAS 1.2 Jy galaxy catalogue [3] with limiting depth of $100h^{-1}\mathrm{Mpc}$ ($H_0 = 100\,h$ km s^{-1} Mpc^{-1}). In Kerscher et al. [6] we discuss how to correct for boundaries. There we also look at a deeper sample, and compare with a simulation. We additionally use the next neighbour statistics and the fluctuations of galaxy counts σ^2 as statistical tools and confirm our results.

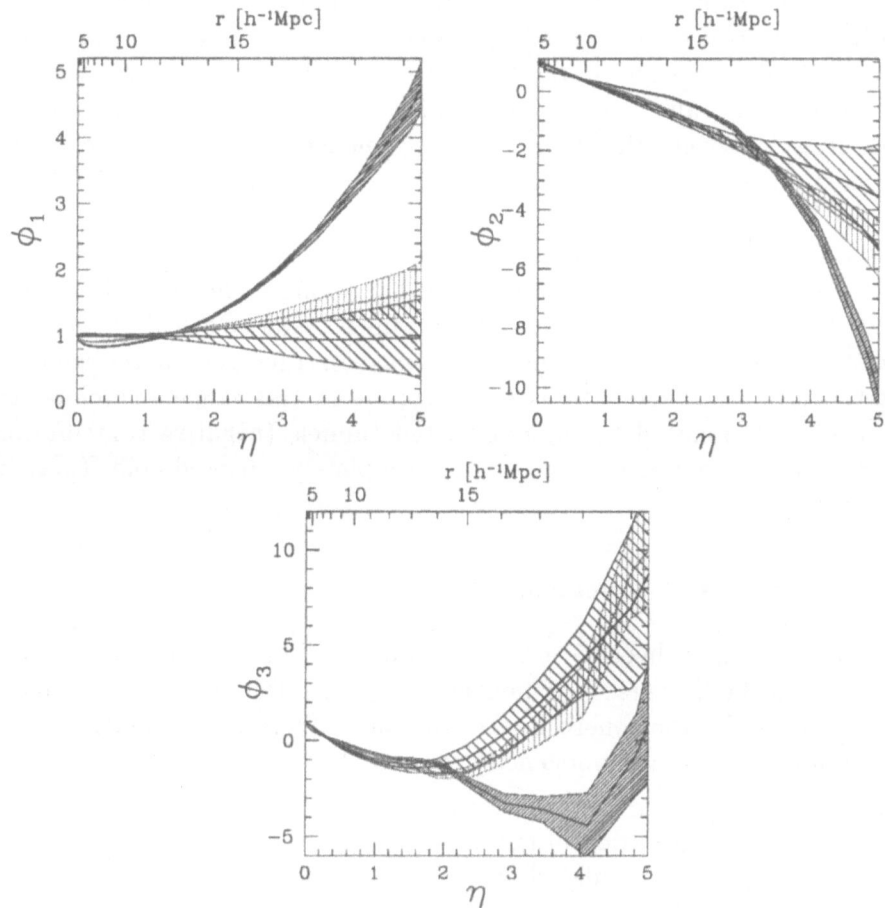

Figure 1. Minkowski functionals ϕ_μ of a volume limited sample with $100h^{-1}$ Mpc depth; the dark shaded areas represent the southern part, the medium shaded the northern part, and the light shaded a Poisson process with the same number density. The shaded areas are the 1σ errors as explained in the text.

In Figure 1 we show the Minkowski functionals of the southern and northern parts together with the functionals of a Poisson process with the same number density. The northern part contains 352 galaxies, and the southern part 358 galaxies. Thus, the sample shows only small fluctuations in the number density. The errors of the Poisson process were calculated using twenty different realizations. To estimate the error of the data we calculated the Minkowski functionals of twenty subsamples 90% of the galaxies, randomly chosen from the volume limited subsample (in Kerscher et al. [6] we discuss error estimates and selection effects in detail).

In both parts of the 1.2 Jy catalogue the galaxies cluster on scales up to $10h^{-1}$Mpc significantly stronger than in the case of a Poisson process resulting in the lowered surface functional ϕ_1, integral mean curvature functional ϕ_2 and the Euler characteristic functional ϕ_3. Moreover, we realize that the northern and southern parts differ significantly. The northern part shows less clustering. Probably the most conspicuous features are the enhanced surface area ϕ_1 in the southern part on scales from 12 to 20 h^{-1}Mpc and the kink in the integral mean curvature ϕ_2 at 14 h^{-1}Mpc. This indicates that dense substructures in the southern part fill up at this scale (i.e. the balls in these substructures overlap leaving no holes). This is probably the signature of the Perseus–Pisces supercluster. On scales from 15 to 20 h^{-1}Mpc ϕ_2 is negative indicating concave structures. In the southern part the Euler characteristic is still negative in this range, therefore, the structure is dominated by interconnected tunnels (negative contributions to the Euler characteristic) and not by completely enclosed voids (positive contributions).

4. Discussion and conclusions

Recently, there have been several contradicting claims that the distribution of galaxies in the Universe is homogeneous (see e.g. Davies [7]) or fractal (see e.g. the article by Pietronero in this volume). We recall the mathematical definition of statistical homogeneity (see e.g. [11]):

For point processes the *almost–sure–homogeneity* is defined as the stationarity of the distribution of the point process under translations, i.e. the statistical properties calculated from an ensemble of realizations must not change under translations of the points. Isotropy is the stationarity of the distribution under rotations. In looking at only one realization we are not able to make statements about the statistical homogeneity. A statistically homogeneous point process may show large fluctuations in statistical quantities estimated from finite volumes in one realization. Fluctuations on the same length scale as the sample or even larger will not be detected. According to this definition certain fractal distributions are statistically homogeneous.

The volume limited subsamples we look at show fluctuations in the number density of galaxies of less than 1%. Recently Davis [7] used the whole galaxy catalogue with flux limit 1.2 Jy, but no volume limitation, and looked at galaxies in different redshift intervals, projected onto the sphere. He found strong anisotropies in the number density in the nearby region, but less in the regions further apart. Except from the subjective judgement made by examining the sample by eye, the quantitative argument is based on the number density only. In looking at the flux limited sample in the whole redshift range we find a difference of 15% in the number density between northern and southern part; the northern part incorporates 2848 galaxies, the southern part only 2465. But this may be attributed to luminosity effects. Another test on fluctuations of the number density in the flux limited 1.2 Jy catalogue is used in Kaiser et al. [5]. With statistics like σ^2, the correlation function, the power spectrum, or the correlation integral one estimates the two–point distribution only. Already here, both parts differ (see [6]). The discrepancies between north and south are enhanced by taking into account higher moments. Indeed correlations of arbitrary order enter into the Minkowski functionals [8] which reveal significant differences (see Figure 1).

Since almost identical selection effects in northern and southern parts together with the fact that the number densities of the galaxies in the volume limited subsamples analyzed are almost equal in both parts, we can draw the following conclusion:

A significant departure from a Poisson process and different clustering properties in the northern and the southern parts are detected on the scale of $200h^{-1}$Mpc (the diameter of the volume limited sample with depth of $100h^{-1}$Mpc). As shown in [6] the detected differences are less significant, but do not disappear on the scale of $400h^{-1}$Mpc. Similar anisotropies have already been reported by Rowan–Robinson et al. [10] in looking at the two–dimensional distribution of IRAS galaxies around the northern and southern galactic pole. Recently, Yess et al. [12] find indications for structures which percolate the whole sample of the 1.2 Jy catalogue.

One interpretation of the discrepancies described above could be based on the following reasoning: we assume that both the northern and the southern parts are typical for our Universe, i.e. our position does not entail any peculiarity compared with other randomly chosen positions. Then, this means that the (ensemble) variance of the statistics applied to the 1.2 Jy galaxy catalogue is at least of the order of the difference between the results for the northern and southern parts. In other words the process of large scale structure admits large fluctuations in the clustering properties on scales of $200h^{-1}$Mpc. The physical objects causing these fluctuations may be superclusters [1] or superlarge–scale structures [2]. Another interpretation in terms of a fractal distribution of galaxies is advocated by Pietronero and coworkers (see e.g. their articles in this volume).

The fluctuations detected on scales up to $200h^{-1}$Mpc (see Figure 1) imply that cosmological simulations which *enforce* homogeneity on the scale of a few hundreds of h^{-1}Mpc and suppress fluctuations on larger scales by using periodic boundary conditions cannot reproduce the large–scale fluctuations indicated by the present analysis of the 1.2 Jy catalogue, this is confirmed in [6] by comparison with a simulation.

It remains an open question, whether it is possible to explain these observed fluctuations in the clustering properties on scales up to $200h^{-1}$Mpc with linear theory and with the usual COBE normalized Gaussian initial density fields.

Acknowledgements

We would like to thank Thomas Buchert and Herbert Wagner for fruitful discussions and the allowance to publish common work prior to submission. MK acknowledges support from the Sonderforschungsbereich SFB 375 für Astroteilchenphysik der Deutschen Forschungsgemeinschaft and together with JS from NATO for support during their stay in Erice.

References

1. A.J. Connolly, A.S. Szalay, D.C. Koo, A.K. Romer, B. Holden, R.C. Nichol & T. Miyaji, *Superclustering at Redshift z=0.54*, Ap. J. Lett. **473** (1996), 67.
2. Andrei G. Doroshkevich, Richard Fong, Stefan Gottlöber, Jan P. Mücket & Volker Müller, *The formation and evolution of large- and superlarge-scale structure in the universe -I: General Theory*, submitted to MNARS.
3. Karl B. Fisher, John P. Huchra, Michael A. Strauss, Marc Davis, Amos Yahil & David Schlegel, *The IRAS 1.2 Jy Survey: Redshift Data*, Ap. J. Suppl. **100** (1995), 69.
4. Hugo Hadwiger, *Vorlesungen über Inhalt, Oberfläche und Isoperimetrie*, Springer Verlag, Berlin, 1957.
5. N. Kaiser, G. Efstathiou, R. Ellis, C. Frenk, A. Lawrence, M. Rowan-Robinson & W. Saunders, *The large–scale distribution of IRAS galaxies and the predicted peculiar velocity field*, Mon. Not. R. Astron. Soc. **252** (1991), 1–12.
6. Martin Kerscher, Jens Schmalzing, Thomas Buchert & Herbert Wagner, *Fluctuations in the 1.2 Jy Galaxy Catalogue*, to be submitted.
7. Davis M., *Is the Universe Homogeneous on Large Scales?*, To be published in "Critical Dialogues in Cosmology" ed. N. Turok, astro-ph/9610149.
8. Klaus Mecke, Thomas Buchert & Herbert Wagner, *Robust morphological measures for large-scale structure in the Universe*, Astron. Astrophys. **288** (1994), 697.
9. Klaus Mecke & Herbert Wagner, *Euler characteristic and related measures for random geometric sets*, J. Stat. Phys. **64** (1991), 843.
10. M. Rowan-Robinson, D. Walker, T. Chester, T. Soifer & J. Fairclough, *Studies of IRAS sources at high galactic latitudes - I. Source counts at $|b| > 60°$ and evidence for a north–south anisotropy of cosmological significance*, Mon. Not. R. Astron. Soc. **219** (1986), 273–283.
11. D. Stoyan, W.S. Kendall & J. Mecke, *Stochastic Geometry and its Applications*, John Wiley & Sons, Chichester, 1987.
12. Capp Yess, Sergei F. Shandarin & Karl B. Fisher, *Percolation Analysis of a Wiener Reconstruction of the IRAS 1.2 Jy Redshift Catalog*, Ap. J. **474** (1997), 553.

TESTING COSMOLOGICAL MODELS USING DAMPED LYMAN-ALPHA ABSORBERS

JEFFREY P. GARDNER AND NEAL KATZ
University of Washington
Department of Astronomy, Seattle, WA 98195, USA

LARS HERNQUIST
University of California
Lick Observatory, Santa Cruz, CA 95064, USA

AND

DAVID H. WEINBERG
Ohio State University
Department of Astronomy, Columbus, OH 43210, USA

1. Introduction

Systems producing absorption in the spectra of distant quasars offer an excellent probe of the early Universe. At high redshifts, they easily outnumber other observed tracers of cosmic structure, including both normal and active galaxies. Mounting evidence that the high column density absorbers are young galaxies links relatively pristine baryonic matter to highly evolved objects at the present day ([1-3] and references therein).

Previous attempts to discriminate cosmological scenarios from the properties of damped Lyman-α (DLA) absorbers have relied on simplifying assumptions. [4] and [5] used a Press-Schechter [8] analysis to predict the abundance of dark matter halos and assumed that some or all of the gas within these halos would cool and become neutral. [6] and [7] proceeded in a similar fashion, but computed the halo mass function from dissipationless N-body simulations. In both methods, ad hoc assumptions are required to estimate the amount of gas that can cool and give rise to DLA absorption.

Katz *et al.* [9] (hereafter KWHM) used gas-dynamical simulations to demonstrate that damped absorption results from concentrations of dense gas embedded in more extended and more massive dark matter halos, with masses and abundances comparable to those of galaxies. Owing to computational expense, the KWHM simulation does not resolve halos with

D. N. Schramm and P. Galeotti (eds.), Generation of Cosmological Large-Scale Structure, 273–278.
© 1997 *Kluwer Academic Publishers.*

virial-radius circular velocities below $v_c \approx 100$ km s^{-1}, although higher resolution simulations of localized regions by Quinn *et al.* [10] (hereafter QKE) indicate that halos down to $v_c \approx 35$ km s^{-1} can host damped absorbers. The KWHM results, therefore, serve as an upper limit.

Since a full simulation is not technically feasible, we employ a two-step correction method to the KWHM data to account for the unresolved systems and then extend those results to predict the DLA absorption distribution in other cosmologies. Hence, we endeavor to use the numerical data to the fullest extent possible to construct a less equivocal cosmological testing ground. Here we shall summarize our methodology, results, and our confidence in them. We encourage the reader to examine the full treatment in Gardner *et al.* [11] (hereafter GKHW) and [12].

2. Methods

Our primary simulation, the same as that used by KWHM, follows the evolution of a periodic cube whose edges measure 22.22 Mpc in comoving units. This region was drawn randomly from a CDM universe with $\Omega = 1$, $h \equiv H_0/100$ km s^{-1} Mpc$^{-1} = 0.5$, baryon density $\Omega_b = 0.05$, and power spectrum normalization $\sigma_8 = 0.7$. A uniform background radiation field was also imposed to mimic the expected ultraviolet (UV) output of quasars. We also employ two simulations that have the same initial conditions and numerical parameters as QKE and identical cosmological and UV background parameters to KWHM. These comprise smaller, 10 Mpc periodic volumes which are evolved using a hierarchical grid of particles in the initial conditions, a technique which allows us to achieve high-resolution locally while preserving the cosmological context of the calculation. Further details of the simulations are given in KWHM or GKHW and references therein.

We use the approximation developed by Press & Schechter [8] to "fill" the gap in the KWHM data for halos of $v_c < 100$ km s^{-1} by requiring the approximation to accurately reproduce the simulated data above this threshold. If we assume an isothermal sphere model for the Press-Schechter halos, their mass and circular velocity are related by a simple expression. In what follows, for comparison with our simulations, we use the CDM transfer function given by [13].

We match the Press-Schechter mass function to the KWHM data, allowing us to extend their halo mass function to lower masses than the simulation cutoff. We then project each halo and measure the area subtended by HI column densities $N_{HI} > 10^{20.3}$ cm^{-2}, yielding a DLA cross-section. Using the halos in both the high resolution and KWHM simulations, we fit for $\alpha_z(v_c)$, the typical DLA cross-section of a halo of circular velocity v_c, by using a power law with an exponential cutoff to reproduce the absence of absorbers in halos with $v_c < 37$ km s^{-1}. As an example, Figure 1a shows the data and fit for $z = 2$. Redshifts 3 and 4 are similar (*cf.* GKHW). By

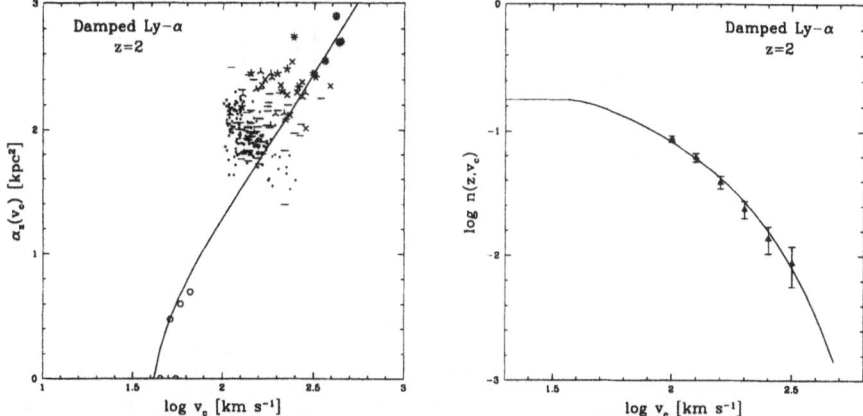

Figure 1. A (left): Comoving absorbing area in kpc^2 vs. circular velocity v_c in km s^{-1} for simulated halos at $z = 2$. The points lying at $v_c \geq 100$ denote halos in the 22.22 Mpc simulation, where the number of vertices in the data point corresponds to the number of gas concentrations in the halo, with the solid points representing halos containing a single absorber. The open circles to the lower left denote halos in the 10 Mpc simulation. The solid line shows the fitted smooth relation. B (right): Incidence of DLA absorption at $z = 2$. Curves show $n(z, v_c)$, the number of absorbers per unit redshift arising in halos with circular velocity greater than v_c. The y-intercepts show the incidence of absorption produced by all halos. Points with $N^{1/2}$ error bars show numerical results from the 22.22 Mpc simulation.

convolving $\alpha_z(v_c)$ with the Press-Schechter mass function we can now use the Press-Schechter formalism to calculate DLA absorption, specifically the observable quantity $n(z)$, the number of DLA absorbers per unit redshift interval along a random line of sight. It is also possible to obtain $n(z, v_c)$, that is the $n(z)$ one would observe if one saw only absorption in halos with circular velocities greater than v_c.

As criteria for the fit of $\alpha_z(v_c)$, we not only require that χ^2 is minimized when matched against the ensemble of simulated halos in DLA cross-section vs. v_c halo, but also that $\alpha_z(v_c)$, when convolved with Press-Schechter, *reproduces the absorption distribution of the simulation.* Figure 1b compares $n(z, v_c)$ of the 22.22 Mpc simulation with that computed using Press-Schechter and our fitted relations $\alpha_z(v_c)$ for $z = 2$. Again, redshifts 3 and 4 are similar (*cf.* GKHW). The curves fit these points quite well — they are, of course, constructed to do so, but the agreement shows that our full procedure, including the details of the Press-Schechter calibration and fitting for $\alpha_z(v_c)$, is able to reproduce the original numerical results in the regime where halos are resolved. This demonstrated ability of the fitting method to reproduce the distribution of absorption in the simulation dispenses with the need to resort to more complex approaches. We must emphasize that our method achieves our goal of characterizing the aggregate absorption as predicted by the KWHM and QKE simulations, and may therefore be used with confidence.

276

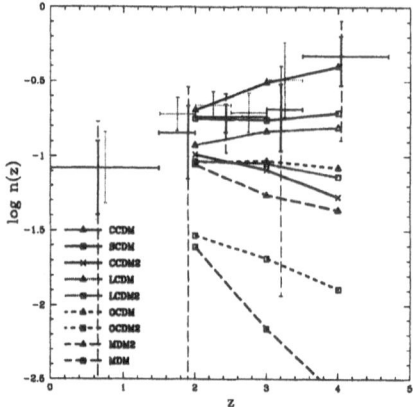

Figure 2. The rate of incidence $n(z)$ of damped Lyman-α systems for various cosmological models versus observational data. The solid error bars are reproduced from [14], with the vertical dashed curve indicating 2σ error. The dotted error bars are values taken from [15]. The heavy curves connecting $z = 2, 3, 4$ are values calculated for each model. The model names correspond to the names in the accompanying table, where their parameters are listed.

3. Results and Discussion

For the halos in the KWHM simulation, there are two competing effects that determine the trend between absorption cross section and v_c (Figure 1a). Higher mass halos have deeper potential wells, so concentrations of cold gas contract further, explaining the downward trend in cross section with circular velocity exhibited by points with a fixed number of vertices. This behavior is in contrast to most semi-analytical models of DLA's [4, 5] in which the cross-section of the halo increases with mass. However, more massive halos tend to harbor more than one concentration of gas, increasing their absorption cross section. The overall trend in Figure 1a is that halos of higher circular velocities have, on average, larger absorption cross sections.

Once $\alpha_z(v_c)$ is known for the KWHM cosmology, one can make the plausible assumption that $\alpha_z(v_c)$ is similar from one model to another, then combine $\alpha_z(v_c)$ from the KWHM simulation with Press-Schechter halo abundances for other models to predict $n(z)$. While it is less secure than the full resolution-corrected numerical treatment of GKHW, it is an improvement over existing semi-analytic calculations of DLA abundances [4-7] which usually assume that *all* gas within the halo virial radius cools and becomes neutral, and which either assume a form and scale for the collapsed gas distribution or compare to observations only through the atomic gas density parameter Ω_g, which is sensitive mainly to the very highest column density systems (GKHW).

Figure 2 shows the resulting rate of incidence for various cosmological models in comparison with observed data. These data are also tabulated and discussed in detail in [12]. The model producing the most DLA absorption is $\Omega_0 = 1$ CDM normalized to COBE ("CCDM"), where we assume scale-invariant fluctuations and $h = 0.5$ [16]. The model we call "SCDM," being the model used in KWHM and GKHW, has an identical power spectrum but is normalized to small-scale structure. In this case, $n(z)$ is lower due to the decrease in power, but still matches observations within current uncertainties. It is possible to construct a CDM model that fits both cluster masses (which requires $0.5 \lesssim \sigma_8 \lesssim 0.7$ for $\Omega_0 = 1$) and COBE data by further adjustments, such as tilting the primeval power spectrum, i.e. $n < 1$, or raising the baryon fraction. CCDM2 [17] is an example of such a model and is COBE-normalized with $n = 0.8$ and $\Omega_b = 0.1$. Unfortunately, these changes reduce the amount of small-scale power relative to SCDM for $z > 2$ resulting in an $n(z)$ which is in significant disagreement with observations for $z = 3$ and $z = 4$.

Our "LCDM" model [16], COBE-normalized with $\Omega_0 = 0.4$, $\Omega_\Lambda = 0.6$ and $h = 0.65$, generally falls within the 1σ errors of the Storrie-Lombardi et al. [14] data except at $z = 4$. Many of the systems which contribute to this data point have not yet been confirmed by high-resolution spectroscopy, however, so the value could decrease with future observations. Hence LCDM could be considered acceptable. LCDM2 is an identical model with the lower normalization favored by Cen et al. [18]. The lessened power of this model invariably leads to a lower $n(z)$. Actually achieving this low normalization while being consistent with COBE would require a tilt or lower h, which in turn would reduce small scale power, further exacerbating the problem.

OCDM [19] ($\Omega = 0.4$, $\sigma_8 \sim 0.7$) shows a COBE-normalized open model similar to LCDM but lacking a cosmological constant, which reduces its power and hence $n(z)$ with respect to LCDM, and skirts the lower end of the observational data. However, given the uncertainty in our estimate of the absorption and in the $z = 4$ observational data, it might still be considered marginally acceptable. The amount of small scale power implied by the COBE normalization in open models is a rapidly decreasing function of Ω_0. Consequently, if $\Omega_0 = 0.3$ (OCDM2 [19]), σ_8 is reduced to near 0.5 causing $n(z)$ to be much too low.

The most popular alternative to lowering Ω_0 is to add hot dark matter. The first incarnations of this model had $\Omega_\nu = 0.3$, which we reproduce here as "MDM" [20]. These models severely suppress structure formation at high redshift, and hence drastically underproduce DLA absorption. Largely in recognition of this problem, $\Omega_\nu = 0.2$, shown as MDM2 [20], has become the favored version. While more palatable than MDM, this model still falls 2.8σ short for $2 < z < 3$ and 2.5σ below observations at $z = 4$. The most optimistic reading would be that mixed-dark matter can, for the present, barely squeak by the combined constraints of cluster abundances on one hand and DLA's on the other.

The more definitive manner to evaluate the different cosmological models is, of course, through numerical gas simulations. As a method to correct for finite resolution, the GKHW technique should be quite reliable, and it can be applied to other cosmological models once the appropriate simulations are available for calibrating $\alpha_z(v_c)$. We are indeed in the process of producing a suite of simulations to address this issue but such a parameter study is quite expensive and it will be a few years before it is completed. Until such a time, we have endeavored to use existing numerical data to their fullest extent to provide what we feel is the best currently possible test of cosmological models against damped Lyα observations.

References

1. Briggs, F.H., Wolfe, A.M., Liszt, H.S., Davis, M.M., and Turner, K.L. (1989) The spatial extent of the z=2.04 absorber in the spectrum of PKS 0458-020, *ApJ*, **341**, 650.
2. Djorgovski, S.G., Pahre, M.A., Bechtold, J., and Elston, R. (1996) Identification of a Galaxy Responsible for a High-Redshift Damped Ly-alpha Absorption System, *Nature*, **382**, 234.
3. Fontana, A., Cristiani, S., D'Odorico, S., Giallongo, E., and Savaglio, S. (1996) The optical identification of a primeval galaxy at $z \geq 4.4$, *MNRAS*, **279**, 27.
4. Mo, H.J. and Miralda-Escudé, J. (1994) Damped Lyman-alpha systems and galaxy formation, *ApJ*, **430**, L25.
5. Kauffman, G. and Charlot, S. (1994) Constraints on models of galaxy formation from the evolution of damped Lyman alpha absorption systems, *ApJ*, **430**, L97.
6. Ma, C. and Bertschinger, E. (1994) Do galactic systems form too late in cold + hot dark matter models?, *ApJ*, **434**, L5.
7. Klypin, A., Borgani, S., Holtzman, J., and Primack, J. (1995) Damped Lyman-alpha systems versus cold + hot dark matter, *ApJ*, **444**, 1.
8. Press, W.H. and Schechter, P.L. (1974) Formation of galaxies and clusters of galaxies by self-similar gravitational condensation, *ApJ*, **187**, 425.
9. Katz, N., Weinberg, D.H., Hernquist, L., and Miralda-Escudé, J. (1996) Damped Lyman-alpha and Lyman-limit absorbers in the cold dark matter model, *ApJ*, **457**, L57 (KWHM).
10. Quinn, T.R., Katz, N., and Efstathiou, G. (1996) Photoionization and the formation of dwarf galaxies, *MNRAS*, **278**, L49.
11. Gardner, J.P., Katz, N., Hernquist, L., and Weinberg, D.H. (submitted) The population of damped Lyman-alpha and Lyman limit systems in the cold dark matter model, *ApJ* (GKHW).
12. Gardner, J.P., Katz, N., Weinberg, D.H., and Hernquist, L. (in preparation) Testing cosmological models against the abundance of damped Lyman-alpha absorbers, *ApJ*.
13. Bardeen, J.M., Bond, J.R., Kaiser, N., and Szalay, A.S. (1986) The statistics of peaks of Gaussian random fields, *ApJ*, **304**, 15
14. Storrie-Lombardi, L.J., Irwin, M.J., and McMahon, R.G. (1996) APM $z > 4$ survey: distribution and evolution of high column density HI absorbers, *MNRAS*, **282**, 1330.
15. Wolfe, A.M., Lanzetta, K.M., Foltz, C.B., and Chaffee, F.H. (1995) The large bright QSO survey for damped Lyman-alpha absorption systems, *ApJ*, **454**, 698.
16. Bunn, E.F. and White, M. (in press) The four-year COBE normalization and large-scale structure, *ApJ*, Astro-ph 9607060.
17. White, M., Viana, P. T. P., Liddle, A. R., and Scott, D. (in press) Cold dark matter models with high baryon content, *MNRAS*, Astro-ph 9605057.
18. Cen, R., Miralda-Escudé, J., Ostriker, J.P., and Rauch, M. (1994) Gravitational collapse of small-scale structure as the origin of the Lyman-α forest, *ApJ*, **437**, L9.
19. Gorski, K.M., Ratra, B., Stompor, R., Sugiyama, N., and Banday, (submitted) COBE-DMR-Normalized open CDM cosmogonies, *ApJ*, Astro-ph 9608054.
20. Ma, C.P. (1996) Linear Power Spectra in Cold+Hot Dark Matter Models: Analytical Approximations and Applications, *ApJ*, **471**, 13.

EVOLUTION OF THE TWO-POINT CORRELATION FUNCTION: ZEL'DOVICH APPROXIMATION VS. SCALING ANSÄTZE

C. PORCIANI

SISSA, Scuola Internazionale di Studi Superiori Avanzati
via Beirut 2-4, I-34014 Trieste, Italy

Abstract. We study the evolution of the mass autocorrelation function in the Zel'dovich approximation and we compare the results with the predictions of the scaling ansatz for the growth of gravitational clustering formulated by Jain, Mo & White [7].

1. Introduction

Nowadays, thanks to new observational resources, we are able to detect a direct signature of redshift dependence in the observed correlation function of galaxies. However, from the theoretical point of view, even neglecting the problems related to galaxy formation and biasing (and so limiting ourself to consider the evolution of the mass density field), we do not dispose of a complete theory for the formation of large-scale structure. For this reason, also approximated schemes for following the behaviour of mass clustering are of great interest. Hamilton *et al.* [6], analysing a large set of numerical simulations, suggested that, in hierarchical models for structure formation, the non-linear mass correlation function, $\xi(r, z)$, can be related to the linear one through a simple scaling relation. Recently, various authors proposed refined versions of this scaling procedure [7],[9]. Here, we investigate whether these scaling ansätze agree with the predictions of the most used approximation schemes for studying the dynamical evolution of mass density fluctuations. In particular, we will consider the mildly non-linear regime of perturbation growth in a CDM model by comparing the correlation function obtained using the scaling ansatz by Jain, Mo & White (hereafter JMW) [7] with its analogue achieved through the truncated Zel'dovich approximation [11],[5]. Actually, it is well known that Eulerian perturbation theory may break down once the mass variance becomes sufficiently

D. N. Schramm and P. Galeotti (eds.), Generation of Cosmological Large-Scale Structure, 279–282.
© *1997 Kluwer Academic Publishers.*

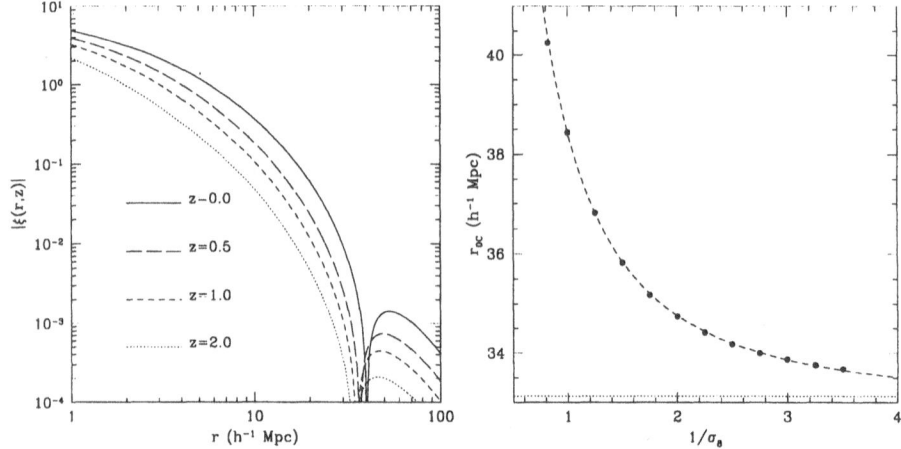

Figure 1. Left panel: redshift evolution of the mass autocorrelation function for a *COBE* normalized standard CDM linear power spectrum evolved through the Zel'dovich approximation. Right panel: the first zero crossing radius of $\xi(r)$ vs. $1/\sigma_8$ (the circles denote the result obtained using the Zel'dovich approximation, the dashed line is the fitting function given in the text while the dotted line shows the prediction of linear theory).

large, while the Zel'dovich approximation, expecially in its truncated form, is able to reproduce fairly well the outcomes of N-body simulations even in the mildly non-linear regime.

A more detailed and complete description of this work is given in [10].

2. Clustering evolution in the Zel'dovich approximation

In this section we briefly analyse the non-linear evolution of the mass autocorrelation function by describing the growth of density fluctuations through the Zel'dovich approximation (more details and mathematical formulae can be found in [3],[8],[10]). The results relative to a *COBE* normalized standard CDM model are shown in the left panel of Fig. 1. Apart from very small scales which are affected by the truncation procedure, we note that $\xi(r, z)$ steepens with decreasing z. Moreover, the first zero crossing radius of $\xi(r, z)$ increases as time goes on (see also [3]). The same behaviour characterizes the evolution of clustering in second-order Eulerian perturbation theory [4],[1]. In the right panel of Fig. 1. we plot the displacement of the first zero crossing of ξ as a function of the linear σ_8. This shifting can be properly approximated by the function:

$$r_{0C}(\sigma_8) - r_{0C}^{lin} \simeq 5.3\,\sigma_8^{(1.5+0.1/\sigma_8)}\,h^{-1}\mathrm{Mpc} \qquad (1)$$

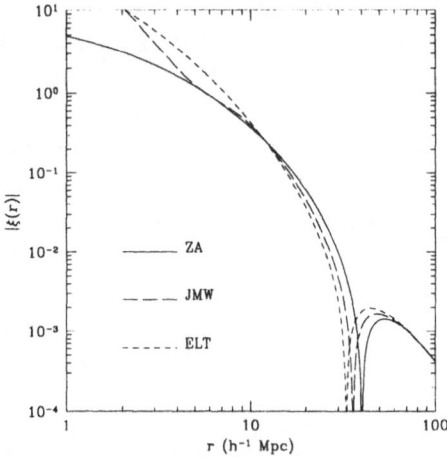

Figure 2. The mass two-point correlation function corresponding to a *COBE* normalized CDM linear spectrum evolved to $z = 0$ by using: the Zel'dovich approximation (ZA), the scaling ansatz of Jain, Mo & White (JMW) and Eulerian linear theory (ELT).

where we denoted by r_{0C} the scale at which the correlation function crosses for the first time the zero–level and by r_{0C}^{lin} its linear analogous.

3. Comparison with the scaling ansatz

We used the scaling ansatz formulated by JMW [7] to evaluate the mass autocorrelation function deriving from a *COBE* normalized standard CDM linear power spectrum. In Fig. 2, the result obtained at $z = 0$ is compared with the corresponding ones achieved through the Zel'dovich approximation and Eulerian linear theory. The agreement between the scaling procedure and the Zel'dovich approximation is remarkable on mildly non-linear scales $(4\,h^{-1}\mathrm{Mpc} \lesssim r \lesssim 20\,h^{-1}\mathrm{Mpc})$ and on linear scales $(r > 60\,h^{-1}\mathrm{Mpc})$. For example, at $r = 5\,h^{-1}\mathrm{Mpc}$, the Zel'dovich approximation underestimates the correlation of JMW by 2%, linear theory overestimates it by 82% while the accuracy of the JMW fit is about $15-20\%$. However, we find that in the interval $20\,h^{-1}\mathrm{Mpc} \lesssim r \lesssim 60\,h^{-1}\mathrm{Mpc}$ the Zel'dovich approximation predicts more non–linear evolution than the scaling ansatz.

While it is easy to ascribe the discrepancy revealed on small scales to the effect of shell crossing, it is much more subtle to understand the disagreement noted on quasi-linear scales. Actually, all the quantitative details of the scaling ansatz are obtained by requiring the resulting $\xi(r)$ to reproduce the linear behaviour for large comoving separations r and, simultaneously, to approximate properly the correlation function extracted from N–body simulations on small scales. However, in order to achieve a detailed description of non–linear scales, JMW used a small box to perform their simulations. Therefore, imposing the match to linear theory on large scales, without having any constraint from numerical data on quasi–linear scales, could seriously alter the accuracy of the scaling ansatz. This implies that the JMW version of the scaling ansatz could be improved on large scales. Our conclusion is shared by Baugh & Gaztañaga [2], who tested the JMW scheme for the evolution of the power spectrum against the results of 5 N–body simulations performed within a $378\,h^{-1}$Mpc box. Indeed, their results show that the JMW formula gives a relatively poor description of the large–scale behaviour even though the agreement between the spectra remains always within the quoted 20 % accuracy.

References

1. Baugh, C.M. and Efstathiou, G. (1994) A comparison of the evolution of density fields in perturbation theory and numerical simulations- I. Non-linear evolution of the power spectrum, *MNRAS* **270**, 183–198
2. Baugh, C.M. and Gaztañaga, E. (1996) Testing ansatze for quasi-nonlinear clustering: the linear APM power spectrum, *MNRAS* **280**, L37–L43
3. Bond, J.R. & Couchmann, H.M.P. (1988) $w_{gg}(\vartheta)$ as a probe of large scale structures, in A. Coley, C.C. Dyer and B.O.J. Tupper (eds.), *Proc. Second Canadian Conference on General Relativity & Relativistic Astrophysics*, World Scientific, Singapore, 385–394
4. Coles, P. (1990) Second-order evolution of Cold Dark Matter perturbations, *MNRAS* **243**, 171–176
5. Coles, P., Melott, A.L. and Shandarin, S.F. (1993) Testing approximations for non-linear gravitational clustering, *MNRAS* **260**, 765–776
6. Hamilton, A.J.S., Kumar, P., Lu, E. and Mathews, A. (1991) Reconstructing the primordial spectrum of fluctuations of the Universe for the observed non-linear clustering of galaxies, *ApJ* **374**, L1–L4
7. Jain, B., Mo, H.J. and White, S.D.M. (1995) The evolution of correlation functions and power spectra in gravitational clustering, *MNRAS* **276**, L25–L29
8. Mann, R.G., Heavens, A.F. and Peacock, J.A. (1993) The richness dependence of cluster correlations, *MNRAS* **263**, 798–816
9. Peacock, J.A. and Dodds, S.J. (1996) Non-linear evolution of cosmological power spectra, *MNRAS* **280**, L19–L26
10. Porciani, C. (in press) Evolution of the two-point correlation function in the Zel'dovich approximation, *MNRAS*, astro-ph 9609029
11. Zel'dovich, Ya.B. (1970) Gravitational instability: an approximate theory for large density perturbations, *A & A* **5**, 84–89

THE SUBSTRUCTURE IN CLUSTERS OF GALAXIES

KĀRLIS BĒRZIŅŠ

Ventspils International Radio Astronomy Center, Akademijas laukums 1, Riga, LV-1524, Latvia[1], and Astronomical Observatory, Niels Bohr Institute for Astronomy, Physics, and Geophysics, University of Copenhagen, Juliane Maries Vej 30, DK-2100 Copenhagen Ø, Denmark[2].

1) E-mail: kberzins@acad.latnet.lv 2) E-mail: berzins@astro.ku.dk

Abstract

A new statistical technique – the *spline method* (hereafter SM) – is introduced in studies of substructure of clusters of galaxies and applied to the *ESO Nearby Abell Cluster Survey* (hereafter ENACS) dataset (see [8], [9]) of Abell 119. We find galaxy clusters to be dynamically unrelaxed systems, i.e., with substructure presented. The algorithm of the substructure finding is discussed using the A119 sample. Some of the potential wells of the subclusters seem to be correlated with x-ray maximums.

1   Introduction

Many statistical tests have been proposed in studies of galaxy clusters (e.g. [13], [2], [1]). At the beginning almost all of them use histograms to divide the dataset into the separate groups. However, the results (subjectively) are strongly dependent on the binsize of the histogram. We find the spline method to be a very good statistical tool producing the parameter independent spline histogram containing all information about the dataset. In Section 2 we will introduce the spline method. Then in Section 3 we will apply the statistical tests on the ENACS dataset of Abell 119 showing the algorithm of the substructure finding. Finally we will give some conclusions in Section 4.

2   The Spline Method

Lets assume that we have measured a set of n discrete values within the interval (a,b), for example radial velocities of galaxies: $X = x_1, x_2, ..., x_i, ..., x_n$ where $x_i < x_{i+1}$, $a < x_1$ and $b \geq x_n$. Without any loss of generality of the method we assume that the realization of each x_i is $f(x_i) = 1/n$ where all data points are

D. N. Schramm and P. Galeotti (eds.), Generation of Cosmological Large-Scale Structure, 283–288.
© *1997 Kluwer Academic Publishers.*

284

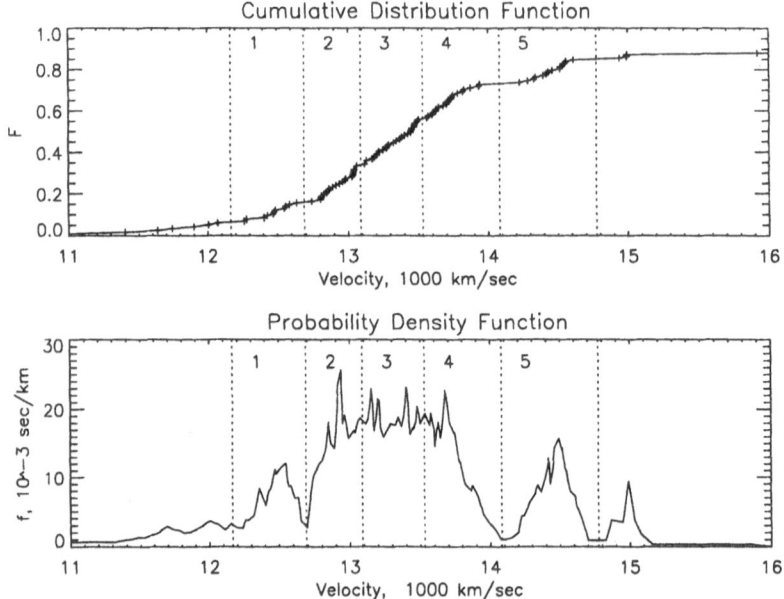

Figure 1: (Top) The cumulative distribution function calculated for the ENACS dataset in field of view of A119. (Bottom) The spline histogram corresponding for the distribution function at the Top figure.

equally weighted. From the given points X we can easily construct the step-type *cumulative distribution* (hereafter distribution) function $F(x) = \sum_{x_i \leq x} f(x_i)$. The continuous differentiable distribution function (if one exists) can be approximated by natural splines $\tilde{F}(x)$ what go through each of $F(x_i)$ points (e.g. [12]) with $\frac{d\tilde{F}^2(x)}{dx^2}|_{x=x_i} = 0$ (see Fig. 1 Top). This transformation is unique.

By definition of the continuous distribution function, we can write:

$$\tilde{F}(x) = \int_{-\infty}^{x} \tilde{f}(v)dv, \qquad \tilde{F}(\leq a) = 0, \qquad \tilde{F}(\geq b) = 1 \qquad (1)$$

where $\tilde{f}$ is the approximation of the continuous *probability density* function $\tilde{f}$. We are using eq. (1) to calculate $\tilde{f}$. In result we obtain the *spline histogram* what is parameter independent (see Fig. 1). The differentiable $\tilde{f}$ contains all information about the given X. The SM provides us with information about bimodality of a dataset. From the geometrical considerations, it is clear that each point of the spline histogram contains some information about the whole distribution. Therefore the edges of the spline histogram may not be suitable for the analysis. The shape of the spline histogram appears to be stable if we add any perturbations to the positions of the given points. We can use the following criteria to define the borders between subclusters:

$$I = n\frac{\int_{v_s}^{v_s+\Delta v/k} \tilde{g}(v)dv}{\int_{x_1}^{x_n} \tilde{g}(v)dv} = n\int_{v_s}^{v_s+\Delta v/k} \tilde{g}(v)dv \tag{2}$$

where the change of distribution is expected to be between v_s and $v_s + \Delta v/k$, and Δv is the constant what is normalized to respect of the whole dataset, e.g., the average separation between galaxies in the sample. Since there is one galaxy in average in Δv in the whole dataset and the density is somewhat higher inside the cluster, we need to correct it by coefficient $k \geq 1$ what is simply an average velocity separation between galaxies in the whole dataset over the one of the cluster member galaxies. $I \leq 1$ indicates a separation between two different groups (there is in average less than a galaxy within a velocity interval $\Delta v/k$ there). Clearly, the criteria (2) will not be able to say anything about the merging groups.

Small deviations from the real density function are present in the spline histogram due to the fact that the observed data points deviate from the real distribution. The artificial maximums and minimums may appear due to the sudden change of distributions thus showing the bimodality of the dataset. These artificial features can be easily found comparing the density function with the corresponding separations between the galaxies. Thus the SM provides us with information about the bimodality of the dataset even for the cases of merging groups of galaxies if the distributions of the both are different.

3   The Substructure of Abell 119

We define the substructure of the galaxy cluster to be the separate dynamical and/or mass density systems. As application for the substructure finding procedure we will use the ENACS dataset of A119 containing 117 galaxies with redshifts and positions in line of sight complete to $m_R = 17.4$ [7]. Abell 119 is well studied before by different authors in optical and in x-rays (e.g. [4], [2], [3], [6]).

The spline histogram of A119 is shown in Fig. 1. It is unlikely that the velocity distribution of galaxies in clusters would be anisotropic. By help of eq. (2), we have divided A119 in three parts (denoted as 1, 2+3+4 (main), and 5 in Fig. 1). The group at around 1500 km/sec contains all together only 3 galaxies therefore we have defined these galaxies as the cluster infall galaxies. The shape of the density function of the main clump is clearly something more boxy than Gaussian, pointing out the unrelaxed stage of this subclump. The most possible reason for the unrelaxed central region of a cluster of galaxies is the presence of the substructure. Dividing the dataset of the main clump randomly into two parts, we compare them using Kolmogorov-Smirnov (hereafter K-S) statistic (see [12]). The K-S returns result that the two data subsets are drawn from the same distribution with probability $P = 0.40$ giving us a clue about bimodality of the distribution.

Assuming Gaussianity, by help of the KMM mixture-modeling algorithm (see [1]) we have separated the groups 2 and 4 and their merging region 3. Afterwards, forgetting about Gaussianity, we have maximized the result by the K-S statistics.

Table 1: Centers of subclusters of A119 as calculated (i) by the SM maximizing the K-S probabilities P that each galaxy from the corresponding group is taken from the same distribution, (ii) by the SM and the 1 dimensional (redshift) KMM, and (iii) by the SM and the 3 dimensional KMM. N is the number of galaxies in the subcluster. $\bar{v}$ and σ_v is the average velocity and the velocity dispersion in km/sec.

group	N	RA	DEC	$\bar{v}$	σ_v	P	Method
1	11	$0^h56^m16.9^s$	$-1°07'36''$	12467	119.8	0.65	
2	25	0:56:25.5	-1:12:59	12962	120.9	0.98	SM
3	17	0:56:16.4	-1:15:22	13323	92.4	0.77	+
4	25	0:55:57.8	-1:13:54	13671	140.6	0.98	K-S
5	14	0:56:26.7	-1:08:33	14440	114.9	0.75	
1	11	(0:56:16.9)	(-1:07:36)	12474	119.9	0.65	
2	25	(0:56:25.5)	(-1:12:59)	12969	121.0	0.98	SM
3	22	(0:56:14.6)	(-1:13:45)	13365	105.4	0.88	+
4	20	(0:55:55.1)	(-1:15:20)	13709	114.1	0.68	1D KMM
5	14	(0:56:26.7)	(-1:08:33)	14440	114.9	0.75	
1	11	0:56:16.3	-1:07:42	12470	119.8	0.65	
2	26	0:56:25.8	-1:11:59	12975	126.3	0.70	SM
3	23	0:56:20.1	-1:14:51	13388	111.1	0.57	+
4	18	0:55:46.4	-1:15:11	13716	107.9	0.83	3D KMM
5	14	0:56:26.7	-1:08:33	14440	114.9	0.75	

The results are shown in Table 1. The radial velocity dispersions characterize the dynamical stage of the system. Clearly, there is no any physical meaning to calculate σ_v for the whole cluster containing many dynamical subsystems. We want to stress that the K-S statistics is very sensitive indicator to even small deviations from the distribution. Nevertheless, we obtain the result confirmed as high as 98% confidence level for groups 2 and 4. The galaxies of each subcluster are scattered around all observed field of view (see Fig.10 in [8]) in the sky projection (see Fig. 2). In fact, one could expect even larger scatter. The results returned by the KMM tests are slightly different (see Table 1).

There are quite large deviations between positions of x-ray and optical maximums there (see Fig. 3). That is because of the projection effects and because the galaxy count density is only the approximation of the mass density. In fact our calculations are limited due to the limited sky coverage of the ENACS dataset. Nevertheless, it seems that the main x-ray source is coupled with the merging group 3, and the North-West x-ray maximum correlates with the group 5.

4   Discussion and Conclusions

In this paper we have introduced the SM and its application to clusters of galaxies. In fact, the SM is the special case of the *adaptive kernel technique* (e.g. [10], [11]). The advantage of the SM is its simplicity and that it uses only very few basic assumptions about the dataset. Another interesting methods for substructure studies are the wavelet analysis (e.g. [5]) and the S-tree method (e.g. [6]).

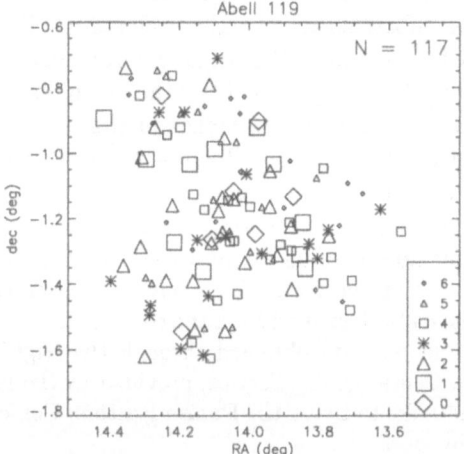

Figure 2: Sky map of A119 by the ENACS dataset. The different symbols are galaxies corresponding to the different subclusters as defined in Fig. 1 (0 and 6 are the foreground and background galaxies, respectively).

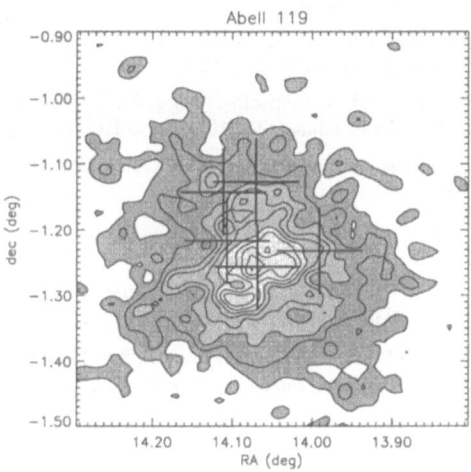

Figure 3: The x-ray (0.2-3.5 keV) contour map of the central part of A119 by Einstein observatory. The contours range between 10% and 90% with steps of 10% and also shown is 99% of the peak level there. The crosses correspond to the number density centers of the subclusters by the ENACS sky limited sample.

We have introduced the simple algorithm based on the SM by help of which we are able to find the substructure of galaxy clusters. The studies show that galaxy clusters can not be treated as the single dynamically relaxed systems. The spline histogram contains all information about the given dataset showing different interesting features (e.g. the four narrow spikes in the A119 spline histogram are thought to be the caustics of subclusters 2 and 4 where density goes to infinity).

To be able to solve completely the problem of the substructure of galaxy clusters we need the data covering the large sky area of the whole cluster and we need to generalize the 1 dimensional spline method to 3 dimensions.

Acknowledgements. I am very grateful to my supervisor Bernard Jones for the many valuable ideas. I have also benefited from the comments of Jesper Sommer-Larsen. This research has made use of data obtained through the High Energy Astrophysics Science Archive Research Center Online Service, provided by the NASA-Goddard Space Flight Center. I also would like to thank the Danish Rectors' Conference whose financial support made possible my work.

References

[1] Ashman, K.A., Bird, C.M., and Zepf, S.E. (1994), *Detecting Bimodality in Astronomical Datasets.* AJ, **108**, 2348-2361

[2] Bird, C.M. (1994), *Substructure in Clusters and Central Galaxy Peculiar Velocities.* AJ, **107**, 1637-1648

[3] Davis, D.S. (1994), *X-Ray Substructure in a Flux-Limited Catalog of Clusters of Galaxies.* Ph.D. Thesis, University of Maryland

[4] Fabricant, D.G., Kurtz, M., Geller, M., Zabludoff, A., Mack, P., and Wegner, G. (1993), *A Study of Rich Cluster of Galaxies A119.* AJ, **105**, 788-796

[5] Gambera, M., Pagliaro, A., and Antonuccio-Delogu, V. (1996), *A 3-D Wavelet and Segmentation Analysis of the Coma Cluster.* Within these Proceedings

[6] Gurzadyan, V.G., and Mazure, A. (1996), *Revealing Galaxy Associations in Abell 119.* Sissa preprint astro-ph/9610056

[7] den Hartog, R., and Katgert, P. (1996), *On the Dynamics of the Cores of Galaxy Clusters.* MNRAS, **279**, 349-388

[8] Katgert, P. et al. (1996), *The ESO Nearby Abell Cluster Survey. I. Description of the Dataset and Definition of Physical Systems.* A&A, **310**, 8-30

[9] Mazure, A. et al. (1996), *The ESO Nearby Abell Cluster Survey. II. The Distribution of Velocity Dispersion of Rich Galaxy Clusters.* A&A, **310**, 31-48

[10] Pisani, A. (1993), *A non-Parametric and Scale-Independent Method for Cluster Analysis - I. The Univariate Case.* MNRAS, **265**, 706-726

[11] Pisani, A. (1996), *A non-Parametric and Scale-Independent Method for Cluster Analysis - II. The Multivariate Case.* MNRAS, **278**, 697-726

[12] Press, W.H., Teukolsky, S.A., Vetterling, W.T, and Flannery, B.P. (1992). *Numerical Recipes in FORTRAN.* 2nd ed. Cambridge University Press, New York

[13] Yahil, and A., Vidal N.V. (1977). *The Velocity Distribution of Galaxies in Clusters.* ApJ, **214**, 347-350

ENTROPY PRODUCTION IN THE EARLY UNIVERSE

A new mechanism

P. DI BARI

INFN-Laboratori Nazionali del Gran Sasso
Strada Statale 17 bis, 67010 Assergi (L'Aquila), Italy

1. Introduction

In this introduction we want to expose briefly the role of entropy in the early universe and in particular the importance of searching for new mechanisms of entropy production. In the following sections we will describe a mechanism [1] that shows how entropy is produced during early annihilations and under which conditions the production is not negligible.

Entropy is not a fundamental quantity as e.g. energy density, but it provides an auxiliary function useful to describe the physical processes in the early universe. This is evident when thermodynamical equilibrium applies (quasi-static expansion) because entropy is conserved and thus it represents a *constant of motion*. This is a traditional result [2] that can be shown easily by applying to the cosmological expansion the usual thermodynamical relation $d(\rho R^3)/dt = T dS/dt - p dR^3/dt + \sum_i \mu_i dN_i/dt$ and imposing the energy-momentum tensor conservation equation $d(\rho R^3)/dt = -p dR^3/dt$. Then necessarily:

$$\frac{dS}{dt} = -\frac{1}{T}\sum_i \mu_i \frac{dN_i}{dt} \qquad (1)$$

But as we are assuming the equilibrium also in the chemical reactions, the right sum is zero and hence $dS/dt = 0$. Entropy conservation allows for the straightforward derivation of the $R-T$ relation. If all species are in radiative equilibrium then $S \propto g_R(TR)^3$, where $g_R = const$ are the *effective degrees of freedom*, and simply $TR = const$, otherwise, the definition of g_R is generalized to the case in which some species are not ultrarelativistic by introducing a function $g_S(T)$ that can be easily calculated [3] and in this case $RT \propto [g_S(T)]^{-1/3}$. The RT factor is important because the particle numbers of any species in radiative equilibrium, and in particular the photon number N_γ, is $\propto (RT)^3$. Thus entropy conservation provides the way to know how the photon number changed during the early universe history. During the annihilations of some particle species the degrees of freedom decrease and photons are produced because the disappearing species releases its entropy to the radiative plasma including photons. Sometimes in the literature this is called entropy production while it is only entropy exchange. Photon production is an important quantity to be known because it is needed to calculate the relic abundances $(n/n_\gamma)_0$ of particle species or of the barionic

D. N. Schramm and P. Galeotti (eds.), Generation of Cosmological Large-Scale Structure, 289–294.
© *1997 Kluwer Academic Publishers.*

number. In fact if at some time t_f a particle or charge number N_f is frozen, we are often able to calculate the abundance $(n/n_\gamma)_f$ at that time and then the relative relic abundance can be expressed as $(n/n_\gamma)_0 = (n/n_\gamma)_f(N_\gamma^f/N_\gamma^0)$. This expression shows how photon production dilutes the abundance from freezing until the present. For this reason it is convenient to define a *dilution factor* $f \equiv N_\gamma^0/N_\gamma^f$. Now it is easy to understand the effect of an entropy production: it gives a further contribution to photon production and hence to f: $f = f_g \cdot f_S$ (where $f_g = g_S^f/g_{S0}$ and $f_S = S_0/S_f$). We point out that the more recent the entropy production is the more effective it is, because it dilutes all densities previously frozen.

2. Entropy production calculation through the L.W. equation

Let us consider the classical problem of the freezing of particle abundances during annihilations. Some particle species h can annihilate eventually into photons through a lighter particle species l: $h + \bar{h} \longrightarrow l + \bar{l} \longrightarrow \ldots \longrightarrow \gamma$'s. Assuming that the elastic reactions rates are strong enough to support the kinetic equilibrium (thus a temperature is defined for the h and it is kept equal to photon temperature), the h distribution function will assume an equilibrium form: $f_h(\vec{p}, t) \simeq \tilde{f}(\vec{p}, t) = \{e^{\beta(t)[E_{\vec{p}} - \tilde{\mu}_h(t)]} + 1\}^{-1}$ where $\tilde{\mu}_h$ is commonly called the *pseudo-chemical* potential to underline that it does not obey, in general, the chemical equilibrium condition $\mu_h + \mu_{\bar{h}} = 0$. In this way the entropy production rate can be obtained easily from thermodynamics using (1), and in our specific case becomes:

$$\frac{dS}{dt} = -\frac{(\tilde{\mu}_h + \tilde{\mu}_{\bar{h}})}{T} \frac{dN_h}{dt} \qquad (2)$$

This expression can be obtained from statistical mechanics as well. It can be noticed that in order to have $dS/dt \neq 0$ two conditions must be satisfied: the first one is departure from chemical equilibrium $(\tilde{\mu}_h + \tilde{\mu}_{\bar{h}} \neq 0)$ and the second is a decreasing particle number $(dN_h/dt \neq 0)$ which means the presence of annihilation processes not balanced by pair production processes. If we want to perform a quantitative calculation of the entropy production, we must use some model to calculate $\tilde{\mu}_h$ and dN_h/dt. The simplest way to describe the annihilations is represented by the special Lee-Weinberg equation:

$$\frac{d\bar{N}}{dy} = A_0 \left[\bar{N}^2(y) - \bar{N}_{eq}^2(y)\right] \qquad (3)$$

$(y = T/m, \bar{N} = N_h/N_{in}, N_{in} = N_h(T \gg m_h), A_0 = 0.055(g_h/\sqrt{g_\rho^{in}})\sigma_0 m m_{PL})$. In the modern literature this equation is usually written using the variable $Y = n/s$ (s is the entropy density) where $Y \propto N$ if entropy is conserved. But as we are relaxing entropy conservation assumption we must use a variable like $\bar{N}$. In this equation particle-antiparticle symmetry is assumed. This is not restrictive because it means only that the departure from equilibrium and the consequent freezing must happen for temperatures when $(N_h - N_{\bar{h}})/N_h \ll 1$. Otherwise it is well known [4] that departure from equilibrium would be greatly delayed and therefore entropy production would be negligible. Thus we can set $\tilde{\mu}_h = \tilde{\mu}_{\bar{h}} = \tilde{\mu}$ and the chemical equilibrium condition becomes $\tilde{\mu} = 0$. Moreover, it is an implicit approximation of the equation that $\tilde{z} \equiv \tilde{\mu}/m = y \ln[\bar{N}(y)/\bar{N}_{eq}(y)]$. In Fig. 1 we show the solutions $\bar{N}$ of Lee-Weinberg equation for different values of A_0 and the correspondent $\tilde{z}$.

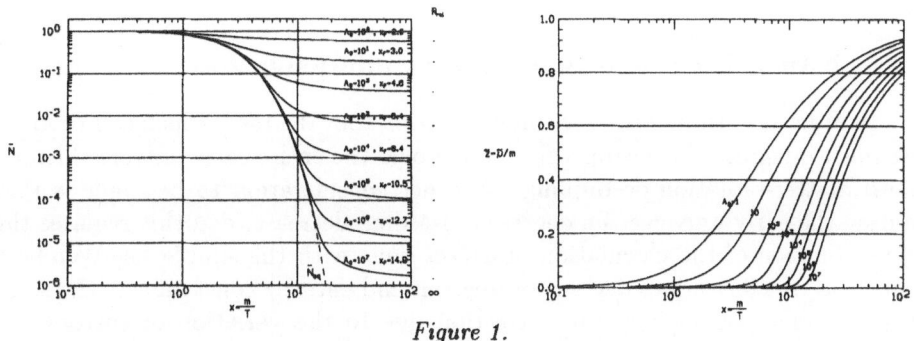

Figure 1.

Finally we can write the entropy production rate through $\bar{N}$ obtaining:

$$\frac{1}{S_h^{in}} \frac{dS(y)}{dy} = -\frac{135\xi(3)}{4\pi^4} \frac{\tilde{z}(y)}{y} \frac{d\bar{N}(y)}{dy} \qquad (4)$$

where $S_h^{in} = g_h S_{in}/g_S^{in}$. By integration one can obtain the entropy production $\Delta S(y, y_{in})$. The results are shown in Fig. 2.

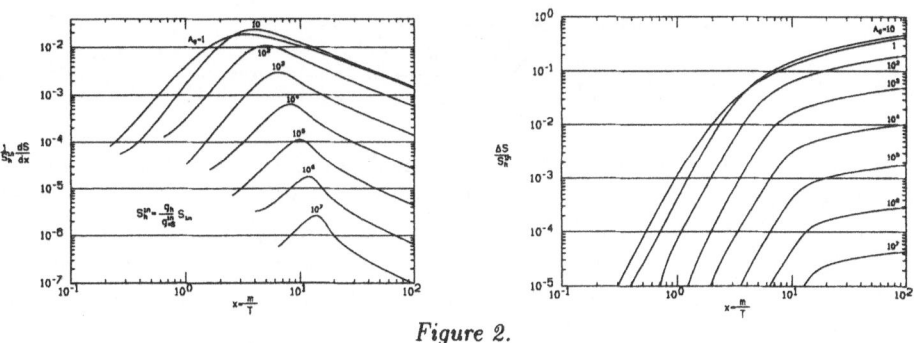

Figure 2.

The rates present a maximum because entropy production is zero before the system goes out of equilibrium (i.e. $\tilde{\mu} = 0$) and after it tends again to zero because the annihilation rate decreases ($dN/dt \to 0$, while $\tilde{\mu}$ reaches its asymptotical value m). Moreover, one can notice the existence of a value for A_0 for which entropy production is maximum. In fact both for $A_0 \to 0$ and for $A_0 \to \infty$ entropy production also tends to zero: in the first case because particles would not annihilate at all, though the system is out of equilibrium, and in the second because the system never goes out of equilibrium. Thus there must exist a maximum. It is reached when $A_0 = 3.65$ and $x_f = 2.5$, where we defined x_f as that value for which $\bar{N} - \bar{N}_{eq} = 0.5\bar{N}_{eq}$. The maximum production of entropy occurs for a freezing during which h-particles are semi-relativistic.

In the *asymptotical limit* $(x = m/T \to \infty)$ all expressions become simpler: $d\bar{N}(x)/dx \longrightarrow -A_0\bar{N}_0^2/x^2$, $\tilde{z}(x) \longrightarrow 1$ and $dS(x)/dx \to \propto A_0\bar{N}_0^2/x$: hence asymptotically the entropy production $\Delta S = \int dx(dS/dx)$ does not stop but it increases logarithmically, thanks to the relic annihilation.

3. Estimation of maximum entropy dilution factor

In order to estimate the maximum dilution that the mechanism can produce we must integrate the entropy production over the entire early universe history, until matter-radiation decoupling. Another consideration to be made is that during the early universe the degrees of freedom decrease, and this requires the generalization of the calculation of dN/dt done with the simple Lee-Weinberg equation which assumes degrees of freedom and entropy constant. At the same time we can also include the correction due to the variation of entropy. It is possible to introduce these modifications only in the asymptotical regime (which gives the major contribution to the entropy production). Hence during this phase the decoupled equations (3) and (4) can be replaced by two coupled equations:

$$\frac{d\bar{S}}{dy} = -\frac{135}{4}\frac{\xi(3)}{\pi^4}\frac{g_h}{g_S^{in}}\frac{1}{y}\frac{d\bar{N}}{dy} \qquad \frac{d\bar{N}}{dy} = A_0 \cdot \frac{\bar{g}_{Si}}{\sqrt{\bar{g}_{\rho i}}}\frac{1}{\bar{S}(y)}\bar{N}_0^2 \tag{5}$$

that yield the solution:

$$\bar{S}_{dec} = \bar{S}_1 = \bar{S}_* \sqrt{1 + 2\frac{a}{\bar{S}_*^2}\frac{g_h}{g_S^{in}}\sum_{i=*,N,\ldots,1}^{1}\frac{\bar{g}_{Si}}{\sqrt{\bar{g}_{\rho i}}}\ln\frac{x_{i-1}}{x_i}} \qquad a = \frac{135}{4}\frac{\xi(3)}{\pi^4}A_0\bar{N}_0^2 \tag{6}$$

If the second term in the square root is much smaller than 1 then, expanding at first order, the logarithmic behaviour is restored. Until now we supposed that all components are coupled to photons and therefore $f_S = \bar{S}_{dec}$. If there is a decoupled component it must be considered that some fraction of the entropy can be given to the decoupled component and thus does not contribute to the dilution.

Now we must find the maximum value that the expression (6) can assume. It depends on two parameters: A_0 and T_{in} (g_h can be set equal 2). We have already indicated that $A_0 = 3.65$ yields the maximum value for $A_0\bar{N}_0^2$. To find the appropriate T_{in} there are two opposite considerations: to get the greatest temperature range requires annihilations as early as possible, but to have the greatest fractional weight of h-particles ($\equiv g_h/g_S^{in}$) requires annihilations that happen when as few particle species as possible have survived. The second consideration suggests to take a time interval such that g_S^{in} is minimum, and the first suggests the insertion of annihilations as early as possible inside this interval. Thus the best situation is when freezing starts soon after electron-positron annihilations so that $m \lesssim m_e/2 \simeq 0.25 MeV$ and from (6) $f_S = 1.43$ is obtained. The second possibility is that freezing starts soon after muon annihilations $(m_h \lesssim m_\mu/2 \simeq 50 MeV)$ and in this case $f_S = 1.31$ is obtained.

4. Application: Mev τ-neutrino

In this section we want to consider a realistic case that produces a dilution factor as much as possible close to the order of magnitude found earlier. This time we must substitute $\sigma_0 = const$ with the thermally averaged cross section $\langle \sigma_{ann} v_{Mo} \rangle (y)$, that can be calculated with a single-integral formula [5] valid for any value of x_f. Unfortunately it is not realistic an hundred Kev particle that undergo the semirelativistic freezing $(x_f \simeq 2)$ we need, though further investigations are in progress considering Majoron or Axino. An Mev τ-neutrino is a particle that would freeze with the right x_f and it is light enough to have a good fractional weight. This then seems to be the best application of the mechanism we described and also because of the possible consequences that it could have on its mass constraints. In last six years constraints from nucleosynthesis [6, 7], SN1987A [8], have been greatly improved, reducing the possibilities for an MeV neutrino from an astrophysical point of view. At the same time the laboratory experiment lower limit has been lowered to 24 MeV [9] and thus in future years there could be an important test for astroparticle physics. This is why we think that it is worthwhile to continue investigations on MeV τ neutrino. Here, however, we want mainly to provide an application for the general mechanism we described.

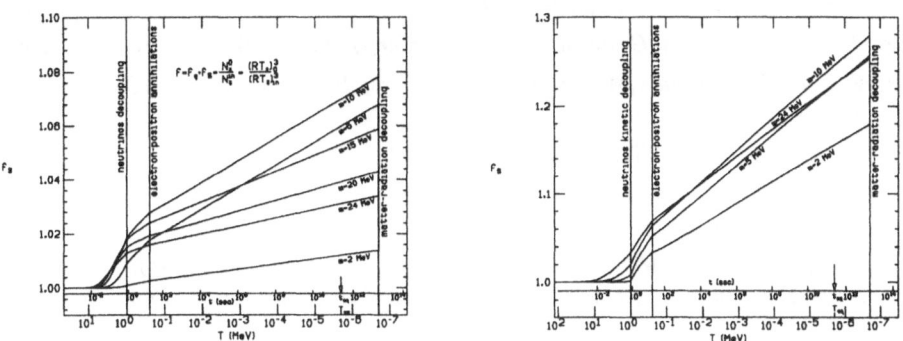

Figure 3. Entropy dilution factor: MeV Dirac ν_τ (left); MeV Giudice ν_τ (right).

We performed the calculations for a Dirac neutrino and for a neutrino with a magnetic moment $\mu \gtrsim 2 \cdot 10^{-9} (m_\nu/10 MeV) \mu_B$ [10]. In the first case it must be considered that ν_τ can annihilate into $e^+ e^-$, $\nu_\mu \bar{\nu}_\mu$, $\nu_e \bar{\nu}_e$ and that for $T \lesssim 1 MeV$ the neutrinos are decoupled: this reduces of a factor $\simeq 1/3$ the dilution factor from the maximum and the result is $f_S = 1.01 \div 1.08$ for masses $m = 2 \div 24 MeV$ (see Fig.3 left). In the second case the annihilation through photon exchange in $e^+ e^-$ is dominant. In this case the factor $1/3$ is avoided and the dilution factor can approach the maximum value permitted ($f_S = 1.1 \div 1.3$ for masses $m = 2 \div 24 MeV$ as shown in Fig.3 right).

5. Conclusions

In the introduction we clarified the difference between the simple entropy transfer from the annihilating particle species to the radiative plasma and an effective entropy production. We presented here the possibility to have a significant entropy production in the late phase of the early universe if the annihilating particle species undergoes a semirelativistic freezing. This mechanism can dilute all relic abundances in the early universe by a factor 1.5 as maximum value. In the last section we presented an application that realizes this possibility: an $MeV\ \nu_\tau$. In the general cases of nonrelativistic and ultrarelativistic freezing, entropy production is a negligible effect as usually assumed.

This talk is based on the work [1]. P. Di Bari has been fortunate to have V.Berezinsky as the supervisor in his "Tesi di Laurea". He wishes to thank A.D. Dolgov for valuable suggestions and M. Lusignoli for his helpful role during the work of thesis. He also thanks all people and the director of Gran Sasso Laboratories P. Monacelli for their warm hospitality during the last two years and Lori Gray for her help with the English.

References

[1] Berezinsky,V. and Di Bari,P. "Entropy production during early annihilations". to be submitted to *Astroparticle Physics*.

[2] Weinberg,S. (1972) *Gravitation and Cosmology*. John Wiley & Sons, New York.

[3] Kolb,E.W. & Turner,M.S. (1990) *The Early Universe*. Addison-Wesley, New York.

[4] Zel'dovich,Ya B. (1965) *Adv.Astron.Astrophys.*, 3,241.

[5] Gondolo,P. and Gelmini,G. (1991) "Cosmic abundances of stable particles: improved analysis". *Nucl.Phys.B*, 360,145–179.

[6] Kolb,E.W., Turner,M.S., Chakravorty,A. and Schramm,D.N. (1991) "Constraints from primordial nucleosynthesis on the mass of the τ neutrino". *Phys. Rev. Lett.*, 67,533.

[7] Kainulainen,K., Fields,B.D. and Olive,K.A. (1996) "Nucleosynthesis limits on the mass of long lived tau and muon neutrinos". *CERN-TH/95-335, UMN-TH-1417/95,HEP-PH/9512321*.

[8] Burrows,A., Gandhi,R. and Turner,M.S. (1992) "Massive Dirac neutrinos and SN 1987A". *Phys. Rev. Lett.*, 68,3834.

[9] Buskulic,D. et. al. (1995) "An upper limit for the τ neutrino mass from $\tau \to 5\pi(\pi^0)\nu_\tau$ decays". *Phys.Lett.B*, 349,585.

[10] Giudice,G.F. (1990) "tau neutrino as dark matter". *Phys.Lett.B*, 251,460.

PRE–BIG–BANG, GRAVITONS AND COSMOLOGY

M. GALLUCCIO
Osservatorio Astronomico di Roma
Via del Parco mellini 84, 00136 ROMA, IT

1. Introduction

The most appealing candidate for a description of physics at the Planck scale is superstring theory.

The tree–level superstring action (low energy limit) is:[1]

$$S = -\frac{1}{2l_{st}^{D-2}} \int \mathrm{d}^D x \sqrt{-g} \mathrm{e}^{-\phi} [R + \nabla_\mu \phi \nabla^\mu \phi + V(\phi)] , \qquad (1)$$

where $D \geq 4$ (spacetime dimension).

The simmetries of the vacuum solutions of the string theory have been studied in cosmology to find non singular models [2]. Then their properties may give useful information for the comprension of the physics of the early universe. The fundamental requirement for such a model is the existence of the so called "branch–changing" solutions [3] which smoothly interpolate between a contracting Universe (*pre–big–bang*, PBB) and an expanding one (*post–big–bang*), related to one another by a duality transformation. This feature is quite general. In fact, "bouncing" solutions are present also if we add the first order string loop expansion, that includes the case where the spatial hypersurfaces have non–zero curvature, [4], or a phenomenological potential for the dilaton [5].

The knowledge of the exact vacuum solutions (a least asymptotically) for the metric, is sufficient to suggest the evolution of this one during the high energy stringy era. This approached procedure is not complete, because we don't know the all order theory (in particular the exact evolution of the dilaton field, and of the inner space) but we may get some important information, by the detection and the misure of a sthocastic gravitational signal from the big bang itself. Infact in the more simple scenario the spectrum of gravitational radiation is peaked around a frequency in the realm

D. N. Schramm and P. Galeotti (eds.), Generation of Cosmological Large-Scale Structure, 295–302.
© 1997 *Kluwer Academic Publishers.*

of the new generation of interferometric detectors (LIGO, VIRGO). After the conformal transformation to the EF one gets [6] for $t \ll 0$ and $D = 4$:

$$a_{EF}(t_{EF}) \propto (-t_{EF})^{1/3}, \quad \phi_{EF}(t_{EF}) \propto - \left(\frac{2}{\sqrt{3}}\right) \ln(-t_{EF}) . \tag{2}$$

Deflation (accelerated contraction) appears as the EF expression of superinflation in the SF, PBB scenario. This means that as a general feature the universe contracts, reaches a minimum size and after that reexpands. The detailed behaviour of the scale factor $a(t)$ depends on the model. It is possible, however, to extract the qualitative effects from general considerations, independently from the knowledge of the exact solutions for $a(t)$ and $\phi(t)$. Obviously, the era around the minimum between collapse and expansion, that is commonly referred to as *stringy*, the solution can be approximated with:

$$a = a_{min} + \left(\frac{t}{t_0}\right)^2 , \quad t \simeq 0 . \tag{3}$$

2. Graviton spectrum

The wave equation for the gravitational waves can be reduced:

$$\ddot{h}_k + 3H\dot{h}_k + \omega^2 h_k = 0 , \tag{4}$$

where a dot means derivation wrt standard time. The present ratio of spectral energy density to the critical density, (at the classical level) is given by the following expression:

$$\Omega_g^c(\omega) := \frac{1}{\rho_{cr}} \frac{d\rho_g}{d\ln\omega} = \frac{\pi c^2}{4G\rho_{cr}} \omega^2 |\Delta h_k|^2 ;$$

here $|\Delta h_k|^2 := k^3 h_k^2$ is the spectral amplitude, $\omega(t) = k/a(t)$ and $h_k(t)$ is the amplitude of the linear wave of wave vector k that adiabatically decreases.

When the primordial waves cross the horizon (the Hubble radius) their amplitudes stay constant. So they grow respect to the other that cross the horizon later, and then the initial quantum spectrum changes [7]:

$$\Omega_g \propto \omega^{4+\alpha-\beta}, \quad \alpha := \sum_i^{1,n} \frac{2q_i}{1-q_i}, \quad \beta := \sum_i^{1,n} \frac{2p_i}{1-p_i} \tag{5}$$

where $a(t) \propto t^{q_i}$ are the scale factor for the cosmological phases in which the waves exit the horizon and $a(t) \propto t^{p_i}$ are the ones for the waves that reenter the horizon.

In some case (decelerated contraction, deflation, etc...) the wave amplitude grow instead to being frozen once outside the horizon, and the previous formula must be modified [8]: we must add a term like

$$\alpha' := -2 \sum_{i}^{1,n} \frac{3q_i - 1}{1 - q_i} \qquad \text{for the modes that exit} \qquad (6)$$

$$\beta' := +2 \sum_{i}^{1,n} \frac{3p_i - 1}{1 - p_i} \qquad \text{for the ones that reenter.} \qquad (7)$$

3. Spectrum from the stringy era

Now we can use this derivation of gravitational spectra to determine the signal from the PBB cosmological model, and in particular from the stringy era . We can divide the spectrum in four branch:

1. the nodes exit the horizon during the dilaton era, and reenter during matter dominated one.
   $q_1 = 1/3$ ($\alpha = 1$); $p_1 = 2/3$ ($\beta = 4$) $\qquad \Omega_g \propto \omega^1$
2. the modes exit during the dilaton era but reenter in the radiation one.
   $q_1 = 1/3$ ($\alpha = 1$); $p_1 = 1/2$ ($\beta = 2$) $\qquad \Omega_g \propto \omega^3$
3. the modes exit during the dilaton era ($t < 0$), reenter the first time in the stringy era ($t < 0$), exit again at $t > 0$ in the stringy era and reenter in the radiation one. In this case (decelerated contraction) the amplitude grow outside the horizon, and we must add the term considered before.
   $q_1 = 1/3$ (dilaton), $q_2 := 2$ (string $t > 0$); $\alpha = -3$
   $p_1 = 2$ (string $t < 0$), $p_2 := 1/2$ (radiation); $\beta = 2$ and $\beta' = -10$
   $\Omega_g \propto \omega^{4+1-2-10} = \omega^{-7}$

Summarizing:

$$\omega_o < \omega < \omega_e \qquad \Omega_g \simeq \Omega_\gamma \left(\frac{H_r}{m_P}\right)^2 \left(\frac{\omega_e}{\omega_M}\right)^3 \left(\frac{\omega}{\omega_e}\right) z_{st}^{7/2} z_{infl},$$

$$\omega_e < \omega < \omega_M \qquad \Omega_g \simeq \Omega_\gamma \left(\frac{H_r}{m_P}\right)^2 \left(\frac{\omega}{\omega_M}\right)^3 z_{st}^{7/2} z_{infl},$$

$$\omega_M < \omega < \omega_1 \qquad \Omega_g \simeq \Omega_\gamma \left(\frac{H_r}{m_P}\right)^2 \left(\frac{\omega}{\omega_M}\right)^{-7} z_{st}^{7/2} z_{infl},$$

$$\omega_1 < \omega < \omega_r \qquad \Omega_g \simeq \Omega_\gamma \left(\frac{H_r}{m_P}\right)^2 \left(\frac{\omega}{\omega_M}\right)^{-2} (z_{st} z_{infl})^{1/2}, \qquad (8)$$

where $\omega_e = 10^{-16} \text{Hz}$.

The full spectrum is shown in Fig.2. Three features emerge: 1) the growth due to deflation at small ω; 2) the existence of a sharp peak at

ω_M; 3) the descent due to power-law inflation at large ω. Concerning 2), we underline that the maximum of the spectrum $\Omega(\omega_M)$ is higher than expected. This is due to the fact that the perturbations that exit the horizon in the stringy era are amplified more than those coming from the deflationary era. Furthermore, a comment must be made about the oscillations showing up in Fig. 2: they arise because waves of different ω and different phases remain subhorizon for different time lengths. Finally, the total, deflation+power-law, duration N_T must be at least of 60 e-folds.

This is a three parameter model: $N_{st} = \ln z_{st}$, the number of e-folds of the stringy era, $N_{infl} = \ln z_{infl}$, the number of e-folds of the power-law era; H_r, the value of the Hubble parameter at the end of inflation and the starting of the radiation dominated era. The shaded area in the Fig. 3 shows the region of the parameter space (z_{infl}, H_r) for which the peak of the spectrum falls in the range of detectability of LIGO/VIRGO interferometers [9] [10], for all z_{st} values.

In addition, we must require that the graviton energy density does not exceed the nucleosynthesis bound [11], $(\Omega_g h_{50}^2 < 5 \times 10^{-5}$ and that the inflationary era solves the monopole problem (limit from the nucleosynthesis, $N_{infl} \geq 10$, i.e. $z_{infl} \geq 5$). Furthermore the low energy limit of string theory and GUT models constrain H_r to be in the range $10^{-10} < H_r/m_P < 10^{-6}$, that means $10^{-16} < T/GeV < 10^{-14}$ [12]. Finally, the value of z_{st} is such that the maximum of $H = H_{st} \simeq l_{st}^{-1}$ is comparable to the string value $10^{-2} < H_{st}/m_P < 10^{-1}$. The region defined by the above constraints has non zero intersection with the shaded region.

4. Conclusions

We have shown that the full stringy phase ($t \simeq 0$) is fundamental for detecting a primordial signature of string cosmology. Our model has three free parameters: we find the region of parameter space which satisfies all the known constraints. In this region both the frequency and the amplitude of the peak may fall within the realm of the new interferometric detectors. A typical value for the peak is $h_{50}^2 \Omega_M \simeq 5 \times 10^{-5}$ at the frequency $\omega_M \simeq 10$ Hz.

Acknowledgements

I would to thank F.Occhionero and M.Litterio for a very fruitful and enjoyable collaboration.

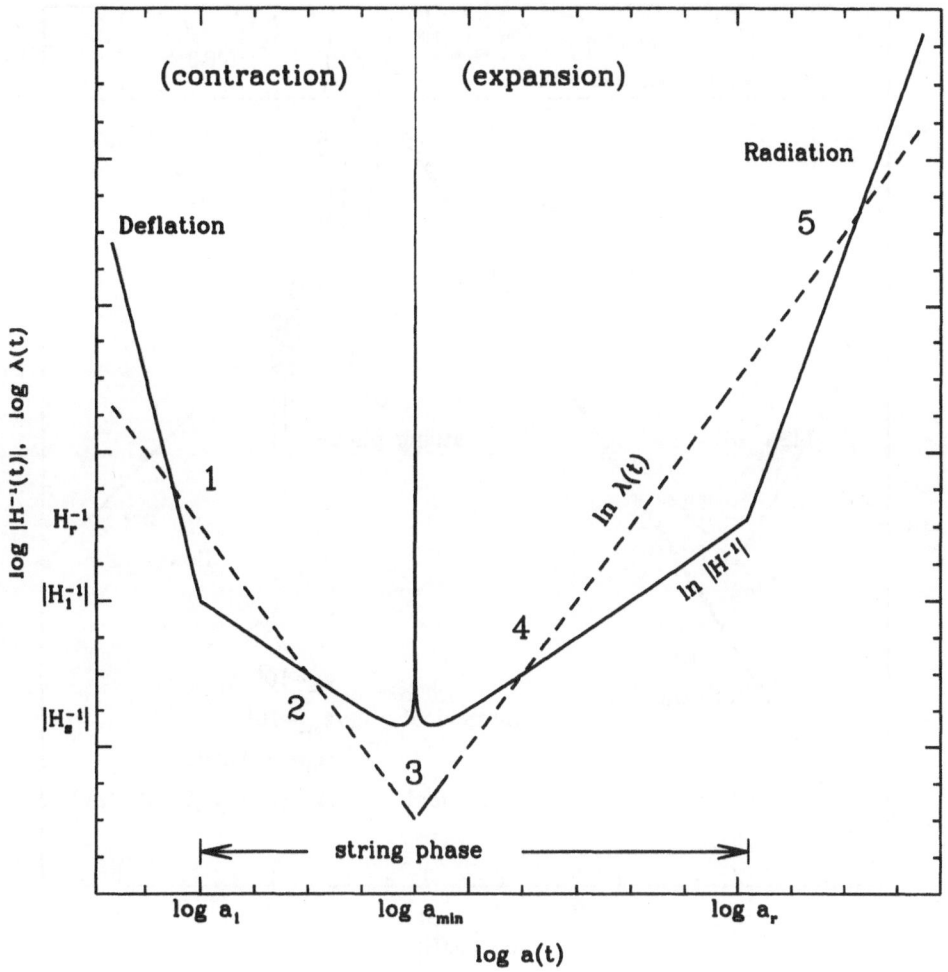

Figure 1. A typical gravitational wave generated during the PBB deflationary presents the features described in the test.

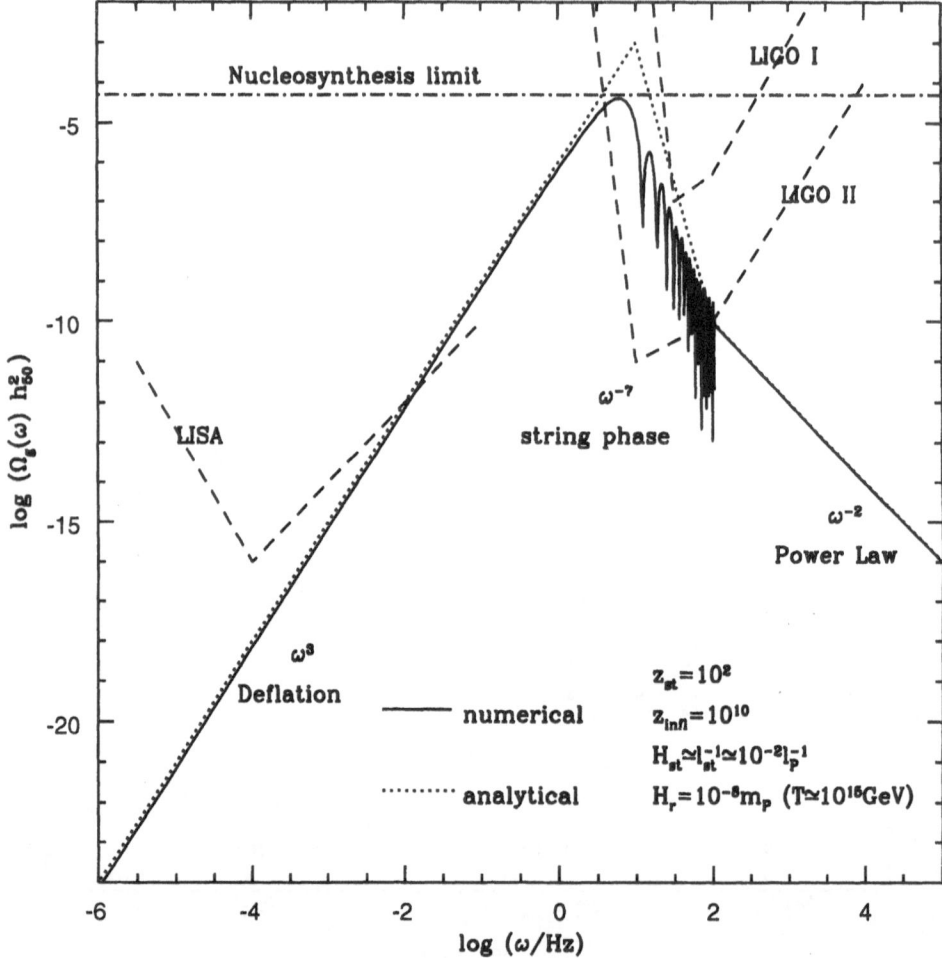

Figure 2. The graviton spectrum of the stochastic background, with the structured peak due to the stringy era, superimposed to the known observational bounds and to the expected sensitivities of the LIGO's.

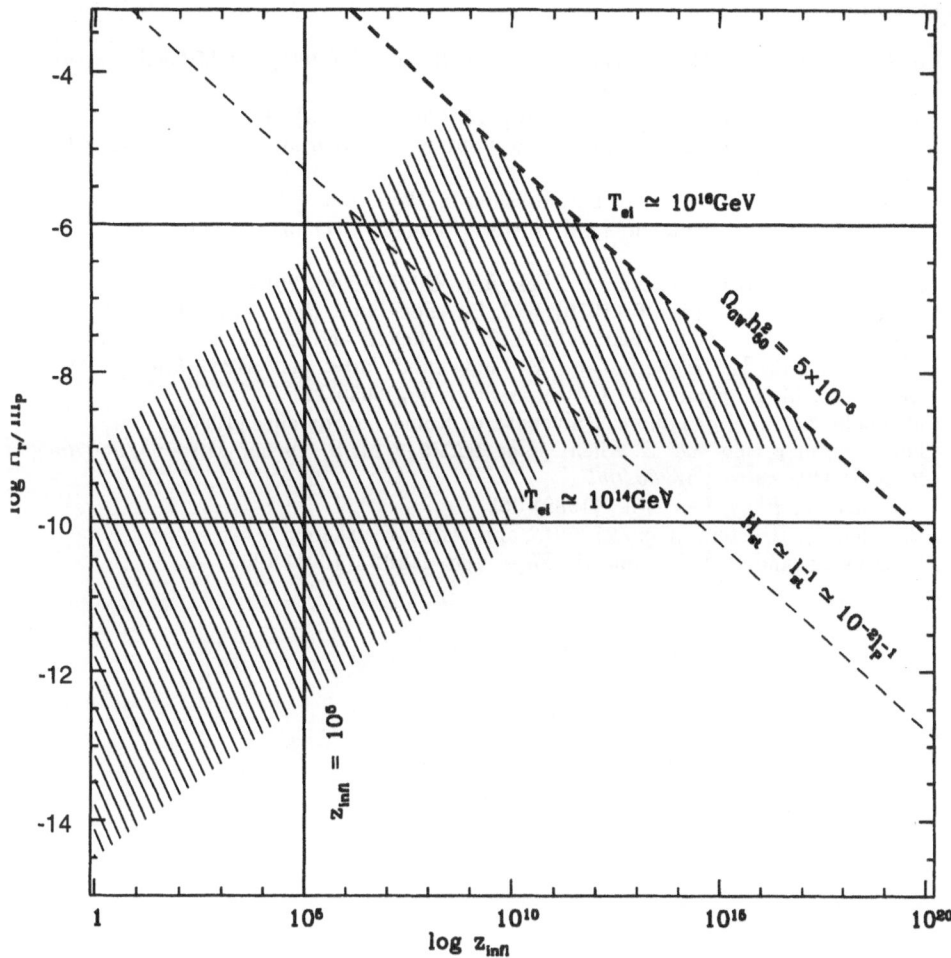

Figure 3. The shading shows the parameter region for which the structured peak could be detected by LIGO. A sensible model falls in the area $z_{infl} > 10^5$ (monopoles), $10^{-10} < H_r/m_P < 10^{-6}$ (GUT), and $\Omega_g h_{50}^2 < 5 \cdot 10^{-5}$ (nucleosynthesis). The final constraint $10^{-2} < g_s < 10^{-1}$ gives the region between the dashed lines.

References

1. K.A. Meissner and G. Veneziano, *Phys. Lett.* **B** 267 (1991), 33 M.Gasperini and G.Veneziano, *Phys. Lett.* **B** 277 (1992), 256.
2. M. Gasperini and G. Veneziano, *Astropart. Phys.* **1** (1993) 317.
3. G. Veneziano, *Phys. Lett.* **B** 265, (1991) 287 R. Brustein and G. Veneziano, *Phys. Lett.* **B** 329, (1994), 429.
4. R. Easther and K. Maeda, to be published in *Phys. Rev.* **D** *hep–th/9605173*.
5. C. Angelantonj, L. Amendola, M. Litterio, and F. Occhionero, *Phys. Rev. D* **51**, (1995) 1607.
6. M. Gasperini and G. Veneziano, *Phys. Rev. D* **50** (1994), 2519.
7. Starobinskij *JETP Lett.* **30** (1979), 682: B. Allen *Phys. Rev. D* **37** (1988), 2078; V. Sahni *Phys. Rev. D* **42** (1990), 453; L.P. Grishchuk and Y. Sidorov *Phys. Rev. D* **42** (1990), 413; M. Gasperini and M. Giovannini *Phys. Lett.* **B282** (1992), 36; A. Buonanno, M. Maggiore and C. Ungarelli gr–qc/9605072.
8. M. Galluccio, *Tesi di Laurea*, Università di Roma *La Sapienza* (1996).
9. For a recent review see: B. Allen, *gr-qc/9604033*; M. S. Turner *astro-ph/9607066*; R. A. Battye *astro-ph/9604059*.
10. A. Giazotto, *Phys. Rep.* **189**, (1989) 365.
11. B. Allen, R. Brustein *gr-qc/9609013*, and references therein.
12. R. Brustein and P. J. Steinhardt, *Phys. Lett.* **B302**, (1993) 196.

THE IMPLICATIONS OF LUMINOSITY SELECTION EFFECTS IN THE INTERPRETATION OF THE ANGULAR SIZE - REDSHIFT DATA

A. A. UBACHUKWU
Department of Physics and Astronomy
University of Nigeria, Nsukka, Nigeria
and Hartebeesthoek Radio Astronomy Observatory,
P.O. Box 443, Krugersdorp 1740, South Africa

Abstract.

We investigate the implications of the luminosity selection effects in the amount of linear size evolution required to explain the observed angular size - redshift ($\theta - z$) data of extragalactic radio soruces. We show that there is a difference in the nature of the dependence of linear size on radio luminosity between low ($z \leq 0.3$) and high ($z > 0.3$) redshift sources which can be attributed to a change in the luminosity - redshift slope at $z \sim 0.3$. This difference can account for the often quoted steeper size evolution found among radio galaxies than quasars which has been shown to have serious implications for the quasar/galaxy unification scheme in which lobe-dominated quasars are powerful radio galaxies seen within 44.4° to the line of sight.

1. Introduction

It has long been realised that the interpretation of the observed angular size - redshift ($\theta - z$) relation for the extended extragalactic radio sources is complicated by both evolutionary and various selection effects present in most source samples. Attempts to disentangle the effects of luminosity from genuine size evolution have yielded conflicting results (see e.g. [1] for a summary). It is not currently certain whether radio galaxies and quasars undergo similar cosmological evolution. An investigation of whether the observed angular size - redshift ($\theta - z$) relation and the dependence of linear size on radio luminosity and redshift are different for radio galaxies and quasars is

D. N. Schramm and P. Galeotti (eds.), Generation of Cosmological Large-Scale Structure, 303–306.
© 1997 *Kluwer Academic Publishers.*

a key to a better understanding of the quasar/galaxy unification scheme of Barthel [2]. In order to do this, large samples of radio galaxies and quasars carefully matched in radio luminosity and redshift space are required.

We report here the results of a preliminary quantitative analysis of the behaviour of linear sizes of radio galaxies and quasars in different luminosity - redshift planes. The analyses were based on the sample of 540 radio sources consisting of 267 radio galaxies and 273 quasars compiled recently by Nilsson *et al.* [3].

2. Luminosity selection effects and linear size evolution

In order to investigate the contribution of luminosity selection effects on the amount of size evolution required to explain the observed angular size - redshift $(\theta - z)$ data, we first estimated the dependence of the median linear size (D_{med}) on radio luminosity (P) and redshift (z) of the sort: $D_{med} \sim P^q$ and $D_{med} \sim (1+z)^{-x}$ for three different redshift bins. The binnings were as follows: $0.3 \leq z$ (low), $0.3 < z < 1.0$ (intermediate) and $z \geq 1$ (high). It should be noted that radio galaxies dominate at the low redshift bin ($\sim 90\,\%$) while quasars dominate at the high redshift bin ($\sim 80\,\%$). This implies that the results of the analyses at the low redshift bin will be more applicable to radio galaxies than to quasars and vice versa at the high redshift bin. The intermediate redshift bin corresponds to their region of overlap within which comparisons can be made between radio galaxies and quasars with high statistical significance.

Furthermore, we have also carried out a linear regression analysis of the dependence of luminosity on redshift of the sort $P_{med} \sim (1+z)^\beta$ for the three redshift bins. This is to enable us to determine the residual size evolution $n = x + q\beta$ after eliminating the combined effects of the $D \sim P^q$ and $P \sim (1+z)^\beta$ correlations present in the $D \sim (1+z)^{-x}$ relation (Ubachukwu [4]). The results are shown in Table 1.

TABLE 1. The regression parameters

Redshift bin	q	x	β	n
$z \leq 0.3$	0.48	0.6	5.6	3.3
$0.3 < z < 1.0$	−0.26	1.2	3.8	0.2
$z \geq 1.0$	−0.32	1.4	4.1	0.1

3. Discussion

Table 1 shows that there is no significant difference in the nature and amount of the dependence of linear size on redshift and radio luminosity at both intermediate and high redshift bins. The results at these redshift intervals show an inverse correlation between linear size and radio luminosity with a mild linear size evolution $x \sim 1.3$ when luminosity selection effects were not considered. In addition, we found no evidence of any size evolution ($n \sim 0.2$) at the two redshift bins when the luminosity selection effects were put into consideration. On the other hand, the low redshift bin shows discrepant results. There is a strong indication of a direct $D - P$ correlation with a milder size evolution when luminosity effects are ignored. Besides, there is a very steep size evolution ($n \sim 3.0$) when luminosity effects are taken into consideration. Another important difference in the present results is the steeper $P - z$ slope, $\beta \sim 5.6$ for $z \leq 0.3$ and $\beta \sim 4.0$ for $z > 0.3$.

In general there appears to be a continuity in the observed variation of linear size with redshift and radio luminosity from intermediate $0.3 < z < 1.0$ (which corresponds to the region of overlap in the redshift distributions of quasars and radio galaxies in the sample) to the high redshift end where quasars predominate. By contrast, the dependence of D on P and z at the low redshift end where the majority of the sources are radio galaxies, appears disjoint.

Oort *et al.* [5], Kapahi [6] and Singal [7] found a very steep dependence of linear sizes of radio sources with $n \sim 3.0$ and a direct correlation between linear size and luminosity similar to the result obtained here at the low redshift bin. Barthel and Miley [8] found a milder size evolution ($n \sim 1.5$) for their quasar sample with an inverse $D-P$ correlation similar to the behaviour found here at $z > 0.3$. In addition to an inverse $D - P$ correlation, Singal [7] found little or no size evolution ($n \sim 0.3$) for quasars - similar to the present result. However Nilsson *et al.* [3] found an inverse $D - P$ correlation for both radio galaxies and quasars and no evidence for any strong size evolution.

We have shown in the present analyses that there appears to be a difference in the evolution of low redshift radio sources and those found at higher redshifts ($z > 0.3$). Since radio galaxies dominate the source counts at these low redshifts ($z < 0.3$), the observed differences between the evolution of radio galaxies and quasars can be attributed to these low redshift galaxies. Moreover, since there is an indication of a discontinuity in the $P - z$ slope at $z \sim 0.3$, it does appear that these low redshift galaxies may belong to a different class of objects (e.g. Laing *et al.* [9]). This has serious implications for the quasar/galaxy unification scheme (Barthel [2]) in which quasars are expected to be radio galaxies seen end-on. This is fully discussed in a follow-up paper (Ogwo and Ubachukwu, in preparation).

4. Acknowledgements

I would like to thank C. S. Flanagan and G. D. Nicolson for useful comments, M. J. Gaylard for proofreading and typing the manuscript. I would also like to acknowledge the hospitality of HartRAO, where this work was done. The visit to HartRAO was supported by the IAU Commission 38 Travel Grant. Finally, I thank the organizers of this School for full sponsorship.

References

1. Singal, A.K. (1996) Cosmological evolution of extragalactic radio sources. In *Extragalactic Radio Sources*, R. Ekers *et al.* (eds.), Kluwer, Dordrecht, pp. 373–378
2. Barthel, P.D. (1989) Is every quasar beamed ?, *Astrophys. J.*, **336**, pp. 606–611
3. Nilsson K., Vatonen, M.J., Kotilain, J. and Jaakola, T. (1993) On the redshift - apparent size diagram of double sources, *Astrophys. J.*, **413**, pp. 453–476
4. Ubachukwu, A.A. (1995) The effect of luminosity - redshift correlation on the analyses of the angular size - redshift relation for radio galaxies *Astrophys. Space Sci.*, **232**, pp. 23–25
5. Oort, M.J.A., Katgert, P., Steeman, F.M.W. and Windhorst, R.A. (1987) VLA high resolution observations of weak Leiden-Berkeley Deep-Survey (LBDS) sources, *Astron. Astrophys.*, **179**, pp. 41–59
6. Kapahi, V-K. (1989) Redshift and luminosity dependence of linear sizes of powerful radio galaxies, *Astron. J.*, **97**, pp. 1–9
7. Singal, A.K. (1993) Cosmic evolution and luminosity dependence of the physical sizes of powerful radio galaxies and quasars, *Mon. Not. R. Astron. Soc.*, **263**, pp. 139-148
8. Barthel, P.D. and Miley, G-K. (1988) Evolution of radio structure in quasars: a new probe of protogalaxies?, *Nat.*, **333**, pp. 319–325
9. Laing, R.A., Jenkins, C.R., Wall, J.V. and Unger, S.W. (1994) Spectrophotometry of a complete sample of 3CR radio sources: implications for unified models, *Astron. Soc. Pacific Conf. Ser.*, **57**, pp. 201–208

PEAKS ABOVE THE HARRISON-ZEL'DOVICH SPECTRUM DUE TO THE QCD TRANSITION

P. WIDERIN, C. SCHMID AND D. J. SCHWARZ

Institut für Theoretische Physik, ETH-Zürich, CH-8093 Zürich

Abstract.

The transition from the quark-gluon plasma to a hadrons gas in the early universe affects the evolution of primordial cosmological perturbations. If the phase transition is first order, the sound velocity vanishes during the transition and density perturbations fall freely. The primordial Harrison-Zel'dovich spectrum for scales below the Hubble radius at the transition develops peaks, which grow at most linearly with wavenumber. Cold dark matter clumps of masses below $10^{-10} M_\odot$ are produced.

QCD makes a transition from a quark-gluon plasma at high temperatures to a hadron gas at low temperatures. Lattice QCD simulations give a transition temperature $T_\star \sim 150$ MeV and indicate a first-order phase transition for the physical values of the u,d,s-quark masses [1]. The relevance of the QCD transition for cosmology [2], has been discussed before, but the focus was on effects of bubble formation. In [3] and in this talk we look at matter averaged over scales λ much larger than the bubble separation ℓ_b. We show that for a first-order phase transition the sound speed c_s drops to zero on these scales when the transition temperature $T_\star$ is reached, stays zero until the phase transition is completed, and afterwards suddenly rises back to $c_s \approx c/\sqrt{3}$. In contrast the pressure stays positive and varies continuously. Since c_s is zero during the transition, there are no pressure perturbations, no pressure gradients, no restoring forces. Pre-existing cosmological perturbations, generated by inflation with a Harrison-Zel'dovich spectrum, go into free fall. The superhorizon modes (at the time of the transition) remain unaffected. The subhorizon modes develop peaks in $\delta p/\rho$ which grow with wavenumber $k > k_\star$, where $k_\star^{phys} \sim$ Hubble rate H at the end of the QCD transition. The details of this growth depend on the QCD equation of state near $T_\star$. We analyze two cases: First we use the bag model

D. N. Schramm and P. Galeotti (eds.), Generation of Cosmological Large-Scale Structure, 307–310.

[4], which allows a simple discussion of the effects. It gives a maximal latent heat, and produces peaks in $\delta p/\rho$ which grow linearly in k. Next we fit lattice QCD results [1, 5], which indicate a smaller latent heat. This produces peaks in $\delta p/\rho$ which grow as $k^{3/4}$.

The sound speed (for $\lambda \gg \ell_b$), $c_s = (\partial p/\partial \rho)_s^{1/2}$, vanishes during a first-order phase transition of a fluid with negligible chemical potential, since

$$\rho + p = T \frac{dp}{dT} , \tag{1}$$

according to the second law of thermodynamics. Since the energy density ρ is discontinuous in temperature at $T_\star$ the pressure p must be continuous with a discontinuous slope. As the universe expands at fixed temperature $T_\star$ during the phase transition, ρ as a function of time slowly decreases from $\rho_+(T_\star)$ to $\rho_-(T_\star)$, p stays constant at $p(T_\star)$, and therefore c_s is zero during the whole time of the phase transition.

The interaction rates in the QCD-photon-lepton fluid are much larger than the Hubble rate, $\Gamma/H \gg 1$, therefore we are very close to thermal and chemical equilibrium and the entropy in a comoving volume is approximately conserved. Estimates show that supercooling, hence entropy production, is negligible, $(T_\star - T_{\text{supercooling}})/T_\star \sim 10^{-3}$ [6]. Bubble formation during the QCD phase transition is unimportant for our analysis, estimates give a bubble separation $\ell_b \sim 1$ cm [7], while the Hubble radius at the QCD transition is $R_H \sim 10$ km. We shall analyze perturbations with $\lambda \gg \ell_b$.

In the bag model [4] the quark-gluon plasma (QGP) for $T > T_\star$ obeys $p_{\text{QGP}}(T) = (\pi^2/90)g_{\text{QGP}}^* T^4 - B$, where g^* is the effective number of relativistic helicity states, and B is the bag constant. We include u,d-quarks and gluons in the quark-gluon plasma, γ, e, μ, and 3 neutrinos in the photon-lepton fluid, and for $T < T_\star$ we have a hadron gas (HG) of pions, which we treat as massless and ideal. ρ follows via Eq. (1), and s from $s = dp/dT$. The bag constant is determined via $p_{\text{QGP}}(T_\star) = p_{\text{HG}}(T_\star)$. Lattice QCD results [1, 5] indicate a latent heat which is about a fifth of the one in the bag model. As a second model we use this smaller latent heat and fit the lattice data above $T_\star$ by $s/T^3 = C_1 + C_2(1 - T_\star/T)^{1/3}$. The growth of the scale factor during the QCD transition, $a_+/a_- \approx 1.4(1.1)$ for the bag model (lattice fit), follows from the conservation of entropy in a comoving volume.

The evolution of linear cosmological perturbations through the QCD transition is analyzed in the longitudinal sector (density perturbations) for perfect fluids. We choose a slicing Σ of space-time with unperturbed mean extrinsic curvature, $\delta[\text{tr}K_{ij}(\Sigma)] = 0$ (uniform expansion gauge [8]). As evolution equations for each fluid we have $\nabla_\mu T^{\mu\nu} = 0$, i.e. the continuity equation and the 3-divergence of the Euler equation of general relativity,

$$\partial_t \epsilon \;\; = \;\; -3H(\epsilon + \pi) - \Delta\psi - 3H(\rho + p)\alpha \tag{2}$$

$$\partial_t \psi = -3H\psi - \pi - (\rho + p)\alpha , \qquad (3)$$

where $\epsilon \equiv \delta\rho$, $\pi \equiv \delta p$, $\rho \equiv \rho_0$, $p \equiv p_0$, $\vec{\nabla}\psi \equiv$ momentum density, $\alpha =$ lapse function. The general relativistic Poisson equation ($R_{\hat{0}\hat{0}}$-equation),

$$(\Delta + 3\dot{H})\alpha = 4\pi G(\epsilon + 3\pi) , \qquad (4)$$

together with the equation of state closes the system of equations.

Our numerical results for the spectrum of density perturbations from the bag model and the lattice QCD fit are given in Fig. 1. We work with the dimensionless variables $\delta \equiv \epsilon/\rho$ (density contrast), $\hat{\psi} \equiv k^{\text{phys}}\psi/\rho$ ($\sim$ peculiar velocity). We show the enhancement of the amplitude $A_{\text{RAD}} \equiv (\delta^2_{\text{RAD}} + 3\hat{\psi}^2_{\text{RAD}})^{1/2}$ of the acoustic oscillations of the radiation fluid after the transition compared to the amplitude without transition. For cold dark matter (CDM) we show the amplitude $A_{\text{CDM}} \equiv |\delta_{\text{CDM}}|$ at $T_*/10$ compared to A_{RAD} without transition. In both cases we obtain peaks over the Harrison-Zel'dovich spectrum of primordial adiabatic density fluctuations. The modes k (horizontal axis) are labeled by the CDM mass contained in a sphere of radius $\lambda/2 = \pi/k$. The positions of the first peaks are the same for the bag model and for the lattice QCD fit. In both cases the peak-dip structure starts at $M_1^{\text{CDM}} \approx 9 \times 10^{-9} M_\odot$, which corresponds to $k_1 \sim k_*$. For $k \gg k_*$ the peaks grow linearly in the bag model and as $(k/k_2)^{3/4}$ for the lattice fit, where k_2 labels the beginning of the asymptotic envelope indicated in Fig. 1. The value k_2 corresponds to $M_2^{\text{CDM}} \approx 2 \times 10^{-10} M_\odot$. The radiation energy inside $\lambda_1/2$ is $\sim 1 M_\odot$, but it gets redshifted as $M_{\text{RAD}}(a) \sim (a_{\text{equality}}/a) M_{\text{CDM}}$.

The origin of these large peaks in $\delta\rho/\rho$ for $k \gg k_*$, where $k_*^{\text{phys}} \sim H$ at the transition, is easily understood in the bag model. The radiation fluid in each mode makes standing acoustic oscillations before and after the QCD transition with gravity negligible and with equal amplitudes A_{in} of δ and $\sqrt{3}\hat{\psi}$. During the QCD transition the sound speed is zero, there are no restoring forces from pressure gradients, the radiation fluid goes into free fall. But during this free fall gravity is again negligible for the radiation fluid, since $\Delta t < H^{-1}$. The peculiar velocity stays constant and the density contrast grows linearly in time with a slope k. Thus, the final amplitude A_+ grows linearly in k modulated by the phase of the oscillation at the beginning of the transition, which produces peaks in the spectrum. The height of these peaks is $(A_+/A_{\text{in}})_{\text{peaks}} \equiv k/k_1$. The slower rise in k for the lattice fit can be understood by a WKB analysis which is presented in [3].

For CDM we consider any non-relativistic matter which decouples kinetically well before the QCD transition. Candidates for our CDM are axions or primordial black holes. CDM falls into the gravity wells generated during the transition by the radiation fluid.

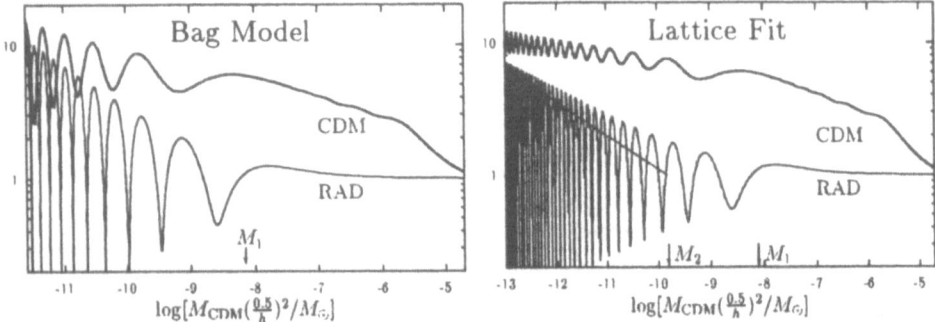

Figure 1. The modifications of the CDM density contrast $A_{\mathrm{CDM}} \equiv |\delta_{\mathrm{CDM}}|(T_\star/10)$ and of the radiation fluid amplitude $A_{\mathrm{RAD}} \equiv (\delta_{\mathrm{RAD}}^2 + 3\hat{\psi}_{\mathrm{RAD}}^2)^{1/2}$ due to the QCD transition. Both quantities are normalized to the pure Harrison-Zel'dovich radiation amplitude. On the horizontal axis the wavenumber k is represented by the CDM mass contained in a sphere of radius π/k.

The implications of these peaks are:

1) CDM clumps with $M_{\mathrm{CDM}} \lesssim 10^{-10} M_\odot$ are produced. They go nonlinear after equality and virialize by violent gravitational relaxation. Assuming a COBE normalized spectrum with tilt $n - 1 = 0(0.2)$ and 3σ peaks the size of $10^{-10} M_\odot$ clumps is ≈ 14 AU (1 AU).

2) Big-Bang Nucleosynthesis will not be affected by the big inhomogeneities of the radiation fluid, because they are wiped out by collisional damping from neutrinos before Big-Bang Nucleosynthesis.

References

1. DeTar, C. (1995) Finite temperature QCD: progress and outstanding problems, *Nucl. Phys. B* (Proc. Suppl.) **42**, 73-84; Iwasaki, Y. et al. (1996) QCD phase transition with strange quark in Wilson formalism for fermions , *Z. Phys. C* **71**, 343-346.
2. Malaney, R. A. and Mathews, G. J. (1993) Probing the early universe: A review of primordial nucleosynthesis beyond the standard big bang, *Phys. Rep.* **229**, 145-219.
3. Schmid, C., Schwarz, D. and Widerin, P. (1997) Peaks above the Harrison-Zel'dovich spectrum due to the quark-gluon to hadron transition, astro-ph/9606125, to appear in *Phys. Rev. Lett.* (Febr.3, 1997).
4. DeGrand, T., Kajantie, K. (1984) Supercooling, entropy production and bubble kinetics in the quark-hadron transition in the early universe *Phys. Lett.* **147B**, 273-289.
5. Boyd,G et al. (1996) Equation of state for the SU(3) gauge theory , *Phys. Rev. Lett.* **75**, 4169-4172 (1995); Beinlich, B., Karsch, F. and Peikert, A. (1996) SU(3) latent heat and surface tension from tree level and tadpole improved actions, hep-lat/9608141.
6. Based on Fuller, G. M. G., Mathews, J. and Alcock, C. R. (1988) The quark-hadron phase transition in the early universe: Isothermal baryon number fluctuations and primordial nucleosynthesis, *Phys. Rev. D* **37**, 1380-1455.
7. Christiansen, M. B. and Madsen, J. (1996) Large nucleation distances from impurities in the cosmological quark-hadron transition, *Phys. Rev. D* **53**, 5446-5454.
8. Bardeen, J. A. (1989) Cosmological perturbations, from quantum fluctuations to large scale structure, in A. Zee (ed.), *Particle Physics and Cosmology* , Gordon and Breach, New York, 1989, pp. 1-64.

THE CRYOGENIC DARK MATTER SEARCH (CDMS)

A. Sonnenschein, D. Bauer, D.O. Caldwell, D. Hale, S. Yellin
Department of Physics, University of California
Santa Barbara, CA 93106

P.D. Barnes, Jr., A. DaSilva, R.J. Gaitskell, S.R. Golwala,
R.R. Ross, B. Sadoulet, D. Seitz, T. Shutt, G. Smith, W.
Stockwell, R. Therrien, S. White
Center for Particle Astrophysics and Department of Physics
University of California
Berkeley CA 94720

P. Brink, B. Cabrera, B. Chugg, R.M. Clarke, B.L. Dougherty,
A. Davies, K. Irwin, S.W. Nam, M.J. Penn
Department of Physics, Stanford University
Stanford, CA 94305

D. S. Akerib, T.A. Perera, R. Schnee,
Department of Physics, Case Western Reserve University
Cleveland, OH 44106

J. Emes, E.E. Haller, W.B. Knowlton, A. Smith, J.D. Taylor
Lawrence Berkeley National Laboratory
Berkeley, CA 94720

B.A. Young
Department of Physics, Santa Clara University
Santa Clara, CA 95053

1. Introduction

Despite the discovery of MACHO dark matter in our galactic halo, much of the inferred halo mass seems still to be missing [1]. If we assume that this missing mass is in the form of the weakly interacting massive particles (WIMPs) predicted by supersymmetry [2], then we expect to find a flux of $10^7/m_\chi$ cm^{-2} in the solar neighborhood or 10^4-10^6 cm^{-2} assuming the most likely mass range $m_\chi \sim 10$ - 1000 GeV and standard models of the galactic halo. These WIMPs have a total cross section for scattering on a nucleon of no more than 10^{-41} cm^2, leading to total scattering rates below 0.1 kg^{-1} day^{-1} of target mass.

These dark matter interaction rates are orders of magnitude higher than the relevant rates for many other types of rare event searches (Eg. solar neutrinos, proton decay) where one looks for a handful of events a year in a ton or more of material. However, detecting this rate of interactions is a formidable problem because the energy deposited is as low as a few keV. All conventional radiation detectors experience significant backgrounds at these energies because of the inevitable low levels of radioactivity present in the materials of the detector itself and its sur-

311

D. N. Schramm and P. Galeotti (eds.), Generation of Cosmological Large-Scale Structure, 311–316.

rounding shielding. Except when operating at great depths below the surface of the earth, cosmic rays also give rise to interactions in the detectors, both directly and by the production of bremsstrahlung, energetic electrons, and neutrons. A compounding problem is that sensitivity to low energy radiation can only be obtained so far in detectors that are highly segmented and consequently it is difficult to achieve large target masses.

An experimental technique capable of reliably distinguishing the interactions of dark matter particles from the interactions produced by radioactivity would be of great value. One must be most concerned with discriminating possible WIMP events from the gamma ray interactions which usually dominate the low energy spectrum. One possible basis for discrimination is the excess in the amount of ionization produced by gamma and electron interactions over that produced in WIMP-nucleon scattering. In a semiconductor target it is possible to measure the energy deposition and ionization production independently by exploiting the vanishing heat capacity of these materials at low (~ 10 mK) temperatures. One can then use the measured ratio of the two quantities to identify candidate WIMP interaction events. In this paper, we describe the setup and initial operation of the Cryogenic Dark Matter Search (CDMS), an experiment based on this discrimination method [3-4].

2. Design of CDMS Experiment

The initial location for the CDMS experiment is a shallow (17 meters water equivalent) tunnel on the Stanford University campus. The relatively small overburden is enough to eliminate the hadronic cosmic ray component and reduce the muon rate to 45 $m^{-2}s^{-1}$. This location was chosen because of its proximity to detector development laboratories. A plan for the extension of the experiment to a much deeper site, the Soudan iron mine (2200 meters water equivalent) is now under proposal. At the shallow site, the experiment will ultimately include 18 detectors incorporating a total of 2-3 kg of target mass. At the deeper site, the active mass could be enlarged to 10-20 kg by removing some of the shielding materials necessary at shallow depth and increasing the number and size of detectors. Figure 1 shows the projected physics reach

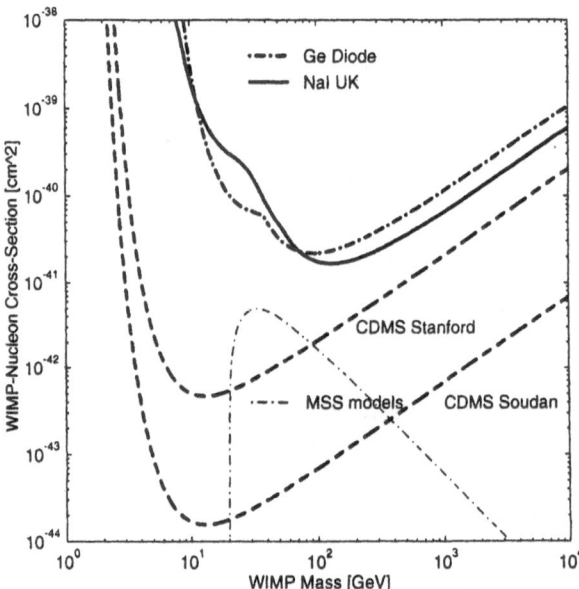

Figure 1. The regions of cross section and WIMP mass parameter space to be explored by CDMS at the current shallow location, assuming a background of 0.03 events/(keV kg day) with a threshold of 2 keV and at a deeper location, assuming 0.001 events/(keV kg day) and the same threshold. Also plotted are existing limits from Ge diode and NaI scintillator experiments.

of the experiment at the two locations. At Stanford, the experiment will just cross in to the interesting supersymmetry region after ~ 100 kg-days of Ge live time.

To compensate for the shallow depth of the Stanford tunnel, we have installed a muon anti-coincidence system which is capable of identifying >99% of the muons which pass through the dark matter detectors. This system is built from 13 pieces of 5 cm thick plastic scintillator surrounding the detectors, cryostat, and radiation shield. (See figure 2) The light from the scintillator is collected with wave-shifter bars into 26 photomultiplier tubes. Muons deposit more than 10 MeV in this scintillator, allowing almost complete separation of the muon events from the large number of events due to radioactivity at lower energy. We measure an integrated rate of 350 muons/s in these counters.

The muons produce neutrons as they pass through the high Z materials of the shield. Neutrons are an important background for this experiment because elastic neutron scattering in our detectors produces the same ratio of ionization to total deposited energy as WIMP scattering. Monte-carlo simulations of cosmogenic neutron thermalization and capture in our shield show that after 100 μs these neutrons have cooled enough for their recoils on our detectors to produce negligible back-

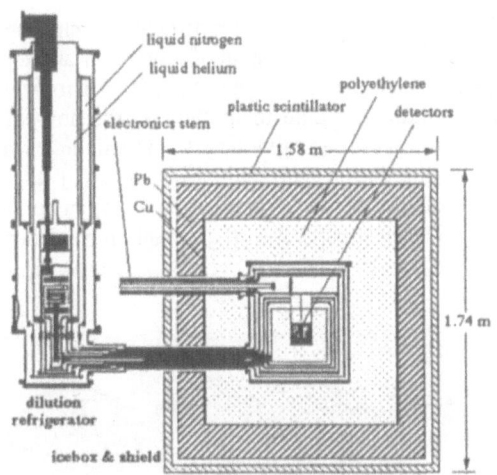

Figure 2. CDMS radiation shield and cryostat.

ground. In anticoincidence with the veto, we expect to see less than 0.05 neutron elastic scattering events/(kev kg day) in the region below 30 keV. Neutron background calculations and flux measurements in a test shield are described in [5].

The radiation shield surrounding the cryogenic detectors is composed of 10 cm of ordinary lead to screen environmental gamma rays, 5 cm of low Pb-210 lead from the Glover mine to screen Bi-210 bremsstrahlung from the outer lead, 25 cm of polyethylene neutron moderator, ~ 2 cm of OFHC copper making up the walls of the cryostat, and an inner 1 cm of very low Pb-210 old lead (<0.12 Bq/Kg). The polyethylene is inside the outer lead wall since neutrons produced in the lead by muons would otherwise dominate the background. There is room inside the cryostat for additional polyethylene to moderate neutrons produced in the copper and old lead. All of these materials were screened for radioactive content by conventional low-level Ge spectroscopy and were assembled under clean room conditions. The shield is sealed against radon gas with aluminized mylar and continuously purged with nitrogen boil off gas.

The cryostat is built almost entirely from electron beam welded OFHC copper and is capable of cooling to 5 mK, but is usually operated at 20 mK. It was designed as a horizontal extension to a commercially available dilution refrigerator so that the complex, and presumably radioactive, dilution apparatus could remain outside the radiation shield and still fit inside the confined space that was available in our tunnel. This system can be cooled in about 5 days and consumes 22 liters of liquid helium a day when running. Its cold volume is 30 cm in diameter by 30 cm high and it has been tested successfully with a heat load equivalent to 42 of our detectors.

Signals from the cryogenic detectors, which are described in the next section, are amplified at the 4 K stage of the cryostat and brought to room temperature on copper-Kapton strip lines which facilitate heat sinking for the several thousand wires that will ultimately be needed. At room temperature, the signal waveforms are digitized at up to 5 MHz in a VXI bus based data acquisition system. Waveforms from all detectors are recorded at the occurrence of a simple threshold trigger. A 16 ms time history of activity in the veto counters is also recorded. All further reduction of the data is accomplished off line. This allows the visual inspection of the small number of candidate WIMP scattering events for artifacts such as microphonic bursts, pileup, etc. and permits the use of optimal noise filtering techniques.

314

3. Cryogenic Detectors

We have developed two types of detectors sharing the same background rejection method. One is based on the measurement of ionization and temperature ("thermal phonons") in germanium crystals, the other on the measurement of ionization and non-equilibrium phonons in silicon. In the future, we plan to implement the non-equilibrium phonon technology on germanium, which is a more sensitive target than silicon for most WIMPs.

For the current germanium detectors [6], the thermal measurement is made with neutron transmutation doped germanium thermistors bonded onto the target crystal by a Au-Ge eutectic layer. The heat capacity of germanium vanishes quickly at low temperatures, becoming ~1 μK/keV at 20 mK for a 100 g crystal. The resistance of the thermistors is a very strong function of temperature. A 1 μK temperature change will produce a resistance change of ~500Ω in a 1 MΩ resistor. A typical "thermal pulse" has a rise time of 2 ms due to the thermal resistance between the bulk crystal and the thermistor and a 30 ms fall time as the crystal cools back down to the temperature of the cryostat through a thin heat sink wire.

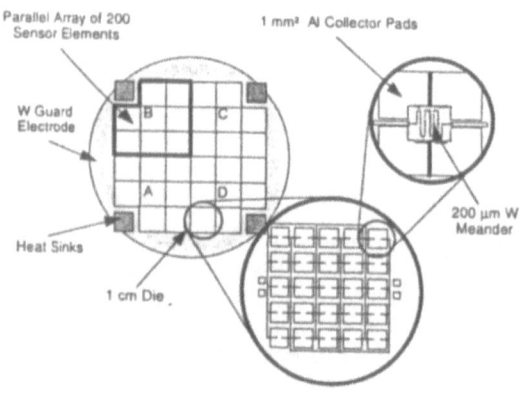

Figure 3. CDMS 100 g Silicon detector. Four sensors (ABCD) are shown patterned on the surface of a 7 cm diameter by 1 cm thick wafer.

In our silicon detectors, athermal phonons produced at the site of a particle interaction are absorbed by thin superconducting aluminum pads deposited on the surface of the crystal by conventional photolithographic techniques (See figure 3). The phonons generate quasiparticles in the aluminum which diffuse into a matrix of thin superconducting tungsten lines and cause heating. This scheme ("quasiparticle trapping") amplifies the phonon signal by draining the energy collected over a large area into a small amount of metal with low heat capacity [7]. The tungsten is biased on its superconducting transition which occurs at ~80 mK in our case. It is held near a stable point in the middle of the transition by electrothermal feedback; a constant voltage applied to the tungsten causes Joule heating (V^2/R) which increases at lower temperatures and decreases at higher temperatures. Excess heat flowing from an interaction in the crystal causes a reduction in the electrical power needed to maintain constant temperature [8]. The current flowing to the tungsten at constant voltage is measured with a high bandwidth SQUID array.

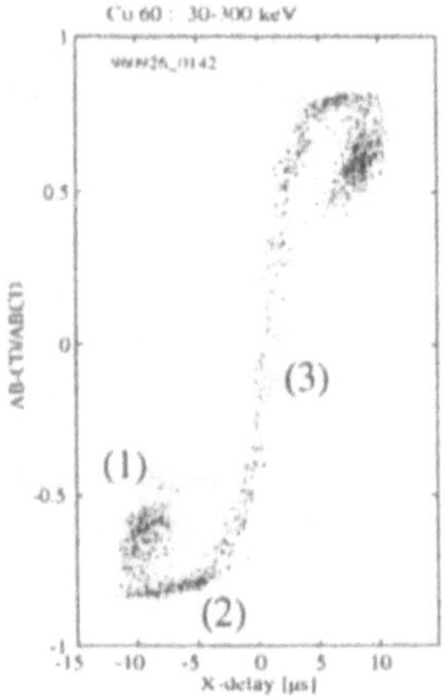

Figure 4. The difference in pulse heights between the sensors on the left side of the detector (AB) and right side of the detector (CD) is plotted against the time difference in reception of the signal at the two sensor groups (i.e., AB and CD are treated as single sensors for this measurement). Events at position (1) occur to the left of the sensors AB; at (2) under sensors AB; and at (3) in the middle of the crystal.

In the sensors on our 100 g detectors, we find a reduction in the time integrated Joule heating equivalent to about 0.3% of the energy of interaction.

The rise times of these signals, 5 µs, are determined by the time required for the quasiparticles to diffuse from the aluminum into the tungsten. Since phonons propagate in the crystal at a velocity of ~0.5 cm/µs, the time differences between arrival at different sensors, together with relative pulse heights, can be used to determine the event location. This allows us to reject events occurring close to the surface of the crystal or employ more sophisticated background subtraction based on the spatial distribution of events. Figure 4 illustrates the reconstruction of event position in the 100 g detector pictured in figure 3. This detector can determine radial position to 10% over the inner 70% of the crystal with somewhat reduced resolution at the edges.

The ionization measurements are similar in germanium and silicon [9]. Charges drift to the surfaces of the crystal under the influence of an electric field created by applying a voltage bias between conductive surface contacts. At low temperature, the electrical impurities are frozen out and no depletion voltage is needed, so it is possible to collect the charge with extremely low electric fields, ~1 V/cm. The surface contacts are divided into two regions on one face of the crystal, a central region and a surrounding annular ring. This provides a limited ability to measure the radial position of the interaction by charge sharing.

Figure 5 shows the response of a 60 g germanium detector to gamma rays and neutrons. Two "bands" appear, one with a high ratio of ionization to heating, due to gamma rays; the other with a low ratio, due to neutrons. In the 10-30 keV range of recoil energy, it is possible to eliminate 99% of the gammas while accepting 90% of the neutrons (or WIMPs) by imposing a cut on the ratio of collected charge to heat. For our silicon detectors, which currently have more noise, a 99% rejection of gamma events can be achieved with 75% acceptance in the 30-60 keV range.

4. Initial Operation

In the summer of 1996, CDMS began taking data with two prototype detectors, a 60 g germanium detector and a 100 g silicon detector. The experiment has now accumulated about 1.6 kg-days of live time for germanium and 0.5 kg-days for silicon. Very preliminary germanium results are shown in figure 6. Features in this spectrum include a 511 keV line from positron annihilation, Pb X-ray fluorescence lines at 75 keV and 85 keV, a copper fluorescence at 8 keV, and a Bi-210 line at 46 keV. There is also a line at 11 keV from Ga-68, which was produced in the germanium by cosmic ray spallation while the detector was above ground. The electronics noise in this detector is equivalent to ~1.5 keV (FWHM) in the charge channels and ~500 eV in the phonon channels, resulting in a 3 keV threshold for gamma rays or 10 keV for nuclear recoils. Spectral lines appear with this resolution when integration is over a short period of time. However, slow temperature drifts in our system currently broaden these lines significantly for long counting periods. This problem can be readily solved in the future by the use of a stable source of heat pulses to recalibrate the pulse amplitudes at frequent intervals.

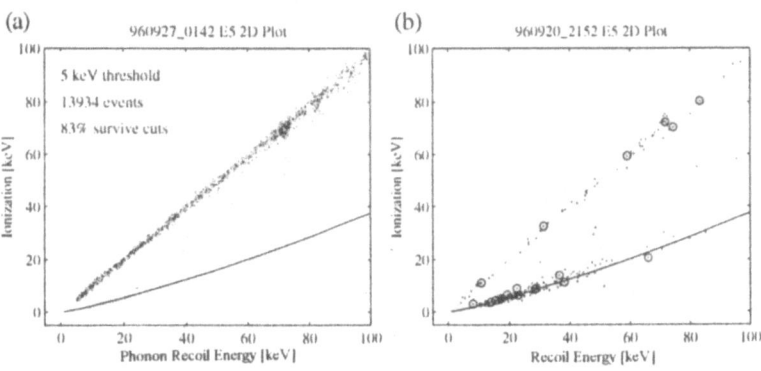

Figure 5. Ionization and phonon signals from a 60 g Ge detector exposed (a) to a Co-60 gamma ray source and (b) to a Cf-252 neutron source showing discrimination. WIMPs would appear on the neutron line. Events between the two lines are due to the existence of regions of the crystal where charge collection is not complete. Circled events were coincident in the Si detector.

316

The muon veto system attenuates the low energy portion of the spectrum by 90% (100 µs veto window), indicating that a large fraction of the low energy background is cosmic ray related. After the veto cut is applied, there are 7 counts/(keV kg day) in the region around 15 keV, about a factor of 2 higher than our ultimate goal for the Stanford phase of the experiment. The spectrum above 3 MeV, resulting almost entirely from direct muon interactions in the crystal, is reduced by a factor of 99.7%, indicating that the muon veto is performing with great efficiency.

After a cut on the ionization/phonon ratio, there are 0.3 counts/(keV kg day) in the region 20-35 keV of recoil energy. The background is limited at this level by the unexpected appearance of a population of events with ionization/phonon ratios partially overlapping those expected for neutrons (or WIMPs) and photons. These events seem to be caused by beta particles from radioactive contamination very near the detector. The betas fall on a "dead layer" of 10 - 30 µm at the surface of the crystal in which charge collection is poor. Although all the materi-

als around the detectors were well screened for bulk radioactive contamination, the technique used was rather insensitive to surface contamination. A prime suspect for the source is the plateout of radon daughters onto the apparatus. In the future, the detectors will be encapsulated by materials known to be low in surface activity. To this end, we are investigating new cleaning and screening procedures, with results that are already encouraging.

For silicon, background results are similar to those reported above for germanium. There are 0.2 counts/(keV kg day) between 30 and 60 keV, based on 3 events. The current generation of these devices does not have sufficient energy resolution to resolve the spectral lines and has a high (30 keV) recoil threshold. However, the silicon detectors do not suffer strongly from the beta emitter contamination problem.

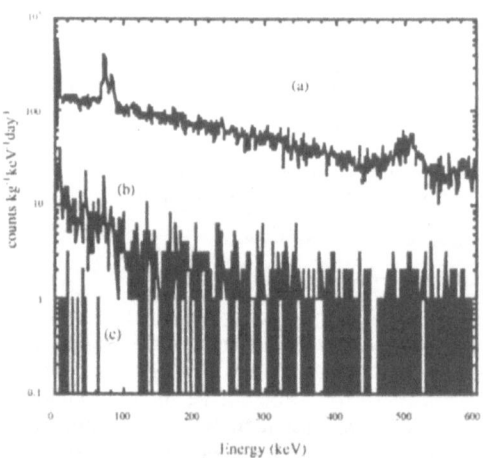

Figure 6. Spectra from 60 g CDMS germanium detector: (a) the raw spectrum; (b) with a 100 µs muon veto; and (c) with a cut on the ionization/phonon ratio.

5. Acknowledgments

This work was supported by the Center for Particle Astrophysics, an NSF Science and Technology Center under Cooperative Agreement No. AST-912005, and by the DOE under Contract No. DE-AC03-76SF00098 and award Nos. DE-FG03-91-ER40618 and DE-FG03-90ER40569. A.S. would like to thank R. Bernstein for organizing a very exciting conference.

6. References

[1] Gates, E., Gyuk, G., and Turner, M.S., *Phys. Rev. Lett.* 74 3724-3727 (1996)
[2] Jungman, G., Kamionkowski, M., and Griest, K. *Phys. Rep.* 267, 195-376 (1996).
[3] Shutt, T. *et al, Nucl. Phys. B* 51B, 318-322 (1996).
[4] Barnes, P. D., Jr., Ph.D. Thesis, U.C. Berkeley (1996).
[5] Da Silva, A. *et al, Nucl. Inst. Meth.* A 354, 553-559 (1995).
[6] Shutt, T. *et al, Phys. Rev. Lett.* 69, 3531, 3452 (1992).
[7] Nam, S.W. *et al, Nucl. Inst. Meth.* A 370, 187-189 (1996).
[8] Irwin, K., *Appl. Phys. Lett.* 66, 1988 (1995).
[9] Penn M. J., Dougherty B.L., Cabrera B., Clarke R.M. and Young B.A., *J. Appl. Phys.* 79 (11), 8179-8186 (1996).

Subject Index